ALLE ZEIT WACH
1842

Time, Mind, and Behavior

Edited by

John A. Michon and Janet L. Jackson

With 69 Figures

Springer-Verlag
Berlin Heidelberg New York Tokyo

Professor Dr. JOHN A. MICHON
Dr. JANET L. JACKSON
Institute of Experimental Psychology
University of Groningen
Kerklaan 30
9751 NN Haren, The Netherlands

ISBN-13:978-3-642-70493-2 e-ISBN-13:978-3-642-70491-8
DOI: 10.1007/978-3-642-70491-8

Library of Congress Cataloging-in-Publication Data. Main entry under title: Time, mind, and behavior. "Result of the International Workshop on Time, Mind, and Behavior, which was held at the University of Groningen in September 1984"–Pref. 1. Time–Psychological aspects–Congresses. 2. Time perception–Congresses. 3. Human behavior–Congresses. 4. Biological rhythms–Congresses. I. Michon, John A. (John Albertus), 1935–. II. Jackson, J. L. (Janet L.) III. International Workshop on Time, Mind, and Behavior (1984 : University of Groningen) BF468.T56 1985 153.7'53 85-26166.
ISBN-13:978-3-642-70493-2 (U.S.)

Media Conversion: fotosatz & design. Berchtesgaden

2126/3130-543210

Preface

This book is the result of the International Workshop on Time, Mind, and Behavior, which was held at the University of Groningen in September 1984. The aim of the workshop was to produce an up to date review of the state of the art in the field of time psychology. The rapid development of a cognitive outlook in experimental psychology has, among other things, underlined the need for a reconsideration of time experience, the coding and representation of temporal information, and the timing of complex responses. Since the publication of Paul Fraisse's classical *Psychologie du Temps* in 1957, time psychology has slowly but steadily drawn an increasing amount of attention, to a point where it now seems to be incorporated into the mainstream of research. At the same time a noticeable tendency for a renewed general interest in time can also be traced in several other disciplines. These two observations supported our belief that it was time for a review of the sort we had in mind.

At the close of 1983 we completed a project supported by the Dutch Organization for the Advancement of Pure Research in which we had studied the coding and retrieval of temporal information. This provided us with a plausible pretense for organizing a workshop. Around Christmas time 1983 we were able to mail a preliminary invitation to a number of our colleagues whom we knew to be currently active in the field. At the same time we started the laborious process of fund raising. To our great pleasure almost all invitations were accepted and the burden of financing the workshop proved less heavy than we had anticipated.

The format of the workshop implied a session on each of the six main topics, a session consisting of a major review paper plus two shorter research papers, each covering a recent piece of research that we knew to be backed by a research program of some substantial size and significance. While editing this book we have come to the conclusion that even although we had to rearrange the contributions to some extent, the format was successful. Also the fact that a slightly different structure than we had originally envisioned has imposed itself on the table of contents is, in our opinion, a positive sign that time psychology has indeed an intrinsic logic and that the reader will be able to extract from the text a consistent picture of what is going on in this area.

While we were in the final stages of editing this volume we learned of the sudden and unexpected death of our colleague Gerard Groos, at the age of 33. Ever since he had agreed to participate in *Time, Mind, and Be-*

havior (and to contribute Chapter 4, with Serge Daan), he had played that remarkable role only the very bright can play: a quiet, unassuming and yet genuine and solid support. In him chronobiology lost one of its truly outstanding scholars.

We are happy to acknowledge the fact that all contributors not only provided us with interesting papers but also complied extremely punctually with the tight schedules we imposed on them. Thus it has been possible for us to complete the editorial work on the manuscript within seven months of the date of the workshop. In hindsight it is surprising that we succeeded in doing this. When we started to arrange for the final revisions we felt that information technology might have arrived at a point where most manuscripts are produced by means of text processors. Consequently we asked authors to submit their manuscripts not only on paper but also to send us a copy of their 'floppy' for further speedy compilation. This has turned out to be a mild disaster. The possibilities of transferring materials from one system to the other are, in our opinion, still extremely poor and in many cases not really worth trying. Computer manufacturers, including the 'big' ones, should be ashamed of themselves!

A more pleasant task is to convey our deeply felt appreciation towards all those who, right from the early inception of the workshop have played an active and indispensible role, large or small, long or brief.

In the first place we thank the team that ran the workshop with us. Ans van Rijsbergen took care of the workshop secretariat in an incredibly efficient way; Harm Hospers, Annemiek Vermeeren, Alma Schaafstal and Jan Maarten Schraagen dealt with every conceivable aspect of the logistics in a smooth, intelligent and effective way. In turn their efforts were made possible by those who, perhaps less visibly, provided their supporting services, Dini Batstra, Herman Hofman, Henk Visscher, Anton Nolle, and Mrs. Eisses and her staff. In its Lord Mayor, Mr. H. Sybesma, we acknowledge the hospitality of the City of Dokkum, one of the historical eleven towns of Friesland which recently, in February, became world news again because of the celebrated Elfstedentocht, the 200 kilometer skating race. A visit to this city was one of the memorable lighter events during the workshop.

Several agencies have contributed financially to the event. We mention them in the order in which their support was obtained: Traffic Research Center (University of Groningen), Gasunie, Department of Experimental Psychology (Vakgroep Functieleer), Groninger Universiteitsfonds, Organization for the Advancement of Pure Research, Royal Netherlands Academy of Arts and Sciences, Organization Committee First International Symposium on Drugs and Driving.

The process of producing the book was greatly facilitated by Ans van Rijsbergen who took care of the often complicated written communications with the authors and helped to keep track of the paper work involved, including part of the ultimate production of the manuscript. She, together with 'De Vries Advies' in Groningen and the Secretariat of the

Traffic Research Center produced a manuscript of such perfection that the transfer to the final printed version could be performed very fast. Harm Hospers took the inestimable responsibility of checking the literature references and compiling the author index.

The most indispensable help and support, however, we received from our long-time life companions, Hetty and Sandy, to whom we dedicate the results of our efforts. They carried much of the burden created by our investment in extra time during holidays, weekends and late evening hours, needed to get the book in press as quickly as possible.

As a reflection of the current concern of behavioral scientists, many of them working in an intellectual context that is dominated by the cognitive and information processing paradigm, *Time, Mind, and Behavior* will, no doubt, provide a time-bound view of the role those scientists think time plays in human experience. At the same time, however, we wish to stress the fact that many of the chapters in our book provide the reader with an outlook on 130 years of empirical study of what is perhaps the most elusive dimension of human experience: Time.

The editors wish to thank the following publishers and authors for permission to reproduce illustrations.

Academic Press, Inc., for figures from the following: From C. L. Lee & W. K. Estes, Order and position in primary memory for letter strings, *Journal of Verbal Learning and Verbal Behavior,* 1977, *16,* 395–418. From A. F. Healy, Separating item from order information in short term memory, *Journal of Verbal Learning and Verbal Behavior,* 1974, *13,* 644–655. From L. R. Peterson, S. T. Johnson & R. Coatney, The effect of repeated occurrences on judgments of recency. *Journal of Verbal Learning and Verbal Behavior,* 1969, *8,* 591–596. From P. Weisberg, Effects of reinforcement history on timing (DRL) performance in young children, *Journal of Experimental Child Psychology,* 1970, *9,* 348–362. From R. Efron, An invariant characteristic of perceptual systems in the time domain. In S. Kornblum (Ed.), *Attention and Performance IV,* 1973.

The American Psychological Association for figures from the following: From G. ten Hoopen, J. Vos & J. Dispa, Interaural and monaural clicks and clocks: Tempo difference versus attention shifting, *Journal of Experimental Psychology: Human Perception and Performance,* 1982, *8,* 422–434. Copyright 1982 by the American Psychological Association. From D. L. Hintzman, J. J. Summers & R. A. Block, Spacing judgments as an index of study-phase retrieval, *Journal of Experimental Psychology: Human Learning and Memory,* 1975, *1,* 31–40. Copyright 1975 by the American Psychological Association. From R. Collard & D. J. Povel, Theory of serial pattern production: Tree traversals, *Psychological Review,* 1982, *89,* 693–707. Copyright 1982 by the American Psychological Association. From J. Greeno & H. A. Simon, Processes for sequence production, *Psychological Review,* 1974, *81,* 187–198. Copyright 1974 by the American Psychological

Association. From F. Restle, Theory of serial pattern: Structural trees, *Psychological Review,* 1970, *77,* 481–495. Copyright 1970 by the American Psychological Association.

Archives de Psychologie for a figure from J. Montangero & J. L. Gurtner, Vitesse-fréquence, vitesse-déplacement et jugements de durée chez l'enfant, *Archives de Psychologie* 1983, *51,* 368–384.

New York Academy of Sciences for figures from J. A. Michon & J. L. Jackson, Attentional effort and cognitive strategies in the processing of temporal information. In J. Gibbon & L. Allen (Eds.), *Timing and time perception,* 1984.

Pergamon Press, Inc., for a figure from K. Honma, C. von Goetz & J. Aschoff, Effects of restricted daily feeding on free running circadian rhythms in rats, *Physiology and Behavior,* 1983, *30,* 905–913.

Presses Universitaires de France for figure from M. Richelle & H. Lejeune, L'animal et le temps. In P. Fraisse (Ed.), *Du temps biologique au temps psychologique,* 1979.

Perceptual and Motor Skills for a figure from N. Stein & R. Landis, Effects of age and collateral behavior on temporally discriminated performance of children, *Perceptual and Motor Skills,* 1978, *47,* 87–94, figure 1. Reprinted with permission of authors and publisher.

Psychonomic Society, Inc., for figures from the following: From A. B. Kristofferson, A real-time criterion theory of duration discrimination, *Perception and Psychophysics,* 1977, *21,* 105–117. From H. Eisler, Applicability of the parallel-clock model to duration discrimination, *Perception and Psychophysics,* 1981, *29,* 225–233. From R. S. Lockhart, Recency discrimination predicted from absolute lag judgments, *Perception and Psychophysics,* 1969, *6,* 42–44.

Yale University Press for a figure reprinted from L. W. Doob, *The patterning of time,* 1971, figure 1, p. 31. Reprinted by permission.

Haren, October 1985

JOHN A. MICHON
JANET L. JACKSON

Contents

List of Contributors

JOHAN F. A. K. VAN BENTHEM, Department of Philosophy, University of Groningen, 9718 CA Groningen, The Netherlands

RICHARD A. BLOCK, Department of Psychology, Montana State University, Bozeman, Montana 59717, U.S.A.

SERGE DAAN, Department of Behavioral Biology, University of Groningen, 9751 NN Haren, The Netherlands

WILLIAM K. ESTES, Department of Psychology, Harvard University, Cambridge, Massachusetts 02138, U.S.A.

PATRIK FERY, Department of Experimental Psychology, University of Liège, B-4000 Liège, Belgium

MICHELANGELO FLÜCKIGER, Department of Psychology, University of Geneva, 1211 Geneva 4, Switzerland

GERARD GROOS †, Department of Behavioral Biology, University of Groningen, 9751 NN Haren, The Netherlands

GERT TEN HOOPEN, Department of Experimental Psychology, University of Leiden, 2312 KM Leiden, The Netherlands

JANET L. JACKSON, Department of Experimental Psychology, University of Groningen, 9751 NN Haren, The Netherlands

MARI RIESS JONES, Department of Psychology, The Ohio State University, Columbus, Ohio 43210, U.S.A.

HELGA LEJEUNE, Department of Experimental Psychology, University of Liège, B-4000 Liège, Belgium

FRANÇOISE MACAR, Institute for Neurophysiology and Psychophysiology, C.N.R.S., 13274 Marseille, France

JOHN A. MICHON, Department of Experimental Psychology, University of Groningen, 9751 NN Haren, The Netherlands

JACQUES MONTANGERO, Department of Psychology, University of Geneva, 1211 Geneva 4, Switzerland

SIEB G. NOOTEBOOM, Institute for Perception, Research, 5612 AZ Eindhoven, The Netherlands

DAVID PARK, Department of Physics, Williams College, Williamstown, MA 01267, U.S.A.

JEAN-JACQUES PERIKEL, Department of Experimental Psychology, University of Liège, B-4000 Liège, Belgium

VIVIANE POUTHAS, Department of Experimental Psychology, René Descartes University, 75006 Paris, France

DIRK-JAN POVEL, Department of Experimental Psychology, University of Nijmegen, 6500 HE Nijmegen, The Netherlands

MARC RICHELLE, Department of Experimental Psychology, University of Liège, B-4000 Liège, Belgium

L. HENRY SHAFFER, Department of Psychology, University of Exeter, Devon EX4 4QJ, United Kingdom

HANS-LEO TEULINGS, Department of Experimental Psychology, University of Nijmegen, 6500 HE Nijmegen, The Netherlands

ARNOLD J. W. M. THOMASSEN, Department of Experimental Psychology, University of Nijmegen, 6500 HE Nijmegen, The Netherlands

Introduction

Chapter 1. Introduction: The Psychology of Time

John A. Michon and Janet L. Jackson

A Perennial Issue

Time is a fascinating subject! This much is evident from the innumerable arguments raised in the course of twenty-five centuries of philosophical debate. The presocratic philosophers already showed a lively interest in the nature of time and formulated many of the questions that are still of fundamental concern to humankind. Is time real or an artifact of the way mortals look at things? Is time a sense impression or is it an idea, that is, a mental construction forced upon us by the innate properties of our minds? The mind-boggling complexities emerging from this debate can be traced in numerous disguises through the ages, up to the present day, and quite a few authors have done admirable jobs by summarizing the different positions or by providing anthologies of indispensable primary sources (e.g. Sivadjian, 1938; Whitrow, 1960; Smart, 1964; Gale, 1968; Sherover, 1975).

This relentless attention may seem strange at first sight. It has been pointed out that a primitive notion of time emerges almost as a necessity and that this notion is, in actual fact, »natural and nearly complete« (Toda, 1978). Human beings, in other words, are not likely to be caught in a web of mystery when they are dealing with the temporal vicissitudes of ordinary, daily life. Proof of this apparent naturalness of time could be the fact that although in most ancient religions we find deities worshipped because they knew how to influence the course of events in time, none appear to have been attributed the ability to alter the flow of time. Even in Christian faith the fact that God made the sun stand still while Joshua waged his battle against the Amorites (Jos. 10:12-13) is not considered a real intervention in the flow of time. As Saint Augustine pointed out, while the sun stood still Joshua's battle was fought and won. Today we would probably add that it was fought and won *in real time*. Since worshipping gods is an early strategy by which humankind sought to exercise control over reality, we may assume that apparently no one really ever felt a strong need for controlling time. The general feeling must have been that time can be adapted to, but that it cannot be controlled and that therefore even »the Gods followed time as given« (Toda, 1978).

Time seems to have acquired its mysterious character only when the notion of spatialized, physical time began to develop. In this context physics is just another attempt at mastering time, but now through inventing clocks, time sharing, and parallel processing systems, rather than through worship. Conventional time as we know it today is to be considered a surplus structure, imposed on an already existing and naively accepted temporal order in the universe: if we cannot change the rate of flow of time, we can at least impose a deliberate structure on the ways events are ›packaged‹ within the confines of the time we find at our disposal.

From Philosophy to Psychology

Upon reflection questions about the conceptual status of time seem to multiply more quickly than the mind can follow. This feeling is certainly what inspired Saint Augustine's well-known lamentation about the incommunicability of the concept, but it also emerges from every one of the results of the detailed philosophical analysis to which time has been subjected since the Middle Ages. Time is, no doubt, a *fundamental* concept. It appears in the definitions of many other concepts and for that reason it may be said to occupy a central position in our cognitive representations of reality. This may be the reason why it seems so self-evident. Time is, no doubt, also a very *old* concept. It appeared early in human experience and most known cultures have some grasp of the concept of time as is revealed by the almost universal existence of time keepers, however crude (e.g. Zern, 1967; Wendorff 1980), and by the existence of temporal expressions in almost any and perhaps every language. Unlike some other ancient concepts, however, the concept of time is *not simple*. Like space and causation, which have also very early roots in human thought, it cannot be decomposed very easily into more simple concepts.

That time is not simple follows from the ease by which we may construct a nearly interminable inventory of problems directly related to time and its passing. Fraser, for instance, lists no less than 300 such problems (1978). Among those problems are all the well-known fundamental philosophical questions about the status of the distinctions before-after and past-present-future, the true ›nature‹ of time, its origin, its reversibility and its continuity. Yet, while such problems offer themselves to the philosophical mind, they are not necessarily, or even primarily, philosophical problems. Neither does philosophical analysis seem to exhaust the set of potentially relevant questions about time. Some are more likely to be raised and answered by physicists, others clearly belong to the domain of psychology. Philosophers sometimes appear to be underestimating the need for scientific analysis of philosophical problems as when they argue, for instance, that »the only empirical research that is needed for the solution of a truly philosophical problem like the meaning of one of our fundamental concepts [such as time] is research into the use of the term in question in daily life« (Zwart, 1976, p.11). Although this invitation to intentionality may lead us some way along the narrow path of insight, it may not lead us very far. The dangers of analysis in the context of common language have been pointed out by so many authors that we will refrain from further commentary (see e.g. Linschoten, 1964; Mandler, 1975; Dennett, 1978; see also Michon (1985) in chapter 2 and, for a different opinion, Park (1985) in chapter 3 of the present volume). We agree, however, with Mandler (1975; p.7) who summarized this position as follows:

»...a philosophical system about knowledge, a system in which people talk about what they believe they know and how they believe they know things, has no logical or psychological one-to-one correlation with the theoretical-empirical attempt to build a psychological theory that explains how the human individual acquires knowledge, perceives the world, engages in ethical endeavors, *and* talks about it.«

As soon as we accept the fact that not every experience is accessible to introspective scrutiny and conscious analysis or, in current terminology, that some information processing is automatic or ›veiled‹, the limitations of a strictly philosophical analysis are immediately clear.

It remains to be decided therefore, not only if we can conceive of time as a unitary concept in conscious experience (much like we conceive of absolute time in classical physics), but also if it is possible to treat the various aspects of time as they are revealed by psychological analysis and experiment from a single, consistent theoretical point of view. If such an integrated view of time in human experience and behavior would turn out to be impossible, we would be left with the alternative to explicate what the several conceptually and functionally independent processes are that such a pluriform concept of psychological time would need to consist of. We would rather prefer, however, to be able to show that all of the various processes that can be distinguished empirically do, in fact, serve one ultimate and very important purpose, namely that of relating the organism to the events that happen in its environment. As Michon (1985) argues in chapter 2 of the present volume, time is the conscious product of the organism's efforts to stay tuned to the temporal contingencies of its environment, and the tuning aspect will be seen to play a very crucial role in this book: the music of the spheres resonates in the rapping of the mind! In this context resonance is not a passive process of adaptation. It is, on the contrary, the result of the struggle of living matter to become independent of the temporal exigencies of the universe. As such it involves the creative structuring of reality in an active sense: humankind creates order, and temporal order in particular!

If we undertake to analyze the complex human notions of time it will be necessary, as Zwart (1976) pointed out, to do so by reducing them to more primitive concepts.

Philosophically speaking this is certainly true although, as we have already argued, it may be difficult to find such concepts. Psychologically, however, an analysis of time is an altogether different task. A scientific psychological analysis must be performed in terms of the processes that underlie what is clearly a very basic characteristic of conscious experience, and in terms of the temporal structure of the environment in as far as it exists independent of the experiencing organism.

The Rise and Fall of Early Time Psychology (1865-1957)

Philosophers have been concerned primarily with the metaphysical and the epistemological status of time as consciously experienced. Psychologists, on the other hand, are dealing with an entirely different set of time-related problems: humans live in an ever-changing world, a fact which turns time into a very real problem. In order to survive it is essential that an organism be capable of correlating its behavior very precisely with the temporal course of events in its environment. Acting five or ten milliseconds too early or too late may demarcate the dividing line between survival and death. This functional requirement has largely been neglected in philosophical analysis, although it is precisely this aspect from which the significance of time for human existence is ultimately derived.

Psychologists want to know what mechanisms and processes are underlying the experience of time. In order to cope with reality the individual creates images and expectancies about the world outside and these representations provide the guiding principles for behavior, the actual rules of conduct. Empirical psychology is concerned with the structure of mental representations: What aspects of reality are perceived directly and ›automatically‹ and what is inferred indirectly by some cognitive process of construction?.

The empirical study of psychological time is well over a century old. It was initiated around 1865. The first study on record was by Ernst Mach, who addressed the question of what we mean when we speak about the ›time sense‹. He was followed by many others – Vierordt, Wundt, Exner, Benussi among them – and as a result Titchener (1905) could qualify the study of time as one of the most important areas of psychological research, no less than »a microcosm, perfect to the last detail«. In Titchener's opinion time reflected all the major questions of psychology. Yet, interest declined rapidly after 1910, partly because of the elusive character of many temporal phenomena, partly also because of the decline of introspective method.

Since that time only a handful of prominent investigators have paid attention to the subject. Only the French have witnessed a very strong and uninterrupted tradition in the psychological study of time which involves many of their great psychologists and psychologically inclined philosophers, including Jean Marie Guyau, Henri Bergson, Pierre Janet, Gaston Bachelard, Henri Piéron, the Swiss Jean Piaget, and in particular Paul Fraisse. Everywhere else time psychology all but disappeared. Thus, in 1964 a reviewer could qualify time psychology as a »venerable tired topic«, attributing the general lack of interest to the fact that none of the students of time had bothered to try explaining temporal experience in terms of other – more basic – psychological mechanisms (Adams 1964). A similar observation was made by Creelman (1962), who correctly observed that whatever remained of Titchener's microcosm had developed »off the main stream of empirical research«. The psychological study of time had indeed become sterile. Experiments were based over and over again on the same petty paradigms, with very little progress in the sophistication of the questions asked or the hypotheses put forward. A topic from the psychophysics of time may illustrate this.

Eisler (1976), whose thoughtful methodological approach to problems of psychophysical methodology deserves mentioning, was able to list no less than 112 psychophysical scaling studies in which subjects were required to judge the durations of brief intervals by means of magnitude and ratio estimation, which is illustrative for the ›overkill‹ of the few research problems that were attacked by time psychologists until quite recently. But, not only were the same experiments repeated again and again in very much the same way, it was also established in most of these experiments that the subjective time scale is practically veridical. That is, an interval that is physically twice as long as a given standard interval will, by and large, also appear twice as long. In an earlier discussion of subjective scales of duration

Fraisse (1957) had already suggested – not without wit and reason – that perhaps the best unit of subjective duration is the second. Nevertheless the search for ›chrons‹ and ›temps‹ went on. Although several exceptions to the general finding that duration is perceived veridically are on record, Eisler's review indicates that no theoretical framework had been developed that might have explained such exceptions. However, without theoretical support to explain how temporal information is stored and processed further studies of the subjective time scale would seem to be entirely useless.

One of us has publicly wondered before why there has been so little interaction between the developments in ›mainstream‹ psychology and time psychology proper, why so little cross-fertilization took place (Michon 1967, 1972). The reason appears to be threefold.

In the first place there is the methodological complexity of time experiments. There is much that can work to disturb such experiments. Subjects may have difficulties obeying instructions, such as the instruction not to count or think during an interval. As a rule the variability of the responses of subjects will be rather large even though it may be possible to reduce the variance of the responses quite dramatically by training a highly motivated subject over a very long period spanning months or even years (Kristofferson 1976, 1984). Since most investigators hate variability in their data as much as very long training sessions, they will be repelled by the study of time. Fortunately there are also those who believe that some of the answers to the question of what time is are revealed by that very variability.

In the second place, little attention has been paid to the fact that time is more than just a dimension in which to express reaction times or the persistence of certain events. The fact that time is information (Michon, 1972; 1985, chapter 2 of the present volume; see also Estes, 1985, chapter 10 of the present volume) has not received proper attention as it constitutes an unusual point of view for many investigators. Yet, it simply implies that simultaneity, order, duration, and instant are meaningful notions that can be judged on the basis of perceived or stored information.

The third factor contributing to the relative neglect of time as a topic for serious psychological investigation is that its ramifications are so tremendous. Once one has embarked on a close analysis of the role of time in human mental activity and behavior, one cannot avoid cutting through a number of well investigated research areas at cross angles and bringing out highly intricate relations. The recent collection by Gibbon & Allan (1984) and the present book bear ample witness of this fact.

Rather than being a microcosmos as was suggested by Titchener, the study of time seems to offer a perspective, a way of looking at behavior, a paradigm, if you like. At least that is what optimists will think. Pessimists on the other hand have insisted that time is not a meaningful descriptive concept, and that, therefore, its adoption in theoretical contexts must necessarily lead to chaos and confusion. Time has proven the pessimists wrong.

The Renaissance: 1957 and After

From about 1960 onward there has been a gradual return of interest in time psychology. This revival was greatly stimulated by Paul Fraisse's epochal monograph (1957; see also 1964, 1967), summarizing a century's work up to 1957. From then on a slow but steady increase in the number of substantive studies can be observed. Other factors of importance in this renaissance were the improved technology of laboratory equipment, and the growing conceptual and methodological sophistication of experimental psychology. Recent studies, more often than not, deal with the problem of time within the framework of some more extensive problem area such as memory, attention, learning theory or psycholinguistics.

In short, the study of time experience seems on its way back to a more prominent position. However, in reviewing the many studies of recent years, we find no reason to believe that we are about to recover Titchener's Microcosm, and the epitheton ›perfection to the last detail‹ still seems a bit farfetched. Yet, it seems to us that we are also far removed from chaos and confusion. The last decade has brought a number of important and interesting developments related to the temporal structure of phenomena in pattern recognition, speech perception, memory retrieval and motor skills many of them to be covered in depth in later chapters. Each of these developments appears to transgress the boundaries of classical time psychology which primarily dealt with the perception, estimation or reproduction of empty or artificially filled intervals. These newer findings are actually embedded in theoretical frameworks of wider importance. Moreover, the advances made in the study of biological rhythms and of the role that temporal factors play in real life task performance have brought a new impulse to classical time psychology (e.g. Aschoff, 1984; Brown & Graeber, 1982). There are also important developments in performance theory, the theoretical and empirical analysis of human behavior in task situations – a task being an integrated pattern of goal directed activities that is subject to certain environmental and structural limitations (e.g. Sternberg et al., 1982; Massaro, 1984; Shaffer, 1985, chapter 15 of the present volume; Thomassen & Teulings, 1985, chapter 17 of the present volume). Finally many things are stirring in cognitive psychology. Important are some of the more recent developments dealing with the role of structural information, epitomized in J.J. Gibson's well known assumption that the world presents information which is picked up rather than constructed from relatively meaningless bits and pieces (Turvey, 1977; Pick & Saltzman, 1978). A related, very important development is coding theory (Jones, 1985, chapter 13 of the present volume; Leeuwenberg & Buffart, 1979; Povel, 1985, chapter 14 of the present volume). It should be possible to map the classical subject matter of time psychology into these areas, thereby uniting the various viewpoints, and thus to show that time psychology is now indeed very much in the mainstream of experimental psychology. Providing such a map has been the motivating force behind our efforts to prepare this volume.

It is not only within the area of psychology proper that we observe a greater effort being spent on temporal problems. The study of time is increasingly drawing attention in a more comprehensive scientific and philosophical context. The number of conferences and books has been increasing steadily over the last 15 years and there is no sign that this trend will soon pass. Time remains a fascinating subject!

Time as Information: Three Fundamental Questions

It is somewhat surprising to find that in most psychological research time is still being treated as if it were applied to a physical system. Subjects are not supposed to have temporal awareness.

We should know better. Although we find that people are subject to several important biological rhythms, they have no innate clocks of a great enough sophistication to help them cope automatically with all the intricacies of temporal structure of human life (Groos & Daan, 1985, chapter 4 of the present volume; Michon, 1985, chapter 2 of the present volume; Richelle et al., 1985, chapter 5 of the present volume). People tend, in fact, to be well aware of the order and location of events in time and of the duration of such events. The skilled execution of many perceptual motor tasks depends on evaluation – be it conscious or implicit – of the temporal properties of events.

In summary, any theory of human performance, and in particular any theory that deals with the interaction between man and a structured task environment, must incorporate the fact that man is aware of or is, at least, using the relations of simultaneity, order, temporal locus and duration when coping with the requirements of the environment.

In short, time is not to be treated simply as a homogeneous dummy parameter describing input-output relations. Time also means information to man. The basic question we should answer is whether time has an independent status in this respect. Is there such a thing as temporal information? The answer is yes, for the following reasons. First we observe that people discriminate temporal information quite spontaneously and directly, even at a very early age (Demany et al., 1977). They react discriminatively to stimuli that differ only in duration or in moment of occurrence and to rhythms that differ only minimally. They can do so with fairly great precision and with a high degree of veridicality. Equally, after practice they are able to time their activities very precisely. And early in this century Pavlov (see Pavlov, 1960; Richelle & Lejeune, 1980) established that time intervals as such may serve to make dogs salivate (something, incidentally, which dog owners have known for centuries if they were feeding their animals at fixed hours of day). However, not all temporal information processing seems to occur in this simple and apparently straightforward way. On the contrary: frequently the processing of temporal information appears to be a highly complex symbolic construction, not in any way something that might affect a time sense, and probably not even analogous to the way we perceive space, but rather more comparable to the way in which the semantic comprehension of a verbal sentence is constructed (van Benthem, 1985, chapter 18 of the present volume; Michon, 1985, chapter 20 of the present volume).

Time appears as an independent property of the flow of information that we experience as reality. There is no fundamental distinction between it and other properties of patterns of information such as size, color, loudness or spatial locations (Michon, 1972). This equivalence raises the following fundamental questions central to time as a topic for psychological study:

(1) In what ways can people process temporal information about events and objects: how do they code, store, and retrieve temporal information, as well as act upon it?

(2) What must the structure of the external information and of the internal processes be in order that the answers to the first question fall into a meaningful pattern?

(3) What relations do exist between, on the one hand, these processes and structures and, on the other hand, the various forms of conscious time experience, including the semantic aspects as expressed in natural language?

Stating these questions does in itself not help much; after all they are no more than rephrasings of questions that have kept time psychologists busy for more than a century. Yet, the situation is crucially different in two ways. The first difference is that the paradigmatic outlook of contemporary cognitive psychology is a powerful stimulant for asking new, relevant questions about temporal experience. The second is that, for the first time since the Big Bang took place in Titchener's microcosm, the echo is heard by experimental psychologists at large. It is our hope that *Time, Mind, and Behavior* will amplify that signal.

The Content of the Book

The remainder of this chapter is devoted to a brief outline of the contents of the book.

Part I - Origins: The Nature and Development of Time (Chapters 2-6)

In chapter 2 John A. Michon offers a review of the present state of thought and insight about psychological time. It first traces the specific aspects that distinguish psychological time from physical time and biological time. Then it discusses the various sources that may be tapped to find out what processes are underlying the temporal experience of humans. There are several such sources, intentionalistic, functionalistic and structuralistic. The analysis of time in narrative provides an example of the intentionalistic approach. It differs in a rather fundamental way from psychonomic analysis which tends to have strong functionalistic overtones. The present state of the art in psychological research is considered from points of view that draw upon the various approaches to time-as-information in the rest of the book. The chapter is in fact a search for conditions to be met when we are to construct *The Compleat Time Experiencer.*

In chapter 3, entitled *Brain Time and Mind Time*, David Park identifies some of the physical constraints that are imposed on the human – or any – experience of time. He points out that physics is struggling with such problems as free will, which, some physicists hope, will eventually be tackled by quantum mechanics. The problem centers around the role of the observer in the universe. The observer in some sense needs consciousness, but physical theory has not been able to handle consciousness at all.

Any form of consciousness presupposes temporal awareness and time is brought into the picture through the geometric concept of spacetime. However, geometrization of spacetime would seem to throw man out of the universe. A crucial aspect which indicates that this is ultimately impossible is the fact that a *now* is required if an observation is to be made. And such is the case if we are to make sense of quantum theory. In quantum mechanics we are dealing with composite states (collections of potential states). In what state a system (particle, body) is, can only be determined by an action of an observer who must meet the requirements of having a now, a memory, and freedom to choose. Some provision for mental phenomena must therefore be made in physics. Otherwise quantum mechanics will forever remain shrouded in mystery.

In chapter 4, *The Use of Biological Clocks in Time Perception*, the late Gerard Groos, and Serge Daan have dealt with the ontogenesis of the temporal ›independence‹ of the organism from its environment, recognizing the special role of biological rhythms, in particular the 24-hour rhythm. Whereas some 24-hour periodicities are innate and triggered automatically, others will emerge only when learned. Examples of the latter are feeding times at regular hours. Groos and Daan have also addressed the question whether intervals other than 24 hours, particularly the shorter ones, can be derived from the 24 hour period. This indeed turns out to be the case: other processes, governing periods perhaps as short as one hour, are demonstrably related with the 24-hour rhythm. But evidence also shows that there may be no relation with intervals in the minute and second range.

Temporal regulations in this range must be based on other processes, and their biopsychological origins are discussed at length by Marc Richelle, Helga Lejeune, Jean-Jacques Perikel, and Patrik Fery in chapter 5. Moving *from Biotemporality to Nootemporality*, terms coined by J.T. Fraser (e.g. 1978, 1982; see also Michon, 1985b, chapter 20 of the present volume), this chapter provides an overview of the relation, across species, between the biological subtratum of temporal behavior and the psychological mechanisms and processes involved. The authors formulate three major questions that specify this relation. The first deals with the degree of independence between the temporal control or regulation of behavior and the well-established biological rhythms discussed in the preceding chapter. Like Groos & Daan, Richelle et al. argue that there appears to be a large degree of independence, although some temporal regulations follow intrinsically the 24 h periodicity. In other words, there appears to be a rather inessential relation in such cases, although data are too scant for a definitive conclusion. The second question raised concerns the comparability of findings across species. Frequently animals are ›asked questions‹ that have no answer in their behavioral patterns and, consequently, they tend to be judged to be more temporally dumb than they might if answers had been elicited from more pertinent behavioral domains. In the third place, the question is asked as to the ontogenetic aspects of temporal regulations: when do certain temporal discriminations and behaviors emerge and are some of them perhaps already present at birth?

Shifting from animal to human behavior this question is picked up by Viviane Pouthas and discussed at some length in chapter 6, *Timing Behavior in Young Children*. Her research contribution seeks to establish at what age temporal discriminations reach a stable level in the course of human development. The method of operant conditioning is used in order to avoid complicated cognitive procedures that young children may not yet have available. The advantage of the paradigm of differential reinforcement of long delays which is used throughout, is that it is ›ecologically relevant‹: after all, being human involves much waiting for things to happen.

It is shown that even very young children can refrain from responding too early, although with considerable variability. Collateral behavior is a way of coping with delays, and such behavior is already found in two-year old children. This collateral behavior can be seen to develop gradually into counting behavior and ultimately into ›chronometric‹ counting.

Part II - Processes: The Perception and Retention of Time (Chapters 7-12)

This part of *Time, Mind, and Behavior* deals first with the metric aspects of temporal information: its amount and the ways of coping with different amounts. In chapter 7, entitled *Time Psychophysics and Related Models* Françoise Macar presents three bodies of core data, plus the models to account for them, in the framework of three basic psychophysical points. Having introduced a number of methodological issues, she first turns to the absolute threshold for duration and the concept of the perceptual ›time quantum‹, which remains one of the puzzling phenomena of time psychology. In the second place the discrimination of duration and the shape of the psychophysical function are taken into consideration. Weber's law is found to hold only under certain heavy constraints, and the nature of the psychophysical function remains somewhat unclear, although the results seem to favor a linear model more than anything else. Thirdly the relation between temporal and nontemporal (modal) information is considered in the light of the question how information about duration is coded. The author's emphasis is on psychological (behavioral) data, although she extends her considerations also into the domain of psychophysiology in a very interesting and promising way.

Whereas in this chapter the discussion is pertinent to single trial or ›static‹ situations, the chapters 8 and 9 deal with the dynamic or sequential approach to the psychophysics of time. The two authors have adopted a somewhat similar approach although they raise quite different issues.

First, Michelangelo Flückiger in *The Effects of Time Pressure on Duration Discrimination* describes a sequential paradigm in which two lights are presented side by side alternating within a fixed cycle with a length of 300 ms. Initially the left-right fractions are exactly 50:50, but then a linear gradient is introduced, away from equality, at various rates. As soon as a subject perceives the irregularity (which appears as something that can be described by the term ›galoping‹) he or she will press a button, which will then reverse the gradient.

The author shows that the rate and the precision of the subjects' behavior are interrelated in a complex way that is, however, not predictable from the static ›single presentation‹ type of experiments that abound in psychophysics. He concludes from his data that temporal information processing is very much a matter of integrating information over time and of filtering out other, irrelevant, temporal cues. The rate of events may introduce ›time pressure‹ when much integration or filtering is required.

A similar experimental approach is adopted by Gert M. ten Hoopen in chapter 9, entitled *The Detection of Anisochrony in Monaural and Interaural Sequences*. It deals with the fact that when pulses in a train are presented alternatingly to either ear (interaurally), the perception of their number or rate turns out to be consistently lower than when presentation is to one ear (monaurally). This effect occurs under a considerable range of circumstances, including the ›stop-reaction task‹ in which a subject is required to react as quickly as possible when an isochronic series of clicks or tones suddenly stops. The author finds evidence for a substantial ›slowing down‹ of mental time.

Chapter 10 by William K. Estes, deals with *Memory* for *Temporal Information*. The paper first discusses six different research paradigms that have given us insight in the ways temporal information is encoded and retained in memory. Convergent research findings are grouped under four major headings. First there is the well known overestimation of short, vs. underestimation of long intervals, probably the oldest experimental result in time psychology (Vierordt, 1868) but looked at in a new light in this chapter. The second theme considers the various ways in which estimates of related vs. unrelated items are produced on the basis of memory contents. The third has to do with the repetition of items: if an item is repeated in the course of a list, its judged recency is greater, that is, it seems more recent. Finally, information seems to get lost by losing its temporal specificity, and this leads to uncertainty gradients influencing temporal judgments.

Models and interpretations are offered for these phenomena. The author considers strength theories, explaining why he is not very fond of them. Encoding theories, which include his own encoding perturbation model do better in his eyes, especially when combined with some kind of search model. The last category discussed, and the least developed thus far, are the memory organization models.

In chapter 11, *Contextual Coding in Memory*, Richard A. Block deals with memory organization from a special point of view. Block holds that remembered duration has to do with the changes in the context in which encoding of information takes place. The more dynamic the context, the longer the remembered duration. Experiments are presented which support this view and the author has shown that Ornstein's (1969) so called ›storage size‹ theory of memory for duration is not tenable in its original form. The qualitative character of the concept of ›context‹ as it appears here and the author's attempts to define the relevant situational parameter do not yet allow very specific predictions: if a certain parameter of the situation does not work, then by definition it is not a contextual change, a potentially vicious circle, which the author progressively succeeds in eliminating. The problems raised in Block's chapter are closely related to those discussed in chapter 12 which ask the timely question *Is the Processing of Temporal Information Automatic or Controlled?* In this chapter Janet L. Jackson considers the coding of temporal information. The view that temporal information is being processed automatically as a byproduct of other information processing has acquired some popularity in recent years. There is, however, only little evidence for it, and most of it is of a rather piecemeal nature. The alternative position, that much temporal information processing is indeed deliberate and related to the effectiveness of rehearsal strategies and attempts at organization is defended. It is shown how and how much certain strategies affect different temporal tasks. The author stipulates that not all temporal tasks appear to tap the same underlying encoding and retrieval processes: different tasks require qualitatively different procedures.

Part III – Patterns: The Structure and Organization of Time (Chapters 13-17)

In chapter 13, Mari Riess Jones considers the *Structural Organization of Events in Time*. Her paper looks at the research on the structural principles of the organization of strings of events from three theoretical points of view. The first she calls rate-relational. It is mostly a psychophysically inclined stance, although it includes some implicit principles of coding, or parsing. Structural aspects are, however, mostly restricted to restatements, but not explanations, of Gestalt principles. The second approach is the structural information theory as it is known from, among others, Simon's (1972), Restle's (1970) and Leeuwenberg's (1969) work. The problem is, as Jones points out, the difficulty of connecting the structural codes to processing principles. This is achieved in principle in the last of the three approaches: the Dynamic Serial Transformation approach which represents Jones' own current position. It maintains that codes, which of themselves never have a unique interpretation, derive an appropriate interpretation from their own temporal structure.

Time, Rhythms and Tension is the topic discussed by Dirk-Jan Povel in chapter 14. Povel develops a model by which the perception of a rhythm as interesting or tension creating, can be explained. The basic idea is that subjects set up an internal beat and that rhythmicity is then derived from the discrepancy between what they expect on the basis of this beat, and the actual rhythm; to the extent that beats are displaced the tension rises. As an illustration of his ›displaced beat hypothesis‹ Povel describes an experiment in which subjects were presented pairs of rhythmic patterns. Of each pair they had to indicate which pattern they experienced as the most rhythmical. The actual choices are largely predicted by the tension scores.

In chapter 15 Henry Shaffer discusses four typical examples of *Timing in Action*: handwriting, musical performance, speech, and typing, and derives a number of generalizations from them. In particular the question of motor programming is discussed, and it is argued that in many cases time need not have a direct representation in the programming of movement. If it plays a role it is mostly in the frame of larger chunks; that is, when movements are connected that themselves have a rather ballistic character.

Expressive timing is discussed separately, and in this part the author suggests that in musical pattern there may be intrinsic temporal and melodic groupings that invite a certain performance, even at first sight. This is an important suggestion, in line with Jones' dynamic serial transformations, and some other related suggestions brought up by Michon (1985a) in chapter 2 and Estes (1985) in chapter 10.

Chapter 16, *A Functional View of Prosodic Timing in Speech* by Sieb G. Nooteboom ties in quite closely with some of the topics raised by Shaffer. It sets out to understand timing in speech on the basis of the plausible functional requirement that the listener should be able to understand what is being said. Timing thus serves the purpose of maximizing the ease of analyzing and coding of the speech signal by the listener, and in order to achieve this the speaker will time his utterances in accord with the expectations he holds about the listener and the quality of the acoustic environment. Speech rate and articulatory precision are to a large extent based on the principle of minimal effort. This allows a number of highly specific predictions about speech timing that, unfortunately, are not always borne out in a straightforward fashion. Additional (but equally plausible) assumptions need to be imported to account for the observed outcome, which demonstrates that the functional approach in its present formulation requires further refinement.

In chapter 17, entitled *Time, Size and Shape in Writing,* Ar J.W.M. Thomassen and Hans-Leo Teulings take a close look at handwriting. On the basis of pseudo-writing movements an analysis is made of the dynamic variables that can potentially serve as the parameters of a motor program that control writing movements at the macro- (word), meso- (letter), and micro- (stroke) levels. Plausible candidates are force, time and distance. Different parameters appear to operate at the three levels.

A general conclusion drawn by the authors, and one that support similar conclusions drawn by Shaffer (1985) in chapter 15, is that *time* is not likely to be coded in motor command structures that control complex movements. Instead temporal precision seems to be a natural outcome of well tuned, smoothly performing output systems.

Part VI – Notions: The Concept and Meaning of Time (Chapters 18-20)

The *Semantics of Time*, and its connections with temporal logic are reviewed by Johan van Benthem in chapter 18. The author first traces two lines of development that are of relevance to the question of temporal semantics: philosophy of science and logic.

The semantics of time is demonstrated to consist of three elements: a grammar for the language in which temporal expressions are stated, a ›mathematical ontology‹ of time (i.e. a structure for representation), and a ›systematic recipe‹ that connects both formal systems. A preliminary conclusion of the author is that language constrains temporal structures but it does not fully determine them; any particular language allows the choice of various ordering patterns. It is not implausible to assume, however, that at a deeper linguistic level further constraints are operating.

Van Benthem is also concerned with the relation between logic and the structure of time: to what extent does logic prescribe temporal structure and to what extent does the actual temporal structure determine the possible logics? An example is the important distinction between point logic and interval logic. These are closely related and can, in principle, be derived from each other although psychologically they do lead to quite different results.

Chapter 19, *The Development of Temporal Inferences and Meanings in 5-8 Year Old Children* by Jacques Montangero deals with the development of temporal reasoning and with the constituent components of the concept of duration as they develop with age.

A conceptual model is proposed which incorporates three different ›frames of reference‹ or ›subsystems‹, *duration* (time interval) being the point of intersection between these frames. Each of the three subsystems comprises three ›meanings‹ or cues that are closely interrelated. Important questions raised by the author include whether the three subsystems of the model are sometimes intermingled so that the various ›meanings‹ get mixed up, causing systematic confusions and errors of judgment.

Eight year old children are able to deal with the three ›meanings‹ within one subsystem, younger children tend to work with pairs only. The study reveals that duration is certainly not a unitary concept, but can be described in terms of partial combinations of the basic triadic model.

The relation between *Temporality and Metaphor* is discussed by John A. Michon in the final chapter 20. This paper takes as its point of departure the complex conceptualization of time as it has been proposed by J.T. Fraser (e.g. 1978, 1982). Fraser has adopted a genetic view of time, which – on a cosmic scale – incorporates five hierarchical forms or levels of temporality, each related to a particular ›state of nature‹. Analysis shows that these five ›states of nature‹ and the corresponding levels of temporality may be conceived as deriving from basic views of the world that people apparently hold. Such views, known as root metaphors (e.g. Pepper, 1942), are interpretations of the ›real world‹ that generate concrete representations of events and event sequences (histories). The levels of temporality as identified by Fraser correspond to particular measurement scales that have been identified in psychophysics as an (exhaustive) set of canonical scale types.

References

Adams, J.A. Motor skills. *Annual Review of Psychology*, 1964, *15*, 181-202.
Aschoff, J. Circadian timing. In: J. Gibbon & L. Allan (Eds.), *Timing and time perception*. Annals of the New York Academy of Sciences, Vol. 423, 1984, pp. 442-468.
Benthem, J. van. Semantics of time. In: J.A. Michon & J.L. Jackson (Eds.), *Time, mind, and behavior*. Heidelberg: Springer Verlag, 1985, pp. 266–278.
Brown, F.M., & Graeber, R.C. *Rhythmic aspects of behavior*. Hillsdale, NJ: Lawrence Erlbaum Associates, 1982.
Creelman, C.D. Human discrimination of auditory duration. *Journal of the Acoustical Society of America*, 1962, *34*, 582-593.
Demany, L., McKenzie, B., & Vurpillot, E. Rhythm perception in early infancy. *Nature*, 1977, *266*, 718-719.
Dennett, D.C. *Brainstorms: Philosophical essays on mind and psychology*. Hassocks, Sussex: Harvester Press, 1978.
Eisler, H. Experiments on subjective duration 1868-1975: A collection of power function experiments. *Psychological Bulletin*, 1976, *83*, 1154-1171.
Estes, W.K. Memory for temporal information. In: J.A. Michon & J.L. Jackson (Eds.), *Time, mind, and behavior*. Heidelberg: Springer Verlag, 1985, pp. 151–168.
Fraisse, P. *Psychologie du temps*. Paris: Presses Universitaires de France, 1957.
Fraisse, P. *The psychology of time*. London: Eyre and Spottiswoode, 1964.
Fraisse, P. *Psychologie du temps*. Paris: Presses Universitaires de France, 1967 (2nd edition).
Fraser, J.T. *Time as conflict: A scientific and humanistic* study . Basel: Birkhaeuser, 1978.
Fraser, J.T. *The genesis and evolution of time: A critique of interpretation in physics*. Amherst: University of Massachusetts Press, 1982.
Gale, R.M. (Ed.), *The philosophy of time: A collection of* essays. Hassocks, Sussex: Harvester Press, 1978. (Reprint of a 1968 edition).
Gibbon, J., & Allan, L. (Eds.), *Timing and time perception*. Annals of the New York Academy of Sciences, Vol. 423, 1984.
Groos, G., & Daan, S. The use of biological clocks in time perception. In: J.A. Michon & J.L. Jackson (Eds.), *Time, mind, and behavior*. Heidelberg: Springer Verlag, 1985, pp. 65–74.
Jones, M.R. Structural organization of events in time. In: J.A. Michon & J.L. Jackson (Eds.), *Time, mind, and behavior*. Heidelberg: Springer Verlag, 1985, pp. 192–214.
Kristofferson, A.B. Low-variance stimulus-response latencies: Deterministic internal delays? *Perception and Psychophysics*, 1976, *20*, 89-100.
Kristofferson, A.B. Quantal and deterministic timing in human duration discrimination. In: J. Gibbon & L. Allan (Eds.), *Timing and time perception*. Annals of the New York Academy of Sciences, vol. 423, 1984, pp. 3-15.
Leeuwenberg, E.L.L. Quantitative specification of information in sequential patterns. *Psychological Review*, 1969, *76*, 216-220.
Leeuwenberg, E.L.L., & Buffart, H.E.J.M. (Eds.). *Formal theories of visual perception*. New York: Wiley, 1979.
Linschoten, J. *Idolen van de psycholoog*. Utrecht: Bijleveld, 1964.
Mandler, G. *Mind and emotion*. New York: Wiley, 1975.
Massaro, D.W. Time's role for information, processing and normalization. In: J. Gibbon & L. Allan (Eds.), *Timing and Time perception*, Annals of the New York Academiy of Sciences, Vol. 423, 1984, pp. 372-384.
Michon, J.A. *Timing in temporal tracking*. Assen: Van Gorcum, 1967.
Michon, J.A. Processing of temporal information and the cognitive theory of time experience. In: J.T. Fraser, F.C. Haber, & G.H. Müller (Eds.), *The study of time*. Heidelberg: Springer Verlag, 1972, pp. 242-258.
Michon, J.A. The compleat time experiencer. In: J.A. Michon & J.L. Jackson (Eds.), *Time, mind, and behavior*. Heidelberg: Springer Verlag, 1985a, pp. 20–52.

Michon, J.A. Temporality and metaphor. In: J.A. Michon & J.L. Jackson (Eds.), *Time, mind, and behavior.* Heidelberg: Springer Verlag, 1985b, pp. 288–296.
Ornstein, R.E. *On the experience of time*. Hammondsworth, Middlesex: Penguin Books, 1969.
Park, D. Brain time and mind time. In: J.A. Michon & J.L. Jackson (Eds.), *Time, mind, and behavior.* Heidelberg: Springer Verlag, 1985, pp. 53–64.
Pavlov, I.P. *Conditioned reflexes*. New York: Dover Publications Inc., 1960 (2nd edition).
Pepper, S.C. *World hypothesis: A study in evidence*. Berkeley: University of California Press, 1942.
Pick, H.L., & Saltzman, E. (Eds.), *Modes of perceiving and processing information*. Hillsdale, NJ: Lawrence Erlbaum Associates, 1978.
Povel, D.J. Time, rhythms and tension: In search of the determinants of rhythmicity. In: J.A. Michon & J.L. Jackson (Eds.), *Time, mind, and behavior.* Heidelberg: Springer Verlag, 1985, pp. 215–225.
Restle, F. Theory of serial pattern learning: Structural trees. *Psychological Review,* 1970, *77,* 481-495.
Richelle, M., & Lejeune, H. (Eds.). *Time in animal behaviour.* Oxford: Pergamon Press, 1980.
Shaffer, L.H. Timing in action. In: J.A. Michon & J.L. Jackson (Eds.), *Time, mind, and behavior.* Heidelberg: Springer Verlag, 1985, pp. 226–241.
Sherover, C.M. *The human experience of time*. New York: New York University Press, 1975.
Simon, H.A. Complexity and the representation of patterned sequences of symbols. *Psychological Review,* 1972, *79*, 369-382.
Sivadjian, J. *Le temps: Étude philosophique, physiologique et psychologique*. Paris: Hermann, 1938.
Smart, J.J.C. (Ed.), *Problems of space and time*. London: MacMillan, 1964.
Sternberg, S., Knoll, R.L., & Zukofsky, P. Timing by skilled musicians. In: D. Deutsch (Ed.), *The psychology of music*. New York: Academic Press, 1982, pp. 181-239.
Thomassen, A.J.M.W., & Teulings, H.-L. Time, size and shape in handwriting: Exploring spatio-temporal relationships at different levels. In: J.A. Michon & J.L. Jackson (Eds.), *Time, mind, and behavior.* Heidelberg: Springer Verlag, 1985, pp. 253–263.
Titchener, E.B. *Experimental psychology,* Vol. 2. London: MacMillan, 1905.
Toda, M. The boundaries of the notion of time. In: J.T. Fraser, N. Lawrence, & D. Park (Eds.), *The study of time III*. New York: Springer, 1978, pp. 370-388.
Turvey, M.T. Contrasting orientations to the theory of visual information processing. *Psychological Review,* 1977, *84*, 67-88.
Vierordt, K. *Der Zeitsinn nach Versuchen*. Dissertation Universität Tübingen, 1868.
Wendorff, R. *Zeit und Kultur: Geschichte des Zeitbewußtseins in Europa*. Wiesbaden: Westdeutscher Verlag, 1980.
Whitrow, G.J. *The natural philosophy of time*. London: Nelson, 1960.
Zern, D. The influence of certain developmental factors in fostering the ability to differentiate the passage of time. *Journal of Social Psychology,* 1967, *72*, 9-17.
Zwart, P.J. *About time: A philosophical inquiry into the origin and nature of time*. Amsterdam: North Holland Publishing Company, 1976.

Part I. Origins: The Nature and Development of Time

Chapter 2. The Compleat Time Experiencer

John A. Michon

Prologue: What is it Like to Be a Time Experiencer?

In a well known article about consciousness Thomas Nagel (1974) raised the fascinating question what it would be like to be a bat. A similar question can be asked with respect to human time consciousness. If our experience of time would turn out to be something that is conceptually unitary, in the sense that it can be defined by a distinct set of related events, attributes, or processes, then it must mean something, however little, to be a Time Experiencer.

If we practice psychology as a ›science of the artificial‹, to use Herbert A. Simon's (1969) phrase, the study of time becomes essentially a matter of designing a temporal interface between the human organism and the external world. The ultimate problem of time psychology then is to demonstrate that the human mind operates in such a fashion that it can cope with the temporal contingencies of its natural and self-created environments and that, at the same time, it does produce the experiential appearances of time which conscious reflection and observation reveal. This would have to include such ›products‹ as the conscious experience of present, past and future, the rate at which time is felt to pass, but also the possibility of such ›pathologies‹ as *déjà-vu* and *stasis*, viz. the apparent freezing of time. And it would also have to deal with the insight that some of these ›products‹ may themselves serve as aids in the coping process. This problem statement leads me to the following working definition of psychological time:

Time is the conscious experiential product of the processes which allow the (human) organism to adaptively organize itself so that its behavior remains tuned to the sequential (order) relations in its environment.

The remainder of this chapter is aimed at clarifying this definition in the context of a general overview of some of the pertinent issues in time, mind, and behavior.

The famed Japanese psychologist Masanao Toda once said: »Obviously, trying to ›define‹ time is a fool's errand. To define a notion is to find for it an equivalent ideational construct made of some other, usually more primitive, notions (...) Any attempt to define time, therefore, is bound to be ridiculous, since nothing in this world even remotely resembles time« (Toda, 1978, p. 371/2). In order to avoid the impression that I am about to embark on a fool's errand, I wish to point out that we need not define time at all in the manner suggested by Toda. Instead time may be defined operationally, namely by what it is that generates our experiences and notions – actual and potential – of time. That is, we may attempt to explain time in terms, not of notions, but of mechanisms and processes, specifically those that provide adequate tuning of the organism to its environment.

The Physical Fundament of Time Experience

Time appears to imply more than is structurally available in reality. It primarily seems to constitute a quality of experience. Indeed there has been a persistent debate among philosophers about the ultimate status of time. Is time a mental construction, an illusion if you like, imposed on reality as human experience it? Or is it (also) a fundamental property of the physical world, independent of who or what is looking at it?

The Reality of Time's Arrow

The debate about the status of time as a property of nature is vacillating between the poles of Becoming and Being which, for brevity's sake, may be qualified as the respective positions pro and con the intrinsic directionality of time. The issue is simply whether or not sequential order is a structural feature of the universe, also known as the problem of ›time's arrow‹. In this debate physicists have, by and large, taken sides with the ideologues of Being, consistently trying to eliminate temporality as an inherent property of the universe from their theories. Not only is Newtonian mechanics, the fundament of classical physics, neutral with respect to the direction of time, but Einstein's relativistic mechanics is so too.

Even the evidence for a genuine break of time symmetry, among other things contained in the second law of thermodynamics, has been attacked. Gibbs (1902), for instance, argued that the intrinsic time-asymmetry of thermodynamics must be an illusion caused by the poor acuity of the observer's sense organs. If we throw a dash of milk into a cup of tea and stir, the result will be a most distasteful looking, opaque liquid that cannot be unstirred anymore. Looking through a microscope, however, we would still observe heterogeneity: milk droplets in an ocean of tea. Although Gibbs' objection has not stood up to criticism, arguments like his have induced Prigogine (1980) and others to propose that a fundamental time asymmetry must exist even at the quantum level. But this proposal, in turn, has met with criticism too. Davies (1981), who otherwise appears to have no difficulty in accepting the idea that the break of time symmetry is real, has qualified Prigogine's work as an attempt »to find a mysterious extra ingredient, absent in hitherto known physics, to make the world inherently asymmetric as it evolves« and this, in Davies' view, »entails putting the asymmetry in by hand« (p. 69).

And so the battle rages on, most recently in a domain known as quantum gravity, where the very large and the very small meet (see e.g. DeWitt, 1983; Freedman & van Nieuwenhuizen, 1985). If the effort to unify gravity with the other three fundamental forces of nature into one geometrical framework would meet with success, physics would once more eliminate time. But the fundamental geometrization of physics would do more. If what happens in the universe can be accounted for directly by the global geometrical properties of spacetime, rather than by the local attributes of objects or events, the need for an observer ›interpreting‹ these objects and events would evaporate at the same time.

This brings me close to some of the points raised by David Park (1985) in chapter 3 of the present volume. Park argues that since physics cannot do without the concept of a conscious observer, for reasons which he explains, psychology *must* play a role in the construction of physical theory, whether physicists like it or not.

It occurs to me, however, that the break in time-symmetry keeps turning up faster than physical theory succeeds in explaining it away. This leaves me with a strong feeling that time must indeed be an inherent property of nature, a fundamental break of symmetry, justifying in an objective way the distinction between earlier and later. Furthermore, ›time's arrow‹ is seen to touch immediately upon the problem of ›The Observer‹ in physical theory, an entity that is required to provide an *interpretation* of what happens and of the sequence in which things happen. Perhaps it is therefore safe to conclude that we must assume sequential order to be an inherent property of reality, but that it takes a conscious observer to recognize it.

Psychological Extras

Assuming that the order of events is indeed physically determined (albeit perhaps under certain local constraints such as the particular inertial frame of reference of the observer), the task of the time psychologist is simply to establish how this ›real‹ order is coded and represented, what mechanisms are involved in the process, and how these produce the phenomenology of time experience discussed earlier. Unfortunately the matter is not simple at all. As Davies (1981, p. 63) pointed out, »if our conception of reality is based on our experience of time ... it is seriously at odds with the external world whose reality we are concerned with ... Psychological time possesses apparent qualities that are absent from the ›outside‹ world of the laboratory. This additional structure consists of an awareness of a now or present moment, and an impression that time passes.«These two qualities, which I shall call *now* and *flow*, are additions to the world as we perceive it in absence of specific physical stimuli that could possibly generate them. As such they seem to constitute a sort of ›minimal set‹ of the experiential modes of time, although this privileged status does not *a priori* exclude other qualities that would turn out not to be reducible to either *now* or *flow*.

The initial problem of time psychology, formulated in the prologue has now been reduced, at least provisionally, to the following goal: establish how ›real‹ order is coded and represented, and what mechanisms are involved in this process, such that *now* and *flow* result as attributes of experience. Instead of directly pursuing this goal, I shall adopt a rather more circumambulant approach. After specifying in more detail what people mean when they say they are experiencing time I shall consider why and how Time Experiencers, such as humans, can function when dealing with an ordered environment. This will lead to the conclusion that a biopsychological explanation in terms of adaptive mechanisms for temporal control does not carry us far enough. A different domain of discourse is required, leading into a consideration of time as information – and humans as information processing systems. Beyond that, however, a third level of theoretical consideration will be needed, one that can accommodate the structural properties of the events that are to be coped with by the individual Time Experiencer.

The Phenomenology of Time Experience

For a theory that can account for time-as-experienced as well as for the tuning processes that generate this experience to be successful, we must first specify the proper domains of discourse for each aspect. For that purpose, time-as-experienced will be considered from an intentional stance, while the underlying processes will be looked at later from a functional point of view.

Intentional Systems

A behaving system is called intentional and someone observing such a system is said to adopt an ›intentional stance‹, if it is possible to ascribe to it rational beliefs, or feelings, or intelligence and if on that basis only it is possible to predict the behavior of that system (Dennett, 1978). Whether or not the system under concern ›really‹ has these beliefs, etc., is immaterial. Moreover, the intentional stance does not require statements about the functions and processes that generate the observed (rational) behavior. Actually, mixing intentional and functional theory is considered extremely bad practice and it should therefore be rigorously avoided (Herrmann, 1982; Michon, 1984). The reason is that the resulting mixture will necessarily be infested with *homunculi*, a pernicious race of question begging entities: »Whenever a theory relies on a formulation bearing the logical marks of intentionality, there is a little man concealed«, Dennett warns us. (1978, p. 12).

In Dennett's view »intentional theory is vacuous as [scientific] psychology because it presupposes and does not explain rationality or intelligence« (1978, p. 15). This does not imply that intentional theory as such is inadmissible. Consistent intentional theory, genuine phenomenological explanation capable of predicting behavior, is perfectly well feasible and quite common in the social sciences, but ultimately it cannot count as scientific psychology for the stated reason. Scientific psychology – or perhaps we should say *psychonomics* – must therefore occupy itself with the elimination of intentional theory by replacing it with functionalistic explanation by adopting what Dennett has called the ›design stance‹ or the ›sub-personal stance‹.

There is a second way of eliminating intentional theory, namely by means of structuralistic explanation; incidentally, this is precisely the way in which physicists have been trying to get rid of time. As we shall see later in this chapter, however, that road is blocked with another type of difficulty: structural theory does not necessarily qualify as psychonomics either (Michon, 1984).

Dennett (e.G. 1981) has come to distinguish two levels of intentional theory. At the first level we find the naive behavioral explanations that are part of ›folk psychology‹ and at the second level we have the expurgated versions thereof, namely those explanations which have passed critical examination by philosophers, social scientists and humanists. I shall briefly consider both versions in the light of our present topic, time experience.

Unexpurgated Descriptions of Temporal Experience

Time figures prominently in experience. Quite a number of temporal impressions that commonly occur in everyday life are fairly dramatic or at least accessible even to naive introspection. The *déjà-vu*, for instance, is experienced occasionally by most people, as is the feeling of ›queer coincidence‹ which C.G. Jung (1960) described in his study on synchronicity, an acausal structural connection between simultaneous events. The powerful impact of rhythm on behavior seems to be universal. Time experience is also drastically affected when people are involved in stressful or blissful events. And there are many other equally conspicuous manifestations of time. In a thoughtful paper Gorman & Wessman (1977) have discussed the rich variety of temporal concepts and expressions that ordinary people use when discussing time in their daily lives. These range from the simple, familiar representations of time as a line or a circle, and time as money or boredom, to such fancy metaphors as ›time is a shooting star‹, an ›everbranching tree‹, a ›Bach cantata‹, or a ›chronic thief‹ (Wessman & Ricks, 1966, pp. 117-120). The phenomenology provided by folk psychology is simply overwhelming, and many authors have indeed argued that experiential time, in their opinion, cannot be a unified concept. In short, there appears to be quite a lot more in everyday experience than just *now* and *flow*.

Artist's Impressions: Time and Narrative

To whom should we turn when it comes to Dennett's expurgated variety of ›folk psychology‹? There seem to be two roads to travel: art and science.

Time is thoroughly anchored in art, not only in the so called temporal arts – music, dance and cinematography – but also in painting (e.g. Baudson, 1984; Goodman, 1984) and literature (Ricoeur, 1983). The significance of the time artists resides in their attempts to *manipulate* time in their self-created universes in ways that are internally consistent and that, consequently, appear plausible to the spectator or the reader. Thus, the appreciation of a work of art resembles the thought experiments *(Gedankenexperimente)* in science, and the requirement of internal consistency qualifies the genuine work of temporal art as an expurgated variety of intentional theory. As an example let us consider time in narrative, the home territory of the time novelist.

A time novelist is someone who composes narrative in which explicit time experiences of the protagonists are crucial determinants of the action and the context. Such authors – among them true celebrities like Samuel Beckett, Marcel Proust, J.B. Priestley, Thomas Mann and John Fowles – can be seen to exercise a threefold command over temporal matters. In the first place the time novelist makes his or her characters experience time in a natural way, but on a magnified scale: they are subject to, but also intensely aware of, day and night, the seasons, their lives passing, and so on. Even their decisions to leave a party or to take the early train are seen in the existential light of time: order, delay, and »how long it takes«. In the second place, in order to evoke such experiences in a character, the author creates conditions in which the experiences are plausible. In other words, the protagonists must be endowed with belief structures and behavioral goals that will closely and rationally accommodate their patterns of activity (or perhaps sometimes deliberately not, as in Gogol's *Diary of a Fool*). Errors destroy the narrative: even naive readers are quite sensitive to ruptures in the fabric of story time. Finally the author creates, by stylistic means, concurrent temporal experiences in the reader and thus, among other things, enhance the reader's identification with one or more characters. Authors can induce feelings in an audience of time passing when the protagonist feels it pass, or time dragging when the story requires it to drag, and so on. In other words, the narrative does not only describe temporal facts and relations, but it wishes to evoke a temporal experience in the reader that will fit the intrinsic, generative structure of the narrative as needed (see also Michon 1985, chapter 20 of the present volume).

In this context it is interesting to mention John Fowles' observation that writers are essentially playing godgames. Such godgames are essentially *time games*, as Fawkner (1974, p. 118) remarked, because from a fundamental dissatisfaction with the world as it is, the writer creates »a past that never was ... Even the simplest and shortest act of literary text, as brief as a *haiku*, is a surreptitious bid for immortality, or freedom from ordinary time«. (Fawkner, 1984, p. 10). If, perhaps, ancient cultures may not have felt the need for gods who were in command of time (Toda, 1978; see also Michon & Jackson, 1985, chapter 1 of the present volume), they seem to flourish in modern literature and theater, where the sacred unity of time, place and action has been abandoned.

Scientists' Expressions: Phenomenological Taxonomies

Although we may have to cope with an abundance of intentional descriptions of naive time experience, psychologists and anthropologists have tried to organize time experience in somewhat more parsimonious terms, aiming for a taxonomy of expurgated folk psychology.

Orme (1969), for instance, reduced the substantial variety of phenomenological time experiences to thirteen more or less independent classes of phenomena. Although his taxonomy incorporates most of the subjective, temporal phenomena, there is no underlying principle which could tie these various manifestions of time together.

Exactly the opposite is found in the beautifully symmetric taxonomy of time which does appear in a book on the *Patterning of Time* by the anthropologist Doob (1971). I cannot even begin to describe or explain the intricate relations that are implied by Doob's taxonomy. Fortunately Doob himself has provided a very compact summary (o.c. p. 407-409) to explain the diagrammatic representation of Figure 1. Faced with such complexity we stand in awe: Doob has certainly succeeded in providing an *orderly* picture. But, although it imposes order, it does not relate in a functional way to the empirical findings it is supposed to accommodate.

Rich and Poor Phenomenologies: Can Time Be Modeled?

Although we are apparently living with an extremely rich temporal world we need perhaps not give up our attempts to organize time experience in somewhat more parsimonious terms. Gorman & Wessman (1977) have indeed proposed that we do not necessarily have to adopt a pluralistic view of time experience: »We should view temporal symbols not as distorted or altered perceptions of a determinate thing called ›time‹ but rather as representations chosen and constructed by individuals as apt expressions of their own life situations and feelings. Not only do the symbols used reflect the cultural context in which the person is operating, it seems even likely that he will use different concepts and symbols depending on particular activities that he is engaged in. Each may require a different approach to temporal information and consequently a different set of cognitive operations. This would turn temporal concepts into state variables, pointing to the type of operation being performed on temporal reality when the person is in a particular state« (p. 238).

Stated in terms that are relevant to our discussion thus far, we are facing the question: How easy is it to construct a functional model of the adaptive tuning process, such that this model will appear as a convincing ›time experiencer‹ to an onlooker adopting the ›intentional stance‹? The answer depends to a large extent on how easily people will ascribe intelligent or rational temporal beliefs and attitudes, such as ›punctuality‹ or ›future-orientedness‹, to such a model.

Generally speaking it is simple to design a functional model for a particular behavioral domain with respect to which observers are prepared to adopt the intentional stance, when the behaviors to be modeled have a *poor phenomenology.* For instance, we feel that people do not differ a great deal in what they are doing when they do sums or crosswords. Such activities have poor phenomenologies, and consequently there can be a fair amount of agreement as to the intelligence or the belief structures that we ascribe to a system that can perform such tasks. In contrast experiences which have rich phenomenologies, such as pain, or dreams, or consciousness, cannot easily be incorporated in models that will elicit the intentional stance in an arbitrary observer. If *your* ›real pain‹ is entirely different from *my* ›real pain‹ how could there ever be a robot to which both you and I are willing to ascribe the suffering of ›real pain‹?.

The question is therefore: how rich is the phenomenology of time experience ›really‹? If it is as rich as the preceding two sections suggest, that is, if my time is very unlike your time, or even worse, if my time *today* is very unlike my time *tomorrow,*

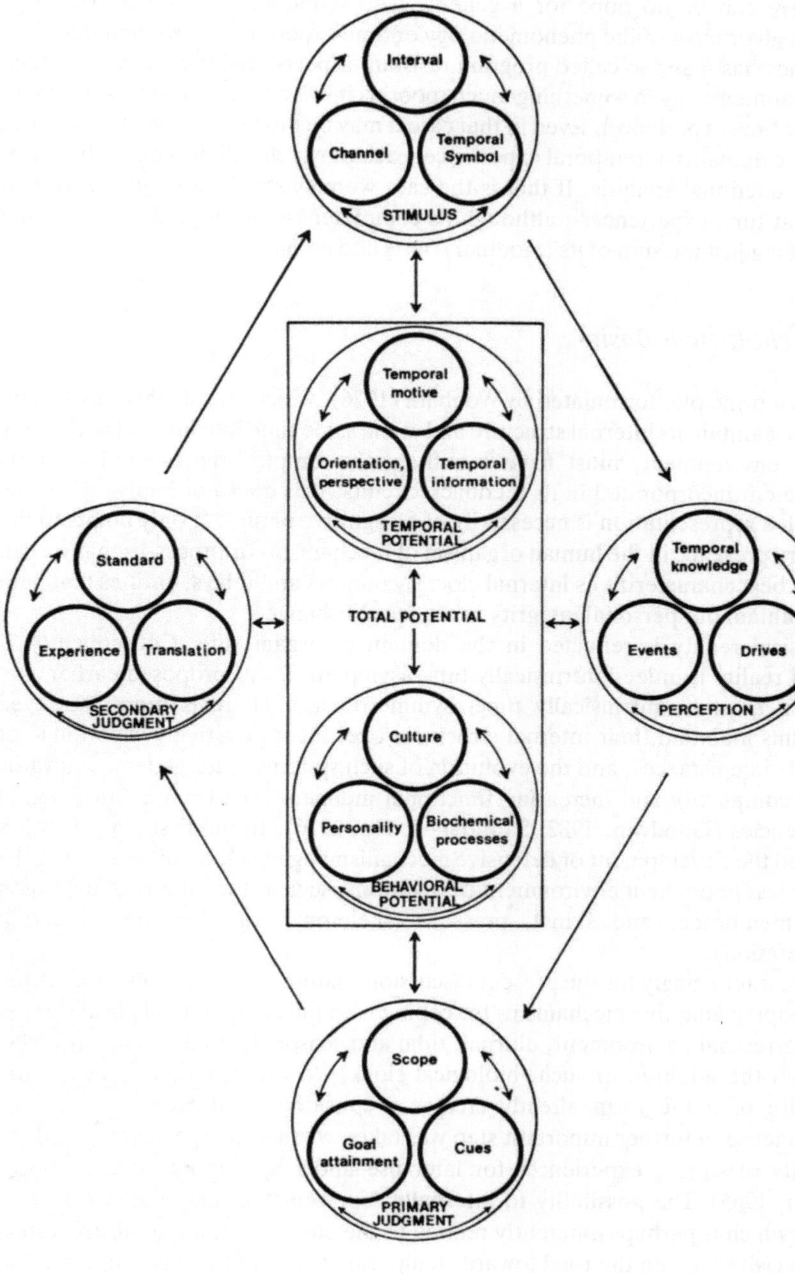

Figure 1. A taxonomy of time.
(From: Doob, 1971, p. 31; reprinted by permission).

then there can be no hope for a general process model that will convincingly generate all or most of the phenomenology of time experience. If we find that time experience has many so called program resistant aspects, and if we cannot reduce that phenomenology to something much ›poorer‹ it will be a hopeless task to design a general time experiencer. Even in that case it may be possible to specify a number of specific domains of temporal experience, each of which will turn out to be accessible to functional analysis. If that is the case we may still be able to construct a ›compleat time experiencer‹, although its completeness will be rather trivial, the whole being just the sum of its (modular) parts and no more.

Biopsychological Basics

There is a principle, formulated by Wonham (1976), which tells us that any system that is to maintain its internal structure and at the same time is to interact adaptively with its environment, must have a sufficiently complex representation of its environment incorporated in its feedback circuits. This does not imply, of course, that such a representation is necessarily of a cognitive nature. It only points to the necessary presence in the human organism of mechanisms or processes that we can perhaps best characterize as internal clocks, counters and relays, entities that help us to maintain our personal integrity in the face of change.

Physical reality is reflected in the domain of organic life. Consequently, if physical reality is indeed intrinsically time-asymmetric, as I proposed earlier, any life form must be intrinsically time-asymmetric too. There is more, however. Organisms maintain their internal structures over a comparatively wide range of external circumstances, and the evolution of such systems tends generally towards greater complexity and increasing functional independence from environmental contingencies (Goodwin, 1982; Saunders & Ho, 1976). In the first place this has provoked the development of defensive mechanisms against local differences in the more or less permanent environmental conditions, such as the salinity of the ocean (outer membranes and skins), pressure (skeletons), or ultraviolet radiation (pigmentation).

More interestingly for the present discussion, nature apparently also succeeded in developing adaptive mechanisms for coping with the comparatively fast changes in the terrestrial environment, diurnal, tidal and seasonal variations in particular. Although the addition of such ›biological clocks‹ to the genetic and behavioral repertoire of a life form already creates a considerable degree of functional independence, a further important step was taken when some species acquired the capability of storing experiences for later use under appropriate circumstances (Richter, 1965). The possibility to internally represent the temporal structure of events (which is perhaps inherently related to the emergence of neural structures) was a decisive step on the road towards temporal control of the environment, that is, towards the ability to negotiate the kinds of temporal contingencies that are not imbedded in such outspoken cyclical patterns of events as the day-night cycle or the solar year.

The process of gradual internalization of environmental variations in a form of ›internal representation‹ creates conditions that allow the optimalization of the organism's functions. At the same time, however, it also creates a new problem. In order to remain in pace with the flow of events in the outer world, a very sharply tuned ›interface‹ with the environment is required. At some point in time, namely the point which we call *now* or *present*, there must be a close-to-perfect correlation between what is happening outside the individual and the representation thereof ›inside‹. Any species that would be incapable of tuning its internal events to those in the outer world would stand as much chance in evolution as the proverbial soluble fish (and pretty much for the same reason).

Let us now turn to a closer consideration of the temporal tuning aspect that, in my view, constitutes the rock bottom for temporal experience. Tuning involves matching events in the outside world events with the corresponding mental events. This, in a very generic and abstract sense, suggests that tuning is in principle a complicated form of time measurement. Time measurement essentially involves two independent series of successive events, the measurement consisting of observing the simultaneity of an event (*p*) in one series with an event (*p*') in the other, followed by another observation of such a simultaneous event pair *q*, and *q*' (Fraser, 1982). Using one of the event series as a reference, the interval between the events *p* and *q* can be expressed in terms of the number or density of events between *p'* and *q'* or *vice versa*. Tuning can be described as the process of keeping track of the correspondence between events in the outside world and the events produced in an internal representation of that world: keeping the two series in synchrony is precisely what tuning is about.

The necessity of tuning lies at the root of the elementary experiences of *now* and *flow*. The interval over which the organism succeeds in directly relating successions of internal and external events determines the width of the *now* or ›specious present‹. Thus, the present may be seen as a dynamic interface, the ›window‹ which interfaces external and mental (internally represented) events. Depending on the state the organism is in and the structure of the event sequence, the width of this window may vary, and the length of the experienced ›present‹ will vary accordingly. At same time, since the matching of external and internal events involves time measurement in the proper sense, it may be assumed that the experience of time *flow* as fast or slow depends on the relation between these two series of events. If the internal events tend to be relatively earlier with respect to their corresponding external events than would be expected on a chance basis, time would seem to flow slowly, and similarly if the internal events would frequently be (too) late relative to their external counterparts, the corresponding experience would have to be one of time flowing fast.

A concomitant aspect of tuning is that it allows the organism to select a time base in synchrony with the ›pulse‹ or ›rhythm‹ of events, thereby freeing it to do other things in between the instants at which perfect coincidence is crucial. This is of great importance in skilled activities such as musical performance, handwriting and speech perception. The chapters by Povel (1985, chapter 14), Shaffer (1985, chapter 15), Nooteboom (1985, chapter 16) and Thomassen & Teulings (1985, chapter 17) discuss this point in considerable detail.

Mechanisms for Temporal Control

The nature and development of the means that made the higher animals and humankind relatively independent of the vicissitudes of a sequential environment are discussed at length in chapter 4 (by Groos & Daan, 1985), chapter 5 (by Richelle et al., 1985), and chapter 6 (by Pouthas, 1985) of this volume. These chapters deal with the biopsychological fundaments of psychological time, and from them we learn that the adaptation of complex organisms, such as humans, to the temporal requirements of their habitats depends on both endogenous and acquired periodicities (biological rhythms or clocks) that are deeply rooted in their physiology and their anatomy. They tend to be phylogenetically ancient, permanent and resistant to external driving except at a narrow range of frequencies. These ›clocks‹ appear to control temporal adaptations to events in the range from one hour upward. Shorter intervals, on the other hand, rely on a rather different type of temporal regulation, for which no predetermined ›clocks‹ are available. Instead, it appears that different species, but also different individuals of a particular species, and even the same individual on different occasions, select a ›time base‹ from a multitude of potentially available, recurrent or nonrecurrent counting and inhibitory delay (waiting) mechanisms.

Much of the temporal regulation of behavior already appears comparatively early in life. In young animals, as well as in human infants adequate temporal control can be observed at an early age, although stable temporal performance is usually acquired fairly late in life. In the human child this state is reached at about 8 years of age (see Pouthas, 1985, chapter 6 of the present volume). This would seem to imply that the selection of an appropriate time base from the available multitude requires a certain degree of maturity plus, no doubt, quite some practice.

Richelle & Lejeune (1980) and their associates have dealt with the various categories of temporal regulation in considerable detail. Broadly speaking they distinguish between physiological and behavioral mechanisms. Within the first category a further division is made between central mechanisms which in part seem to be located in the septal and hippocampal regions of the brain, or are at least observable in the electrical activity of the cortex. Somewhat disappointingly perhaps, there is little evidence for a relation between timing and such conspicuous periodical phenomena as the brain's alpha rhythm. Rather it seems that certain event related cortical responses, viz. the contingent negative variation, also known as the ›expectancy wave‹, correspond somehow to the ways in which a person is coping with the temporal requirements of the environment (see e.g. Macar, 1980; also Macar, 1985, chapter 7 of the present volume).

Among the peripheral physiological mechanisms discussed in Richelle & Lejeune (1980) heart rate plays a prominent role. The relation between the regularity of cardiac activity and mental effort has been thoroughly established (e.g. Mulder & Mulder, 1981). It has also been found that many simple motor responses occur in phase with the cardiac cycle. It remains unclear, however, whether this is in fact due to a specific pacemaker (Zeitgeber) function of the peaks in the cardiac response.

Within the second category of temporal regulatory mechanisms mentioned by Richelle & Lejeune (1980) three subgroups may be distinguished, to wit, proprioceptive mechanisms, collateral behaviors, and (external and internal) temporal cues. Proprioceptive temporal controls make use of the various feedback circuits that can be established in the sensori-motor system, and they play a fundamental role in skilled behaviors such as writing and musical performance. Collateral behaviors, the second subgroup in the category of behavioral mechanisms, are important for temporal control of behavior, although they bear no direct relation to the action that is being timed. Pacing up and down, scratching, finger tapping while waiting for something to happen constitute powerful aids in correct temporal performance. Constraining the subject so as to prevent this collateral behavior has a very negative effect on the temporal precision of the subject's actions, which is overcome only when a new collateral pattern can be developed within the prevailing constraints (Frank & Staddon, 1974; Glazer & Singh, 1971). Rather than being a kind of behavioral clock or a non-verbal way of counting – which pacing or fingertapping may easily appear to be – the received view of the functional significance of collateral behavior is that it serves to inhibit responses. In order that tension will not build up too high while the subject is waiting for something to happen or for an action to be taken, a (partial) release of tension is established by means of unrelated, permissible, so called ›displaced‹ behavior (see Richelle & Lejeune, 1980). As a final subgroup of temporal control mechanisms internal and external temporal cues should be mentioned. These include various forms of implicit or explicit counting, or more elaborate cognitive and environmental clocks and calendars.

The status as ›internal clock‹ of all physiological and behavioral mechanisms remains somewhat ambiguous. Many authors agree that subjects will select whatever mechanism will suit their needs (e.g. Macar, 1985, chapter 7 of the present volume): »multiple time bases are continuously constructed in response to the particular requirements of each situation and replaced by others when they become useless« (Richelle & Lejeune, 1980, p. 165).

Too Many Clocks

The number of different options available to the organism for establishing a time base for its performance in a particular timing task is apparently very large (Gooddy, 1958), and choices appear to be highly opportunistic (Richelle & Lejeune, 1980). Evidently we have reached the limits of a psychobiological explanation of human time experience in terms of ›clocks‹: there are no discernable rules that could explain why one ›clock‹ or ›clock system‹ would be preferred over another. A different frame of reference is needed, one which does enable us to understand how it is that different temporal conditions do elicit specific response patterns and strategies. Such a frame of reference is indeed available: *time as information*. And since we are aware of the fact that humans are a species of information processing systems it is appropriate to see if time experience can be understood as a manifestation of temporal information processing.

Time as Information

Behavior is under the control of time. Something is done simply because it is 1215 h rather than 1900 h, or because someone has waited 20 seconds rather than 10. In music and dance timing quality constitutes the difference between the delightful and the dreadful. Time, in other words, appears to have causative properties deriving from the temporal relations between the organism and its environment. Events that do not distinguish themselves except for their relative temporal position, as for instance in meter and rhythm, are found to have a strong effect on behavior when people are dancing or playing music. The power of rhythmicity can even be observed soon after birth. In short, temporal relations contain information and both humans and animals use this information to guide their behavior.

Fifteen years ago I formulated the ›equivalence postulate‹, explicitly stating the dual role of time: time as information in addition to its usual role as the ordering variable *t* in dynamic equations (Michon, 1972). In this view time has a status as relevant stimulation for the human organism that is not formally different from such extensive and intensive attributes as size, loudness or color. In other words, it is assumed that time is explicitly represented in the mind. At that time I understood this equivalence in a weak sense: stimuli were thought to possess an explicit temporal attribute. Since about 1977 this conception has been replaced by a much stronger equivalence; I now feel more strongly attracted to the idea that temporal information is encoded in a representational system of its own and that the encoding of sequential (or temporal patterns) takes place in a separate representational code which has come under study in the early seventies. Strong equivalence has found expression in the work, among others of Jones (1976, 1985, see chapter 13 of the present volume), and Anderson (1983) who entertains an interesting concept of temporal strings as a separate representational mode, to which I shall return later.

The Information Processing Paradigm

Assuming the independent character of temporal information implies that the general features of information processing theory should also be generally applicable to temporal information processing as well. In recent years psychologists have formulated a rather substantial and internally consistent paradigm, known as cognitive psychology, which not only accounts for an extensive range of mental and behavioral phenomena, but also carries a considerable formal potential since it is closely related to the general theory of computation which also lies at the basis of computer science. The number and variety of models that have been based on the cognitive paradigm is quite considerable, but fortunately the basic features are pretty much the same in all of them. The following capsule review has no other purpose than to introduce some of the basic features of this class of models to those readers who are not sufficiently familiar with this major trend in presentday experimental psychology. Much of the content of *Time, Mind, and Behavior* relies implicitly or explicitly on this or a related model (Figure 2).

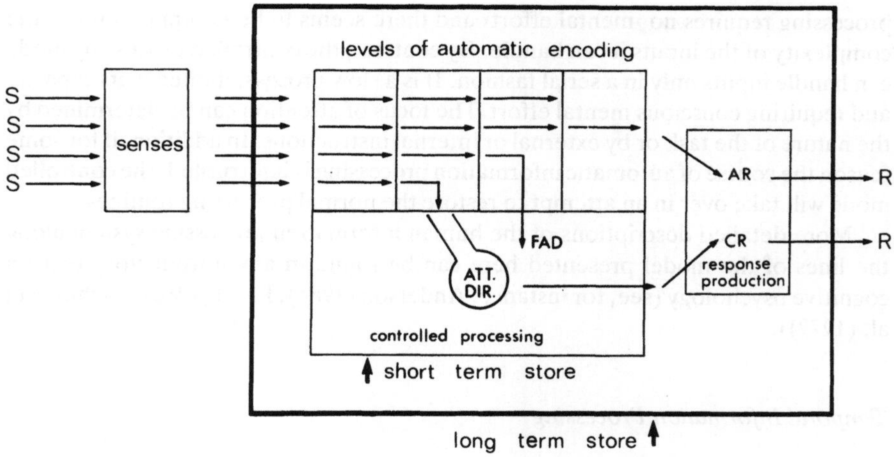

Figure 2. A general view of human information processing. Stimuli *(S)* reach the sensory systems (senses), which include iconic and echoic buffers. Subsequently information enters short-term or working memory, which is the active part of long-term memory. The input information is assumed to be analyzed at progressively deeper 'levels', much of it taking place automatically or habitually and leading to automatic responses *(AR)*. Attention is directed *(ATT.DIR.)* only to a small fraction of the information. This is the information that is processed under deliberate control and that may eventually lead to controlled responses *(CR)*. Since attentional capacity is limited some stimuli will not receive the attention they require, leading to a focal attention deficit *(FAD)*. (From van Zomeren, 1981; reprinted by permission).

Incoming stimuli (*S*) impinging on the receptor surfaces of the senses are subsequently buffered in a sensory register for a brief period of perhaps one second. In this memory buffer some superficial pre-processing filtering may take place. From the sensory register information is next transferred to a ›working memory‹ or ›short term store‹ for a progressively deepening analysis. Working memory is not a structurally isolated component within the memory system, but rather that part within the integral associative memory network which is active at a particular instant. Depending on the task to be performed, the process of analyzing the input information will, after some time, result in an overt or covert response. This may conclude an action cycle, and the information that was involved in it may subsequently be abandoned and ›leak‹ away in about 20 to 30 seconds unless, for some reason, it is kept in the active state through ›rehearsal‹ (silent or aloud). While information resides in working memory it can be consolidated, partly or in its entirety, as a result of repeated use or deliberate associative elaboration and so become a part of long term or permanent memory. In other words, it may leave a more or less robust trace, which can be reactivated associatively on a later occasion. Most information processing models distinguish between two basic processing modes: automatic and controlled (or deliberate). Automatic processing is seen as a fast process that can accept any number of simultaneous inputs and produces outputs almost instantaneously, in a more or less reflex-like fashion. Automatic processing requires no ›mental effort‹ and there seems to be no upper limit to the complexity of the inputs it can handle. By contrast, the controlled processing mode can handle inputs only in a serial fashion. It is a slow process, limited in its capacity and requiring conscious mental effort. The focus of attention can be determined by the nature of the task or by external or internal instructions. In addition, if for some reason the course of automatic information processing is interrupted, the controlled mode will take over in an attempt to restore the normal processing routines.

More detailed descriptions of the human information processing system along the lines of the model presented here can be found in any introductory text on cognitive psychology (see, for instance, Anderson (1985), Reed (1982), Lachman et al. (1979)).

Temporal Information Processing

Let us now turn again to the role that temporal information plays in information processing, fitting it to the general framework outlined in the previous section. Cognitive psychology appears to have come to a point where it accepts that at least three fundamentally different types of internal code or representation are in operation: mental images for coding spatial relations between items, abstract propositions for the encoding of meaning and temporal strings for encoding the order of items in a sequence (Anderson, 1983; p. 45-46). It is, of course, not surprising that temporal information is given a more or less independent status, but it is surprising that it has taken ›main stream‹ psychology so long to accept that fact (some reasons have been indicated by Michon & Jackson, 1985, chapter 1 of the present volume).

What are the characteristics of these temporal strings? Anderson argues that they encode order, but not other aspects of time which, however, »is not to say that we cannot perceive or remember interval properties of a time sequence, but that such properties are not directly encoded in the structure of the temporal string ... Such information can optionally be encoded as attributes of the ordered elements.« (Anderson, 1983; p. 49).

Following this line of thought, the next question we are facing is what, in fact, constitutes the ›temporal attribute‹ encoded in these strings? The simplest assumption consistent with the idea of temporal strings is that the physically given order of events is the rock bottom cue on which temporal information processing is based. However, even if order would turn out to be a *sufficient* temporal stimulus, it cannot be the *only* relevant one. Logical or conventional (viz. proverbial or numerical) order, or even spatial arrangement may well provide (pseudo-)temporal cues that serve in those cases that order cues are lost or not available. If you put your cart in front of your horse, I will see your cart pass sooner than your horse, (at least on the assumption that the horse will persist in moving in its natural direction). If this temporal information is indeed encoded in terms of ordered strings, as is suggested by Anderson, other temporal attributes such as duration or temporal locus (position) being added if need demands, the question becomes how order can be encoded in the subject's mind. To answer this, one should realize that functionally the properties and organization of memory are such that there does indeed exist what would seem a sufficient condition for ordered events to be temporally encoded. If one event is perceived and subsequently retained in memory, the next event to be perceived will necessarily have the earlier event as part of its (remembered) context. Thus the earlier-later relation is represented as a stored-actual or old-new relation which has been used, implicitly or explicitly, by a great many authors from philosophers such as Saint Augustine, Husserl, and Merleau-Ponty, to experimental psychologists such as Janet (1928), Hintzman et al. (1975) and Tzeng et al. (1979) in order to account for the temporal organization of memory.

Recently there has been a tendency to assume that the encoding of earlier-later in terms of old-new is taking place automatically. In this light temporal information processing is seen as an automatic byproduct of the processing of other (categorical, propositional or spatial) information processing modes (Hasher & Zacks, 1979; Reed, 1982). This hypothesis attained a certain popularity, although it was based mostly on invalid assumptions and partly on unreliable empirical data. Moreover it stood almost totally in constrast with everything that was already known about the perception and judgment of time. Michon & Jackson (1984) produced detailed counterevidence. Independently, Zacks et al. (1984) have recently revised their original opinion and now appear to accept the idea that temporal information processing usually involves much hard cognitive work on the part of the subject. A balanced view of the issue is provided by Jackson (1985) in chapter 12 of the present volume. Her study makes abundantly clear that temporal information is indeed derived from various stimulus attributes, both at the lower levels of perceptual motor skill and at the cognitive level. Apart from individual differences in the selection of encoding and retrieval strategies there is a marked influence of the type of processing task the subject is performing: not all tasks tap the same strategy, although some tasks do *suggest* a preferred strategy.

The question how temporal relations are encoded is also addressed by Thomassen & Teulings (1985, chapter 17 of the present volume) and Shaffer (1985, chapter 15 of the present volume). They point out that time is not necessarily encoded *directly* at all in the highly overlearned skills they are studying. The fine and consistent timing that is required, for instance in writing or musical peformance, is likely to be »a consequence of the smooth functioning of the physiological and biomechanical systems involved in the process« (Thomassen & Teulings, 1985, pp. 253–263).

In summary it seems appropriate to accept the idea that time can be, but is not always directly encoded. Especially in overlearned, automatic skills like writing it is not. This strongly suggests that processing temporal information is indeed largely a deliberate cognitive activity, requiring a great deal of attention and strategic flexibility on the part of the individual. When and if temporal information is processed, order may be considered as the predominant stimulus attribute, but it is certainly not the only one. Other, intrinsically non-temporal cues may be given a (quasi-) temporal interpretation. This quality plays an essential role in my further discussion.

Models of Temporal Information Processing

A number of theories has been proposed to account for temporal information processing, although not many have been generated within the confines of the information processing paradigm. It is one of the areas in which time psychology proper and cognitive psychology have not yet merged. The theories and models dealing with the encoding and storage of temporal cues may be subsumed under two labels first proposed by Michon (1967a, 1972) and Ornstein (1969), namely clock theories and event-related theories. Clock theories are based on a general scheme in which a hypothetical internal time base generator is driven by some process of central activation (e.g. Treisman, 1963). The level of activation and, hence, the ›clock rate‹ is determined by the complexity of the external situation and by various physiological and psychological states, such as body temperature or anxiety. The time base emits (quasi-)periodic or random pulses which are then counted or integrated and thus provide the internal reference against which temporal behavior is performed. A detailed account of clock theories is found in Macar's chapter in the present volume (1985, chapter 7). Since in this approach it is impossible to distinguish qualitatively between the various factors that may cause fluctuations in the rate of the internal clock, there has been a steady push towards the second type of theory which makes many more explicit allusions to the role played by the situation (number and complexity of events, contextual stability) and the individual (alertness, expertise, intoxication, etc.). Chapter 11 in this volume, by Block (1985, see also Block, 1974), presents a typical example of this approach, which dates back at least to Guyau (1890; see Michon & Jackson, 1984). Clock theories and event-related theories both acknowledge the necessity of some internal time base, but the latter gain some flexibility since they explicitly avoid the unnecessary extra step of translating event-related states into a state of specific activation that is required in clock theories to account for the fact that a time base generator must somehow be driven. This step appears uninformative given the opportunistic choice of timing mechanism discussed earlier in this chapter.

A Generic Model

The insight that a theory of temporal information processing should be formulated in terms of event-related attributes rather than in terms of clocks or specific activation mechanisms should remind us of the fact that the independence of temporal information has already been claimed. A plausible theory should give an account of this independence. Furthermore, to the extent that temporal information processing is indeed an activity that is under attentional control rather than automatic, such a theory should also take into account that the attentional resources of human beings are limited and that, therefore, explicit temporal control must impede the control of propositional and spatial information.

In 1975 Thomas (Thomas & Brown, 1975; Thomas & Weaver, 1975) proposed a formal model of time estimation which meets these requirements. The model was presented as a summary description of a series of experiments on the perception of brief duration in the 100 ms range, but it was stated in such general terms that it can be transposed unconditionally to a much wider range of intervals (see also Macar, 1985, chapter 7 of the present volume). As such Thomas' model may be considered an important contribution to the theory of temporal information processing. The model can be summarized by the following functional equation,

$$t_{est} = \alpha f(t,I) + (1-\alpha)g^* (I,t),$$

stating that the judged duration of an interval is a weighted function (with weighting parameter α) of the directly encoded temporal information $f(t,I)$, and the remembered temporal information $g^*(I,t)$. In terms of the theoretical position adopted so far in this chapter, directly encoded temporal information should be understood as the temporal cues that a person extracts from a time interval when and as long as it is in progress. Such cues will, as I argued before, mostly but not exclusively be derived from the order of events. Depending on the effort spent in processing temporal cues (t) while the interval lasts, there will be greater or less opportunity to pay attention to other, non-temporal attributes of the stimuli (I). Hence, depending on such factors as instruction, interest, or physiological state, a person can either obtain a good immediate estimate of the duration of the interval or a good encoding of the (non-temporal) content of the interval, but not both. If, on the other hand, a judgment about the duration of an interval is asked after some delay, the interval must first be remembered (that is, located in permanent memory). This involves a reconstruction (g^*) of the duration of this interval on the basis of its non-temporal contents (I) and the originally extracted temporal cues (t) in as far as the latter were at all encoded and stored in permanent memory. Dynamically, judgments about remembered duration depend on the initial division of attention between non-temporal information (I) and temporal information (t), and on the rates of forgetting both kinds of information. It may be assumed that propositional and spatial information, when encoded, will be retained very long in comparision to temporal information, which, supposedly, is extremely volatile. If a remembered interval is judged veridically, therefore, it must be by virtue of a consistent, detailed reconstruction of the non-temporal contents of the interval (from which subsequently a new temporal estimate can be derived). This dynamic balance between the abilities of judging time in passing and in retrospect is represented in Thomas' model by the weighting parameter α. If $\alpha \rightarrow 1$, immediate judgment will be excellent (at least if the content of the interval does provide temporal cues at all) while judgment of remembered time will be poor. If, on the other hand $\alpha \rightarrow 0$, immediate judgment of duration has to be rudimentary while retrospectively the interval may be constructed adequately and a veridical temporal estimate derived from it.

Thomas' model offers a consistent summary of a large number of interrelated facets of temporal information processing. I shall refrain from a detailed discussion of these aspects, only to mention that the model provides an excellent vehicle for the description of *flow*, the subjective rate of time passing, one of the two attributes of time experience that were recognized as basic earlier in this chapter.

It is common experience that time periods which seem to pass quickly tend to become longer and longer in retrospect. An absorbing theater performance, an exciting holiday, or a romantic evening have a content (I), so absorbing that little or no attention remains for the temporal cues (t) that such events might in principle provide. The resulting immediate judgment therefore tends to severely underestimate the actual duration. In retrospect, however, the intense cognitive commitment must result in an overly detailed mental reconstruction of what happened during the original event. The density and the detail with which events have been experienced will generally provide a relatively high quota of temporal cues on which to base a retrospective judgment of the period involved. Immediate and retrospective judgments must therefore necessarily counterbalance each other. More dramatic even than a romantic evening is the near fatal accident. Accounts of people who were nearly killed in a car accident or falling down a precipice frequently amount to the paradoxical statement that it was all over before they really knew what was happening, while at the same time they had all the time in the world to take evasive action. The only way out of this paradox is to realize that the two observations refer, respectively, to immediate and retrospective experience of the duration of such events. We may assume a paroxysmal outburst of brain activity in the endangered organism during the event with no attention for temporal cues whatever, while at the same time generating a very high amount of non-temporal information I.

Some Psychophysical Constraints on Temporal Information Processing

Although Thomas' model covers an important part of the temporal information processing territory, it is formulated as a qualitative model and one remains curious about the actual form of the functions $f(t,I)$ and $g^*(I,t)$. The answer should be derived from the psychophysics of time, actually the oldest branch of time psychology and perhaps the most overworked. Macar (1985, chapter 7 of the present volume) provides a concise summary of the major results in this area. From the available evidence it can be concluded that, at least for the range of intervals that concerns us most – seconds to minutes – the psychophysical function connecting estimated and physical time is following the well known ›power law‹ of Plateau-Stevens (e.g. Stevens, 1975; Eisler, 1975): $t_{est} = at^b$. The various factors that may influence the actual judgment, viz. illness, alertness, or skill, apparently influence the value of both the parameters a (arbitrarily) and b ($0.5 \leq b \leq 1.0$). This amounts to the conclusion that subjective estimates of time are proportional either to true physical time or to the square root of physical time, or somewhere in between, depending on the circumstances and the task imposed on the subject.

Whatever the general shape of the psychophysical function, from the variability of its parameters it should be evident that there is more to time judgments than meets the eye. I shall briefly mention four factors with a quite conspicuous influence on the empirical results.

Volatile Time Constants. The literature on temporal factors in behavior has made us familiar with several time constants that appear to play an important role in the temporal control of behavior. The more prominent among these are the time quantum (for which values between 5 and 100 ms are reported), iconic storage (300 ms), the indifference interval (400-700 ms), the psychological refractory period (initially 250 ms), primary memory (3-5 s), and the range of short-term memory (20-30 s). Some of these mechanisms are discussed in some detail in later chapters of this book.

It is my contention that such time constants will be found operative only in extreme conditions, namely in laboratory experiments and in other unfamiliar conditions in which the subject has no finely tuned temporal response strategies or schemata available and is thrown back on his or her basic processing mechanisms. As soon as more sophisticated task performance is required these time constants stop being constant and take on values that reflect the properties of the task rather than those of the organism. This point is illustrated for instance in the distinction drawn by Jones (1985, chapter 13 of the present volume) between ›rate relational pattern perception‹ and ›coding‹.

The time quantum being set initially at roughly 100 ms, has moved to 50 and later even 5 ms, assuming every conceivable value in the meantime as a consequence of task variables (Michon, 1967a), training variables (Kristofferson, 1976), or structural variables (Warren, 1974). Another fish in the same kettle is the so called psychological refractory period, a very popular research topic in the 1960s. The idea is that upon presentation of a stimulus the processing of that stimulus will absorb the internal processing capacity to such an extent that subsequent information must wait until some capacity becomes available. At one time it was thought that this period was of the order of 250 ms. Then, when data from a wider range of situations became available, it turned out that the 250 ms range was too restricted and other estimates started to appear, until it was decided that the psychological refractory period would easily assume any value between 0 ms and 700 ms, depending only on the nature of the task (Kahneman, 1973).

In summary, although the organism may display certain time constants when brought under extreme task conditions, in most actual situations the evidence for the constancy of such constants disappears and their values appear to be determined only by the requirements imposed on the organism by the task it is performing at the time. It will be evident that this brings us close to what was said in an earlier section about the opportunistic character of the choice of an internal time base.

Linearization of Time. The exponent b in the power relation between estimated and physical time, $t_{est} = at^b$, takes on values between 0.5 and 1.0. Eisler (1976), in a heroic compilation of no less than 112 experimental studies, established that in more than one half of these studies there is a simple linear relation between estimated and physical time. As Michon & Jackson (1985, chapter 1 of the present volume) point out, Paul Fraisse had, already in 1957, reached the conclusion that the best subjective measure for the second is the physical second. Yet, even though one half of the studies support this view, the other half apparently does not, and the question may be raised what is the matter with that second half. In a study reported some time ago (Michon, 1975), I have suggested that there may be two essentially different ways of representing the psychophysical scale of duration, an ›analytic‹, linear scale consistent with Fraisse's contention, the other an ›impressionistic‹ scale in which estimated duration varies with the square root of physical time. (The intermediate values of the exponent *b* should, perhaps, be considered as artifacts from averaging over subjects or as the result of mixed strategies). The impressionistic scale was found over a considerable range of intervals in young children, gradually disappearing with increasing age (Fraisse, 1948); for intervals longer than 20 s in a person with defective memory (Richards, 1973); and in some normal subjects (Clausen, 1950). Furthermore it is normally found for intervals shorter than about half a second (Michon, 1967b).

In summary, adult subjects will always represent time in a linear fashion between 0.5 and 20 s. Both shorter and longer intervals may, however, be represented proportional to the square root of physical time.

These findings suggest an immediate relation to the tuning function of temporal information processing. As I argued before, tuning requires a near perfect correspondence between external and internal events. Such a relation can only be achieved when the internal events run off in ›real time‹. Consequently the internal time scale *must* be linearized over a considerable range in order to make adequate tuning possible. The range between 0.5 and 20 s, being roughly the lower and upper bound of working memory, would serve this purpose perfectly well. This range may be seen as representing the variable temporal window that interfaces us with reality, probably with a lower limit, an ›idling value‹, of 0.5 s, but longer when the input permits. Below 0.5 s information processing is of a highly perceptual nature, fast, parallel and not accessible to cognitive control. Accordingly temporal information could well follow the ›impressionistic‹ square root scale in this ›ultrashort‹ range. At the upper end, beyond 20 s most adults will by and large, but not always, linearize their time scales with the help of clocks and calendars and with a great deal of prior experience in linearizing time. If, however, anchoring points are not available, because long-term memory cannot be accessed in a conventional way for some reason, the impressionistic mode may take over. Only children may have difficulty in linearizing the interval over which working memory is operating; that is why children are so very apt to make timing errors until they are between 8 and 10 years of age.

Partial Order and Hierarchical Organization. A third aspect that deserves brief mention is that it has gradually become clear that temporal information is not necessarily encoded and stored in a simple serial fashion. Although duration is frequently represented as a one-dimensional scale, it is evident that most people do not normally represent their experiences on a fully integrated, single time scale.

On the macroscopic scale of one's personal history there would seem to be at least several more or less independent times, one for each of the major areas of activity; family life, work, community activities, and the public arena as it is reflected in the news media. Connecting these times into one global time scale may be quite difficult and exceed people's cognitive capacity.

At the microlevel it is the phenomenon of perceptual *streaming* which establishes the fragility of the one-dimensional representation of time. Streaming involves the separation of events from a single sequence of tones into two (or more) independent sequences. The best known example is a melody which breaks into two separate ›voices‹ when high and low tones are rapidly alternating. Composers use this phenomenon, especially in their pieces for unaccompanied solo instrument. Between streams no temporal relation can be established: the order of specific tones in two streaming melodies cannot be determined relative to each other. The indeterminacy of temporal relations resulting from streaming can only be solved indirectly if the listener succeeds in establishing a temporal hierarchy among the various ›voices‹ or parts of an event sequence. Hierarchies have been observed in rhythmic keytapping performance (Vorberg & Hambuch, 1978, 1984), and in short-term memory (Estes, 1972; Lee & Estes, 1977, 1981; see also Estes, 1985, chapter 10 of the present volume). Functionally they are of great importance because a hierarchical organization allows the person to incorporate more complicated higher order relations into the way he or she encodes temporal information. As a result longer and more complex event sequences can be coped with and accordingly the tuning process is facilitated. Hierarchical organization plays an important role in the construction of the experienced present, as will be explained later.

Perturbation. Temporal information appears to be very instable, and although it may initially be encoded veridically, forgetting may be very quick. Apparently, however, this does not always happen, which indicates that at least the normal adult person has the means available to compensate for the loss of temporal information when estimating time. Yet, under certain conditions the instability of temporal cues becomes noticeable.

This phenomenon has been studied by Estes (1972, 1985, chapter 10 of the present volume; see also Lee & Estes, 1977, 1981). Estes addressed the issue how the position information of items in a memorized list of letters or words becomes less and less certain over time. By assuming that the initial temporal positions of the items are increasingly perturbed since they are subject to stochastic fluctuations, he was able to give a temporal interpretation of inversions between items, intrusions from other lists and total forgetting of some items. Estes' model supposes that each item's position is subject to a two-directional random walk, a model that is similar to a model proposed by Michon (1967a) to explain irregularities in tapping behavior. Perturbation may account for the quick loss of temporal information revealed by psychophysical studies. It may also account, therefore, for the square root relation that, as pointed out, is occasionally found between estimated and physical duration. A square root relation is actually what one expects on the assumption that the perturbation process has indeed the characteristics of a simple stochastic random walk: in such cases the variance of the estimated length of the interval between two points t_1, and t_2 will vary with the difference $(t_2 - t_1)$.

More important than its ability to explain the loss of temporal information is the fact that the perturbation mechanism confronts us with the limits of the functional information processing approach that I have outlined above. The reason is that with increasing length of retention perturbation would exceed all bounds, and our memories would collapse in chaos. Nothing in the process would seem to stop the loss of information. Yet this is not what we see happen and we must therefore conclude that the structural properties of event sequences contribute in a fundamental way to the coding, storage and retrieval of temporal information.

The Structure of Temporal Information

If we study how people organize their memories it is clear that the processing and retention of temporal information is dependent on the pattern of the input information. Estes (1985, chapter 10 of the present volume) does also hint to this when he suggests that temporal factors in memory should be studied by introducing complexly organized stimuli such as sentences, since »language introduces structures in a hierarchy that constrain perturbation models in meaningful ways«.

In this context it is necessary to distinguish between patterns and codes. A pattern can be defined as a codable array of entities (symbols or events, for instance numbers), while a code is a set of rules that describe patterns in such a way that the code is finite and more compact than the original series. If an array (e.g. a string) cannot be represented in a compact way in any conceivable code it does not count as a pattern; this is the case, for instance, with the decimal fraction of the number $\pi = 3.1415926535...$, whereas the decimal fraction of $1/7 = 0.142857142857...$ represents a very simple repeating pattern with a period of 6 places.

A second basic distinction to be made when we discuss patterns and codes is that between sequential patterns and temporal patterns. The former category is specified only in terms of the order between elements of the sequence, and clearly represents the sort of temporal strings that are supposed to lie at the heart of the representation of temporal information in memory (see p. 34, and Anderson, 1983). Temporal patterns in the proper sense are sequential patterns in which additional (absolute or relative) temporal relations are specified. Musical patterns and more specifically rhythmic patterns, which encode (relative) duration, belong to this category.

The question is how the information processing system does cope with the syntactic structure of these patterns. Considerable attention has been given to this question in the past fifteen years. Several of the chapters of the present volume deal with this matter, particularly Jones (1985, chapter 13), Povel (1985, chapter 14), Shaffer (1985, chapter 15), and to some extent Nooteboom (1985, chapter 16) and Thomassen & Teulings (1985, chapter 17). Rather than summarizing the findings these authors will discuss in great detail, I shall only refer to a few fundamental aspects that are inherent in this approach to temporal information processing.

A first point to mention explicitly is the important distinction drawn by Jones between – what she calls – rate relational and coding theories. This is an important distinction because it focuses on the difference between the approach in which the internal processing stages are considered as the central features of psychological theory an approach that is commonly known as ›performance theory‹ and the approach that is more concerned with the intrinsic structure, the ›affordances‹ (Gibson, 1979), of the information presented to the subject. Coding theory deals with ›competence‹ rather than with performance. The polarity between competence and performance, which originated in the once heated discussion between linguists and psycholinguists (see e.g. Lachman et al., 1979), turns up in various other disguises elsewhere in psychology. There is a general trend in the development of information processing theories in which a particular phenomenon is initially described in terms of an internal, frequently neurophysiological, mechanism. When the search for such a brain mechanism is found to be fruitless, and it usually is, the next stage is a functional description in terms of processing ›stages‹. Subsequently these ›stages‹ tend to become formulated less and less in terms of generic performance terms, such as ›encoding‹ or ›response selection‹ and instead they begin to mimic the structural features of the stimulus domain or domains to which they apply. The processing mechanisms seem to leave the body (and the mind) to become structural ›affordances‹ of the environment (Michon, 1984). Herbert A. Simon (1969, p. 24) epitomized the latter stance when he contemplated the complex behavior of an ant on the beach: »An ant, viewed as a behaving system, is quite simple. The apparent complexity of its behavior over time is largely a reflection of the complexity of the environment in which it finds itself«. This trend from mechanism to contextualism (de Mey, 1982; Michon, 1984) appears in a number of ways also in *Time, Mind, and Behavior*: time psychology is apparently taking part in this general trend, the same trend, incidentally, which I earlier in this chapter diagnosed in physics (with reference to Park, 1985, chapter 3 of the present volume) as the attempt to eliminate the observer from the cosmos and to denounce the reality of time.

Pattern ›Extraction‹

The importance of context in contemporary psychological theory is a product of early applications of statistical information theory to psychological problems. Shannon's theory gave psychologists the insight that the meaning and significance of a stimulus cannot be determined in an absolute sense, but only relative to the set of stimuli to which it belongs. This was borne out most explicitly by Garner (1962, 1974) who, moreover, underscored the psychological significance of what he called ›structural uncertainty‹ (vs. event uncertainty) with a battery of supporting evidence. One revealing phenomenon is colloquially known as ›motorboating‹. It pertains to our ability to perceive the periodic character of a cyclically repeated burst of pure stochastic noise. Although the whole sequence of such bursts is, technically speaking, locally strictly random, people easily extract the hidden autocorrelation and as a result they tend to report a stable perceptual impression, a sound more or less similar to the chugging or throbbing of a ship's engine.

Garner and his associates studied cyclical patterns of high and low tones in order to find out what the structural features of such patterns are that determine the organization of a stable perception (Garner, 1974; see also Jones, 1985, chapter 13 of the present volume). They found two features that largely determine the stability of a cyclical tone pattern: the length of a subsequence that constitutes a ›gap‹ or pause at the end of one cycle and the length of a subsequence of tones that constitutes the starting ›run‹ of a cycle. Ambiguity will result from a conflict between the two features or from the absence of a sufficiently long and therefore dominant ›gap‹ or ›run‹. Garner's research, leading to the ultimate conclusion that stable and perceptually ›simple‹ patterns are those that have a unique description in a given code, has triggered an avalanche of research in an area now known as structural information theory. Jones, in her contribution to the present volume, has outlined the development and the major issues in this fertile domain in as far as they are relevant to the encoding and retention of temporal patterns.

Meaningful Patterns

As I have argued, the encoding of temporal information relies on a mixture of cues: order, causal, spatial, and conventional. The question is in what conceptual structure or structures these cues are eventually imbedded. The authors of the chapters in Part IV of this volume (chapter 18 by van Benthem (1985), chapter 19 by Montangero (1985) and chapter 20 by Michon (1985)) are all concerned in one way or another with just this problem. They make it clear that the classical conceptualization of time as a simple spatial metaphor – an arrow passing from back to front through the body of the person – cannot be maintained. Van Benthem (1985), for instance, stresses the pluriformity of the formal ›mathematical ontologies‹ from which various conceptualizations can be derived, such as, for example, point logic and interval logic. Even though formally the two can be derived from each other, the psychological consequences of a point logic versus an interval logic are quite different. Van Benthem goes even as far as to suggest that the interval representation lies at the root of individual time while the point representation is a cognitive construction, that is characteristic of public time.

It is a long way from a ›mathematical ontology‹ of time representation to a concrete conceptualization. A psychologically important step in that direction was taken by Miller & Johnson-Laird (1976) in their classic book on perception and language in which they developed their ideas of ›procedural semantics‹. Procedural semantics implies that words and sentences serve as prescriptions that tune the organism to certain perceptual or cognitive conditions. People have a sharply tuned ability to decide which temporal relations in a set of verbal assertions are consistent and which are not. Whether a certain temporal expression is accepted as a valid description of a given situation or event depends, according to these authors, on implicit checks about the truth-table values of the constituent elements of the expression. Thus, the sentence »The shop was closed as long as the bomb exploded« is considered incorrect because the expression »as long as« requires a durative proposition on either side; and usually the explosion of a bomb is an instantaneous event. Recent progress towards more sophisticated conceptualizations of time has been made by Allen (Note 1; 1983), who developed a formalism for describing temporal structure on the basis of what the state of the world is before and after a certain event rather than describing the changes during the actual transition. His approach allows the full advantage of the interval logic described by van Benthem (1985, chapter 18 of the present volume).

The Construction of the Present

The experience of *now*, classically known as the specious present, has been considered from several points of view. Some authors (reviewed by Macar, 1985, chapter 7 of the present volume) have chosen the minimum option of the threshold of order perception (20-200 ms) as the possible range for the experience of *now*. Others, including Mandler (1975), Jones (1976), Michon (1978), and Kunst (1978) have taken a much more dynamic view. This view is actually one that is conceptually consistent with the classical view or the present (represented in the work of e.g. Wundt, and James; see Fraisse, 1957, for an overview), but also with much of the work on incremental models of speech understanding and speech production (see e.g. Kempen & Hoenkamp, in press). The basic idea is that the experiential present is initiated at a certain instant, when a sequential pattern of events is beginning to be presented to a subject. On the basis of the first, minimal cues an interpretive context is generated and on that basis a hypothesis (representation) is formed about the possible future course of events. This process of generation and confirmation should actually be seen as the warp and woof of the tuning ›window‹ that interfaces the subject with reality in real time. If, and as long as the anticipatory hypothesis is confirmed by what is happening in the outer world, the ›window‹ can expand and accordingly the experience of *now* becomes more extended (Mandler, 1975). If the hypothesis is not borne out sufficiently well by subsequent events, a new anticipatory hypothesis may be selected *within* the selected interpretive context and tested by backtracking. If this does not meet with success, the chosen interpretation has apparently no further merit and must be replaced. At that point the current *now* is interrupted (segmented), its contents thus far – up to the point where no extrapolation to new events turned out to be feasible – are transformed, from the surface level (at which temporal information is generally encoded) into a deep-level representation, and a new interpretive cycle is started. This idea implies that past or remembered time has a discrete structure and can be represented by strings of (meaningful) chunks that are derived from a highly interactive interpretive tuning process. Although such a view of the *now* as the product of the tuning process is still in its infancy, it seems capable of incorporating many aspects of the information processing approach and the coding approach outlined in the preceding sections.

Selforganization and the Reconciliation of Structure and Function

Structuralistic theories impose their own problems that ultimately disqualify them as psychological theory. If a purely linguistic approach to language eliminates the need for a speaker or a listener, and if geometrical optics is, after all, a theory of visual perception without a perceiver, what would be the merit of a theory or temporal coding as a theory of time experience? There would be no such merit, but at the same time it appears impossible to go very far in that direction. Once more it seems impossible to eliminate the observer from the universe. It has been pointed out that patterns – which include sequential and temporal patterns – are patterns by virtue of the structural context in which they appear. A wedge-shaped symbol V will be different if it is an element of the set (U, V, W, X), than if it belongs to the set (→, V, &, ~). The question that structural theories have always great difficulty in answering is *what*, or *who*, determines that context. This difficulty is a matter of considerable dispute in contemporary cognitive psychology and philosophy of mind and therefore, to find it discussed in several of the chapters of *Time, Mind, and Behavior* is indicative of the timeliness of the book.

Structures, described variously as formal theories, axiomatic theories, mathematical ontologies, etc. are related to a representational mode (or grammar) propositional, imaginal or temporal – which is the vehicle for conceptualization, *through* an interpretation function, a »systematic ›recipe‹ between grammar and mathematical ontology« (van Benthem, 1985, chapter 18 of the present volume). *Point logic* and *interval logic* are examples of such interpretation functions: they specify very different and independent ways of representing temporal relations, and as such they offer independent but not quite unique views of the temporal structure of reality.

This view, however, remains incomplete and arbitrary as long as it is not made clear why some structures appear to invite a certain interpretation rather than another one. In Goodman's (1984) terminology: »humankind makes versions of the world, and true versions make worlds«. But, how do we know which versions, which of our representations are indeed true. In several chapters the authors hint at this crucial question. Thus, Shaffer (1985, chapter 15 of the present volume) points to the fact that the quality of sight performance of musicians suggests that there is a certain necessity to musical pattern. And similarly both Jones (1985, chapter 13 of the present volume) and Povel (1985, chapter 14) appeal to the fact that temporal structures may derive their »appropriate interpretation« and their functional significance from their temporal (rhythmic) aspects.

Summary and Epilogue: What is it Like to Build a Time Experiencer?

In retrospect we can conclude that time psychology has come a long way to becoming a respectable concern for psychologists. The realization that time is a manifestation of the need to exchange information with the environment to enable an organism to stay geared to the course of events in its environment has triggered a substantial and substantive amount of research in the field. Let me summarize the basic task of time psychology stated in the prologue by considering what would be needed to build a special version of the ›compleat time experiencer‹ we have been looking for, namely a robot that would be temporally as independent of its working environment as humans are of theirs. Like humans, robots are operating more or less independently of their environment. Machines have, of course, been known to do jobs under extreme conditions, in heat, cold, poisonous or other adverse circumstances. Classically that does not involve structural adaptation: the structure of a machine normally does not change as a function of the environmental circumstances. Living organisms, however, do have that potential. Even primitive species have managed to cope with variable circumstances. The most fascinating development of this kind has been the ›programmable memory‹, that allowed each individual of a species to learn and benefit from its own private experience. One concomitant of this development was the emergence of symbolic representations that allowed *anticipation* and *advance evaluation* of behavioral options. In higher organisms this independence implies *temporal* independence, both with respect to periodic and singular, progressive processes. This involves tuning to the events occurring in reality. Although clocks may be involved, these clocks require constant tuning and retuning. A robot that is to function truly independently of its environment should enjoy the same temporal independence of the external course of events as people. That is, it should need no time bases that are permanently synchronized with an external clock (although resetting once every day should probably be allowed). For the rest the robot's action should depend only on the representation of what is required by the prevailing circumstances, and on the basis of past experience and future goals. This requires an elaborate representation of time that should allow both overlearned activity sequences to run off more or less uncontrolled in a data-driven fashion and an elaborate ›top down‹ repertoire of cognitive strategies for coping with temporal contingencies that require planning, reflection and communication.

It appears that important progress has been made in the direction of a functional model of time, that is, the design of a model that can explain observed behavioral patterns and subjective phenomenology in terms of underlying processes: biological rhythms, constraints on working memory, thresholds for differences in stimulation, to name a few. Most of these models appear to have used the clock concept in one way or another. I have also considered the need for a complementary approach, based on the idea of structural information theory: not only *order* is given in nature, but certain other temporal relations are also contained in the events that occur in the organism's environment. Rhythmicity, recurrence, etc. are not simply imposed on event sequences by the organism but are ›afforded‹ by the structure of the environment. The question is how these ›affordances‹ are picked up.

Evaluating both ways of coping with temporal experience while avoiding intentionalistic explanations, a fundamental problem has come under consideration in recent years. It is important to mention that this topic is dealt with by several contributors to *Time, Mind, and Behavior.* This clear trend in cognitive theory, concerns *selforganization* as a crucial step in uniting the functional (performance) and structural (competence) approach to information processing. Wonham's (1976) principle, Prigogine's (1980) ›dissipative structures‹ which maintain their internal structure while interacting with their environment, and J.J. Gibson's (1979) ›affordances‹ are complemented, in a fundamental way, by Jones' ›dynamic serial transformations‹, van Benthem's ›systematic recipe‹ for interpreting formal ›mathematical ontologies‹ in terms of (natural) language, Michon's reference to ›generative metaphor‹, and by Park's plea for a conscious ›observer‹ in physical theory. They all imply the deep-felt desire to cope, in information processing theories, with structure and function at the same time. (e.g. Goodwin, 1982; Michon, 1984). Uttal (1978) has phrased this desire in terms of the relation between a wheel and its rotation: we are faced with the functional question how a wheel generates rotation and with the complementary structural question why precisely rotation is the ›natural‹ or ›self-evident‹ product of a wheel. Within the present context of *Time, Mind, and Behavior* all these recent efforts aim at answering one paramount question: How does an event sequence evoke in the observer a certain *interpretation* of context in which the sequence occurs, such that the result is a stable representation of that sequence, that is a *pattern*. Stable representations in this sense are a necessary requisite for adequate tuning performance. But adequate tuning performance is in turn dependent on the interpretation of the dynamic features of the external situation. Like wheels we appear to go round in circles: let us hope that they will not be vicious circles.

Dealing with such a dynamic concept of internal structure requires a formalism, a metamodel if you like, from which any representation of time, and thus any functional basis for time experience, can be derived through a *choice of interpretation*. Only when such a metamodel will be available it will become possible to decide unambiguously whether indeed time is a unitary concept, or perhaps a multiplicity depending on an inhomogeneous set of explanatory concepts. As yet the matter is undecided.

References

Allen, J.F. Maintaining knowledge about temporal intervals. *Communications of the Association for Computing Machinery*, 1983, 26, 832-843.

Anderson, J.R. *The architecture of cognition*. Cambridge, MA: Harvard University Press, 1983.

Anderson, J.R. *Cognitive psychology and its implications*. San Francisco: Freeman, 1985 (2nd edition).

Baudson, M. (Ed.), *Tijd: De vierde dimensie in de kunst*. Brussels: Vereniging voor de Tentoonstellingen van het Paleis voor Schone Kunsten, 1984.

Benthem, J. van. Semantics of time. In: J.A. Michon & J.L. Jackson (Eds.), *Time, mind, and behavior*. Heidelberg: Springer Verlag, 1985, pp. 266–278.

Block, R.A. Memory and the experience of duration in retrospect. *Memory and Cognition*, 1974, 2, 153-160.

Block, R.A. Contextual coding in memory: Studies of remembered duration. In: J.A. Michon & J.L. Jackson (Eds.), *Time, mind, and behavior*. Heidelberg: Springer Verlag, 1985, pp. 169–178.

Clausen, J. An evaluation of experimental methods of time judgment. *Journal of Experimental Psychology*, 1950, *40*, 756-761.

Davies, P.C.W. Time and reality. In: R. Healy (Ed.), *Reduction, time and reality: Studies in the philosophy of the natural sciences*. Cambridge: Cambridge University Press, 1981, pp. 63-78.

Dennett, D.C. *Brainstorms: Philosophical essays on mind and psychology*. Hassocks, Sussex: Harvester Press, 1978.

Dennett, D.C. Three kinds of intentional psychology. In: R. Healy (Ed.), *Reduction, time and reality: Studies in the philosophy of the natural sciences*. Cambridge: Cambridge University Press, 1981, pp. 37-61.

DeWitt, B.S. Quantum gravity. *Scientific American*, 1983, *249* (6), 104-115.

Doob, L.W. *Patterning of time*. New Haven: Yale University Press, 1971.

Eisler, H. Experiments on subjective duration 1868-1975: A collection of power function experiments. *Psychological Bulletin*, 1976, *83*, 1154-1171.

Estes, W.K. An associative basis for coding and organization in memory. In: A.W. Melton & E. Martin (Eds.), *Coding processes in human memory*. Washington, DC: Winston & Sons, 1972, pp. 161-190.

Estes, W.K. Memory for temporal information. In: J.A. Michon & J.L. Jackson (Eds.), *Time, mind, and behavior*. Heidelberg: Springer Verlag, 1985, pp. 151–168.

Fawkner, H.W. *The timescapes of John Fowles*. London: Associated University Presses, 1984.

Fraisse, P. Etude comparée de la perception et de l'estimation de la durée chez les enfants et chez les adultes. *Enfance*, 1948, *1*, 199-211.

Fraisse, P. *Psychologie du temps*. Paris: Presses Universitaires de France, 1957.

Frank, J., & Staddon, J.E.R. Effects of restraint on temporal discrimination behavior. *Psychological Records*, 1974, *24*, 123-130.

Fraser, J.T. *The genesis and evolution of time: A critique of interpretation in physics*. Amherst: University of Massachusetts Press, 1982.

Freedman, D.Z., & Nieuwenhuizen, P. van The hidden dimensions of spacetime. *Scientific American*, 1985, *252* (3), 74-83.

Garner, W.R. *Uncertainty and structure as psychological concepts*. New York: John Wiley & Sons, 1962.

Garner, W.R. *The processing of information and structure*. Potomac, MD: Lawrence Erlbaum Associates, 1974.

Gibbs, J.W. *Elementary principles in statistical mechanics*. New Haven: Yale University Press, 1902.

Gibson, J.J. *The senses considered as perceptual systems*. Boston: Houghton Mifflin, 1966.

Gibson, J.J. *The ecological approach to visual perception*. Boston: Houghton Mifflin, 1979.

Glazer, H., & Singh, D. Role of collateral behavior in temporal discrimination performance and learning. *Journal of Experimental Psychology*, 1971, *91*, 78-84.

Gooddy, W. Time and the nervous system. *The Lancet*, 1958, *nr. 7031*, 1139-1141.

Goodman, N. *Of mind and other matters*. Cambridge, MA: Harvard University Press, 1984.

Goodwin, B.C. Genetic epistemology and constructionist biology. *Revue Internationale de Philosophie*, 1982, *142*, 527–548.

Gorman, B.S., & Wessman, A.E. Images, values, and concepts of time in psychological research. In: B.S. Gorman & A.E. Wessman (Eds.), *The personal experience of time*. New York: Plenum Press, 1977, pp. 217-263.

Groos, G., & Daan, S. The use of biological clocks in time perception. In: J.A. Michon & J.L. Jackson (Eds.), *Time, mind, and behavior*. Heidelberg: Springer Verlag, 1985, pp. 65–74.

Guyau, J.M. *La genèse de l'idée de temps*. Paris: Alcan, 1890 (2nd ed. 1902).

Hasher, L., & Zacks, R.T. Automatic and effortful processes in memory. *Journal of Experimental Psychology: General*, 1979, *108*, 356-388.

Herrmann, Th. Ueber begriffliche Schwäche kognitivistischer Kognitionstheorien. *Zeitschrift für Sprache und Kognition*, 1982, 1, 3–14.

Hintzman, D.L., Summers, J.J., & Block, R.A. Spacing judgments as an index of study-phase retrieval. *Journal of Experimental Psychology: Human Learning and Memory*, 1975, *1*, 31-40.

Jackson, J.L. Is the processing of temporal information automatic or controlled? In: J.A. Michon & J.L. Jackson (Eds.), *Time, mind, and behavior*. Heidelberg: Springer Verlag, 1985, pp. 179–190.

Janet, P. *L'évolution de la memoire et de la notion de temps*. Paris: Chahine, 1928.
Jones, M.R. Time, our lost dimension: Toward a new theory of perception, attention, and memory. *Psychological Review*, 1976, *83*, 323-355.
Jones, M.R. Structural organization of events in time. In: J.A. Michon & J.L. Jackson (Eds.), *Time, mind, and behavior.* Heidelberg: Springer Verlag, 1985, pp. 192-214.
Jung, C.G. Synchronity: An acausal connecting principle. In: *Works of C.G. Jung, Vol. 8*. London: Routledge & Kegan Paul, 1960, pp. 417-531.
Kahneman, D. *Attention and effort*. Englewood Cliffs, NJ: Prentice-Hall, 1973.
Kempen, G.A.M., & Hoenkamp, E. An incremental-procedural grammar for sentence formulation. *Cognitive Science*, 1985, in press.
Kristofferson, A.B. Low variance stimulus-response latencies: Deterministic internal delays? *Perception and Psychophysics*, 1976, *20*, 89-100.
Kunst, J. *Making sense in music: An enquiry into the formal pragmatics of art*. Ghent (Belgium): Communication & Cognition, 1978.
Lachman, R., Lachman, J.L., & Butterfield, E.C. *Cognitive Psychology and information processing: An introduction*. Hillsdale, NJ: Lawrence Erlbaum Associates, 1979.
Lee, C.L., & Estes, W.K. Order and position in primary memory for letter strings. *Journal of Verbal Learning and Verbal Behavior*, 1977, *16*, 395-418.
Lee, C.L., & Estes, W.K. Item and order information in short-term memory: Evidence for multi-level perturbation processes. *Journal of Experimental Psychology: Human Learning and Memory*, 1981, *7*, 149-169.
Macar, F. *Le temps: Perspectives psychophysiologiques*. Brussels: Pierre Mardaga, 1980.
Macar, F. Time psychophysics and related models. In: J.A. Michon & J.L. Jackson (Eds.), *Time, mind, and behavior.* Heidelberg: Springer Verlag, 1985, pp. 112-130.
Mandler, G. *Mind and emotion*. New York: John Wiley & Sons, 1975.
Mey, M. de. *The cognitive paradigm*. Dordrecht: Reidel, 1982.
Michon, J.A. *Timing in temporal tracking*. Assen: Van Gorcum, 1967a.
Michon, J.A. Magnitude scaling for short durations with closely spaced stimuli. *Psychonomic Science*, 1967b, *9*, 359-360.
Michon, J.A. Processing of temporal information and the cognitive theory of time experience. In: J.T. Fraser, F.C. Haber, & G.H. Müller (Eds.), *The study of time*. Heidelberg: Springer Verlag, 1972, pp. 242-258.
Michon, J.A. Time experience and the memory process. In: J.T. Fraser & N. Lawrence (Eds.), *The study of time II*. New York: Springer Verlag, 1975, pp. 302-313.
Michon, J.A. The making of the present: A tutorial review. In: J. Requin (Ed.), *Attention and performance VII*. Hillsdale, NJ: Lawrence Erlbaum Associates, 1978, pp. 89-111.
Michon, J.A. Over de metatheoretische grondslagen van de psychonomie. In: J.G.W. Raaijmakers, P.T.W. Hudson & A.H. Wertheim (Eds.), *Metatheoretische aspecten van de psychonomie*. Deventer: Van Loghum Slaterus, 1984, pp. 18-41.
Michon, J.A., & Jackson, J.L. Attentional effort and cognitive strategies in the processing of temporal information. In: J. Gibbon & L. Allen (Eds.), *Timing and time perception*. Annals of the New York Academy of Sciences, Vol. 423, 1984, 298-321.
Michon, J.A. Temporality and metaphor. In: J.A. Michon & J.L. Jackson (Eds.), *Time, mind, and behavior.* Heidelberg: Springer Verlag, 1985, pp. 288-296.
Michon, J.A., & Jackson, J.L. The psychology of time. In: J.A. Michon & J.L. Jackson (Eds.), *Time, mind, and behavior.* Heidelberg: Springer Verlag, 1985, pp. 2-17.
Miller, G.A., & Johnson-Laird, P.N. *Language and Perception*. Cambridge: Cambridge University Press, 1976.
Montangero, J. The development of temporal inferences and meanings in 5- to 8-year old children. In: J.A. Michon & J.L. Jackson (Eds.), *Time, mind, and behavior.* Heidelberg: Springer Verlag, 1985, pp. 279-287.
Mulder, G., & Mulder, L.J.M. Information processing and cardiovascular control. *Psychophysiology*, 1981, *18*, 392-402.
Nagel, T. What is it like to be a bat? *Philosophical Review*, 1974, *83*, 435-450.

Nooteboom, S.G. A functional view of prosodic timing in speech. In: J.A. Michon & J.L. Jackson (Eds.), *Time, mind, and behavior.* Heidelberg: Springer Verlag, 1985, pp. 242-252.

Orme, J.E. *Time, experience and behavior.* New York: American Elsevier, 1969.

Ornstein, R.E. *On the experience of time.* Harmondsworth: Penguin Books, 1969.

Park, D. Brain time and mind time. In: J.A. Michon & J.L. Jackson (Eds.), *Time, mind, and behavior.* Heidelberg: Springer Verlag, 1985, pp. 53-64.

Pouthas, V. Timing behavior in young children. In: J.A. Michon & J.L. Jackson (Eds.), *Time, mind, and behavior.* Heidelberg: Springer Verlag, 1985, pp. 100-109.

Povel, D.J. Time, rhythms and tension. In: J.A. Michon & J.L. Jackson (Eds.), *Time, mind, and behavior.* Heidelberg: Springer Verlag, 1985, pp. 215-225.

Prigogine, I. *From being to becoming: Time and complexity in the physical sciences.* San Francisco: Freeman, 1980.

Reed, S. *Cognition: Theory and applications.* Monterey, CA: Brooks & Cole, 1982.

Richards, W. Time reproductions by H.M. *Acta Psychologica*, 1973, *37*, 279-282.

Richelle, M., & Lejeune, H. *Time in animal behaviour.* Oxford: Pergamon Press, 1980.

Richelle, M., Lejeune, H., Perikel, J.J., & Fery, P. From biotemporality to nootemporality. In: J.A. Michon & J.L. Jackson (Eds.), *Time, mind, and behavior.* Heidelberg: Springer Verlag, 1985, pp. 75-99.

Richter, C.P. *Biological clocks in medicine and psychiatry.* Springfield, IL: Charles C. Thomas, 1965.

Ricoeur, P. *Temps et récit.* Paris: Editions du Seuil, 1983.

Saunders, P.T., & Ho, M.W. On the increase in complexity in evolution. *Journal of Theoretical Biology*, 1976, *63*, 375-384.

Shaffer, L.H. Timing in action. In: J.A. Michon & J.L. Jackson (Eds.), *Time, mind, and behavior.* Heidelberg: Springer Verlag, 1985, pp. 226-241.

Simon, H.A. *The sciences of the artificial.* Cambridge, Mass.: MIT Press, 1969.

Stevens, S.S. Psychophysics: *Introduction to its perceptual, neural, and social prospects.* New York: John Wiley & Sons, 1975.

Thomas, E.A.C., & Brown, I.B. Time perception and the filled duration illusion. *Perception and Psychophysics*, 1974, *16*, 449-458.

Thomas, E.A.C., & Weaver, W.B. Cognitive processing and time perception. *Perception and Psychophysics*, 1975, *17*, 363-367.

Thomassen, A.J.W.M., & Teulings, H.-L. Time, size and shape in handwriting: Exploring spatio-temporal relationships at different levels. In: J.A. Michon & J.L. Jackson (Eds.), *Time, mind, and behavior.* Heidelberg: Springer Verlag, 1985, pp. 253-263.

Toda, M. The boundaries of the notion of time. In: J.T. Fraser, N. Lawrence, & D. Park (Eds.), *The study of time* III. New York: Springer Verlag, 1978, pp. 370-416.

Tzeng, O.J.L., Lee, A.T., & Wetzel, C.D. Temporal coding in verbal information processing. *Journal of Experimental Psychology: Human Learning and Memory*, 1979, *5*, 52-64.

Uttal, W.R. *The psychobiology of mind.* Hillsdale, NJ: Lawrence Erlbaum Associates, 1978.

Vorberg, D., & Hambuch, R. On the temporal control of rhythmic performance. In: J. Requin (Ed.), *Attention and Performance VII.* Hillsdale, NJ: Lawrence Erlbaum Associates, 1978, pp. 535-555.

Vorberg, D., & Hambuch, R. Timing of two-handed rhythmic performance. In: J. Gibbon & L. Allen (Eds.), *Timing and time perception.* Annals of the New York Academy of Sciences, Vol. 423, 1984, pp. 390-406.

Warren, R.M. Auditory temporal discrimination by trained listeners. *Cognitive Psychology*, 1974, *6*, 237-256.

Wessman, A.E., & Ricks, D.F. *Mood and personality.* New York: Holt, Rinehart and Winston, 1966.

Wonham, W.M. Towards an abstract internal model principle. *IEEE Transactions on Systems, Man, and Cybernetics*, 1976, *6*, 735-740.

Zacks, R.T., Hasher, L., Alba, J.W., Sanft, H., & Rose, K.C. Is temporal order encoded automatically? *Memory and Cognition*, 1984, 12, 387-394.

Reference Notes

1. Allen, J.F. *A general model of action and time*. Technical Report nr. 97, Computer Science Department, University of Rochester, New York, 1981.
2. Zomeren, A.H. van. *Reaction time and attention after closed head injury*. Doctoral dissertation. University of Groningen, 1981.

Chapter 3. Brain Time and Mind Time

David Park

For centuries, scientists and philosophers have struggled with conceptual problems posed by the self, by consciousness and will. Now that theological questions no longer play an important part in this struggle, and with the rise of various exact sciences, mechanistic notions have tended to overshadow the tremendous fact we have to start with: that within limits that seem to us very broad, we have a sense of freedom in choosing what we shall do next. To my surprise these mechanistic notions seem to originate not with chemists and physicists, as one might expect, but within the scientific communities which concern themselves with brain and behavior. In this chapter, as we think about Time, Mind, and Behavior, I shall assume two things. The first is that nobody needs to explain to us what it feels like to be free, and the second is that the task of science is to explain the world in terms which, as far as possible, relate to our experience of it. We interpret the world scientifically to our prescientific selves, and in this process the question what is the exact truth often seems less important than the question how we ought to tell it. This is so much the case that one very rarely hears claims of truth anymore in experimental or theoretical science or even in mathematics, and it is quite remarkable how easily we get along without them. Rather than over the correctness of a theory we tend to argue over the validity of a model, and one of the requirements for validity is that the model should help us feel that we understand.

Let me defend for a moment the claim that human beings are, in the classical sense, free. It is perhaps unnecessary to argue for this at all, but I have a little fear that a physicist might be expected to put forth deterministic arguments concerning the mind, based on deterministic arguments concerning the brain. I want to forestall any such expectations, not by opposing a humanist point of view to one that is supposed to be scientific, but by making scientific claims. It is of course true that the progress of physics could be measured, especially in the early days, by successful demonstrations that A determines B, but that characteristic does more to define the methods and history of physics than it does to chart a path for science as a whole. For the last fifty years atomic theory has done very well without any more than a statistical determinism, and I know that many physicists have seized on this fact to explain freedom of the will. Our experience seems in most cases to have been much the same. We consult our friend the neurophysiologist, who has already been asked the same obvious questions many times, and are told that the smallest neural currents involve billions of electrons and so quantum indeterminism, even though it exists, cannot be important here. Embarrassed, we return to topics we know something about, and it is only after some time that we begin to wonder whether our friend's words perhaps reveal more about the state of his instrumentation than about the state of the thinking brain. In fact, several physicists I know are privately convinced that quantum indeterminacy at the level of very small signals, perhaps of thermal fluctuations, must ultimately be responsible for our freedom.

If the development of physical science does not lead toward a belief in determinism, it is equally true that arguments based on determinism do not seem to explain the activity of scientists. While it is at least possible to argue as Descartes did, but I would not, that cats and dogs function purely as machines as they look after their safety, nourishment, and so on, I do not see how anyone can argue that a scientist's persistent efforts to satisfy what he perceives as speculative curiosity can be explained in terms of the environmental situation. If the task of science is to explain the world in terms which, as far as possible, refer to the world as we know it and to ourselves as we know ourselves, I do not see how any deterministic explanation of mental function can perform this task.

The arguments by which physicists try to explain the material universe have been developed over several centuries and have achieved some notable successes. Physics aspires to accuracy and objectivity, and yet paradoxically it finds itself compelled at important points to introduce an entity calledThe Observer which in some sense represents ourselves, and it endows this observer with consciousness and memory. Physicists have not made the smallest progress toward a physical theory of consciousness; rather, the observer appears in physical formulations as a scientific instrument does: it plays a fundamental role in the language of explanation even if we need not know how it works. But there is a difference: there is somebody who designed the instrument and knows that its operation can be explained scientifically. Can an idealized observer be similarly designed and understood on scientific principles? Later I shall argue that at present it cannot, and if this is true we are indeed far from possessing any physical model that might give insight into the nature of consciousness.

I begin, however, with some remarks on the concept which in this book serves to organize our thoughts on mind and behavior, the concept of time. Is time really there, existing as some process or at least some dimension independent of human cognition, or is it a conceptual device introduced in order to aid our understanding: (Of course one might ask the same question concerning space.) This is a question from the philosophy books, but before it got there it seemed highly important to a number of great thinkers, and the mere fact that it now lives in philosophy books does not mean that it cannot or need not be answered. The best I can do is to mention some current considerations without trying to decide anything.

Space, Time and Spacetime

Space and time, Einstein once said, are modes by which we think and not conditions under which we live (Forsee, 1963). Later he wrote (Einstein & Besso, 1972) »For us who are convinced physicists, the distinction between past, present, and future is only an illusion, however persistent.« Einstein believed in a natural world that exists objectively, independently of ourselves, and so by »modes by which we think«

he meant modes that help us to order and describe and explain sensations which that world gives us. Kant's word was *Anschauungsformen*, often translated as »forms of intuition.« The scientific description of space and time begins, for our purposes, in Einstein's demonstration that spatial and temporal intervals are observer-dependent, and these facts have their beautiful explanation in Hermann Minkowski's concept of spacetime (Lorentz et al., 1923; Taylor & Wheeler, 1966). In the Eighteenth Century it was an obvious idea to describe dynamical processes by means of coordinates in a fourdimensional manifold formed by pasting a time coordinate onto the familiar three coordinates of space. Minkowski's idea went far beyond this: in spacetime the dimension of time is as closely related to the other three as up-and-down is related to North-and-South and East-and-West, and if we think in terms of spacetime, Einstein's observer-dependence of spatial and temporal intervals is no harder to understand than the fact that ›up‹ as determined in Australia is not the same direction as it is in Groningen.

Minkowski's spacetime was a geometrical representation, but a few years later Einstein's General Theory of Relativity (Lorentz et al., 1923) transformed it into physics. Take the idea of a magnetic field which is familiar from elementary physics. A magnet exerts a force on another magnet a few centimeters away. The magnetic field is the agent in the apparently empty intervening space which transmits this force. It contains energy and does work. Similarly the General Theory of Relativity is a theory of the gravitational field. Here also, the field contains energy and does work, but connected with these it has a further remarkable property: it determines the scales that measure the different directions of spacetime. For example, it tells a stick of a certain length aligned in a North-South direction how long it ought to become when taken somewhere else and aligned up-and-down. If we take the very limited view (which is, however, the impression one would get from looking only at the mathematical structure of the General Theory) that spacetime is completely characterized by these scales, we might say that Einstein's field *is* spacetime.

The differential equations which determine the properties of spacetime have a form not very different from those which determine the properties of an elastic medium under stress. In the elastic medium, deformations are produced at the points where it is pushed or pulled; in Einstein's spacetime they are produced wherever and whenever matter and energy are present. The spacetime of Einstein's theory is remarkably like a thing, and if, as Einstein remarked, space and time are modes by which we think, I believe he meant that they are modes by which we think about spacetime, or rather about events in spacetime, for even though spacetime can be represented as a single unified concept, each observer's conception of time is essentially and fundamentally different from his conception of space.

Where, in the theory of events in spacetime, does this essential difference originate? The equations of the theory require four coordinate axes in spacetime, which an observer might choose to name North-South, East-West, up-down, and past-future. In many of the equations an algebraic sign distinguishes the time direction from the three spatial directions. This occurs typically as follows: In an ordinary space spanned by three orthogonal axes labelled X, Y, and Z, the length l of a rod having one end at the origin of coordinates and the other at the point whose coordinates are x, y, and z is given by

$$l^2 = x^2 + y^2 + z^2,$$

and since the length is an intrinsic property of the rod, this formula gives the same quantity whatever directions may have been chosen for the three axes. Similarly, in relativistic spacetime, the length s of the spatio-temporal interval between one event located at the origin of the four coordinates and another at the point whose coordinates are x, y, z, and t is given by

$$s^2 = t^2 - x^2 - y^2 - z^2,$$

and again, different choices of axes corresponding to different observers who might assign numbers to x, y, z, and t all lead to the same value of s. A large class of phenomena in the four-dimensional world are different from what they would be in a world without the change of sign. I mention only the existence of waves, which can originate in a small region of space and spread out into a large one. Since at the fundamental level physics describes the behavior of all kinds of matter in terms of waves, and since our senses of sight and hearing operate with waves, it is necessary to make the distinction of sign in order to explain almost anything we perceive. But not everything; for some theoretical calculations the distinction need not be made. I once recalculated the famous phenomenon of the precession of the perihelion of Mercury, one of the basic observational tests of relativity theory, and found to my surprise that the sign associated with the time axis makes no difference. There are calculations in quantum field theory which are impossibly difficult unless the distinction is ignored, and since the time of Copernicus we have often found that computational simplicity should be regarded as a sign of pragmatic truth. But when need the distinction be made? Whenever we study how events unfold in time. Is the distinction in sign, then, purely a convention adopted in deference to our arbitrary insistence on separating out one dimension of spacetime for separate treatment? It could be argued so, but I think not. Our experiences of space and of time are so very different that I cannot believe that there is anything arbitrary about the distinction: the distinction is already there.

How did it get there? Why does spacetime have three dimensions of space and one of time? Try to imagine what life would be like if we were suddenly taken to a world in which time had two dimensions, analogous to the two spatial dimensions of a table top. Or don't try: you'll fail! Our inner life, when we close our eyes, runs along the single thread of time. We would not even know how to think of a second dimension.

Long ago, Henri Poincaré (1903; 1904; 1913) explored the reasons why we are able to organize our spatial experience by means of concepts belonging to three-dimensional Euclidean geometry. The arguments are subtle, starting from the properties of vision and the fact that our bodies are covered with a single skin which brings us tactile sensations. What I now wish to ask is why we have experiences which, with the addition of a single time coordinate, *can* be organized in this way.

Why is it that a description using the dimensionality 3 + 1 gives an adequate account of our sensations, whereas 2 + 2 or 9 + 7 would not? The explanatory remarks to follow are new and tentative and should not be taken too seriously. They are not part of the main thing I want to say, and I shall make them brief.

Cosmic Interlude

During the past thirty years the most urgent task of theoretical physics has been to understand the multiplicity of particles and fields that can be found in Nature. To provide a mathematical basis for such an explanation every point of the four-dimensional spacetime we know and love has got other spaces, abstract mathematical structures, attached to it. The result has been some very complicated geometry. Recently a new idea has arrrived, called dimensional reduction, or compactification (Davies, 1984; Freedman & van Nieuwenhuizen, 1985). Suppose I start with a piece of paper and roll it into a narrow tube. Viewed from a distance its appearance suggests a line, a one-dimensional figure, but when we look more closely we see that it has thickness as well: the second dimension of the paper is still there, at every point along the line, but curled up so that you cannot see it clearly. By analogy let us suppose now, just for amusement, that out of the primaeval chaos of the Big Bang there emerged for a moment an 11-dimensional spacetime. (The number 11 is not absolutely essential, but the possibilities it opens are especially simple.) Suppose that seven of the spatial dimensions are rolled up into a very small seven-dimensional spheroid, leaving four dimensions spread out. Suppose further that the same general kind of physics goes on in the rolled-up seven dimensions as in the four we know about. Then with a few added assumptions it can be shown that the resulting mathematical structure contains some of the structures elaborated by physicists over the last decades, except that they are no longer so abstract or ad hoc, since they involve merely some more dimensions of the same kind we normally inhabit. Further, if we can understand the rolling-up process, we shall have some insight into why the fundamental fields are as they are and why we are left with exactly four dimensions in which to live our lives. Note that even though the spheroid has very small dimensions it is everywhere. Our own bodies and the things we see around us in spacetime are perhaps manifestations of processes in this ubiquitous little spheroid, and if we look very hard we might some day see signs of it.

There remains enough to be explained to keep people busy for some time: Why eleven dimensions: What made exactly seven of them collapse? Nowadays the second problem has a name – spontaneous compactification – but no accepted solution. It generates a dozen papers every month. If the problems can be solved, there will be substantial gains at several points in physical theory, but we need not go into that here. Let it suffice that (even though from an epistemological point of view it is absurd to say so) spacetime is becoming even more like a thing for physicists, like Newton's absolute space and time which exist, as he wrote, »without regard to anything external,« logically prior to the created universe. But now, since the curvature of spacetime, and perhaps even its dimensionality, are determined by its material contents, the situation is more complicated than Newton imagined. Whatever comes out of this speculation must finally explain one basic fact among others: the world is such that our experience of it can be described in terms of a spacetime of 3+1 dimensions.

In Einstein's words about space and time we recognize echoes of Kant and earlier writers. It is interesting to note how many people have thought this way, but by now there does not seem to be much more that we can learn from them. We must try to put a scientific consciousness into the immensity of spacetime, and for that purpose we turn to that dried-out little creature which physicists call The Observer.

The Conscious Observer

The gateway to a physics in which the observer plays a significant role is quantum mechanics, the theory of atoms which originated with Werner Heisenberg in 1925 and was brought essentially to its present form during the twelve months that followed. From the beginning there were questions of how to interpret the theory's various formulas (Peterson, 1968; Jammer, 1974). The formulas agreed beautifully with experiment but generally failed to give physicists the feeling that they knew what they were doing when they used them. Now, 60 years later, the question of interpretation is still not settled, and eminent modern physicists have admitted they do not understand this theory which they use every day of their working lives. There are conflicting interpretations in the market, but so little progress is made that the whole discussion goes to sleep for long periods. I suspect one of the reasons we have not succeeded better is that all the work has been done by physicists and mathematicians and a few chemists, and that they badly need help from people with a different point of view.

A scientific theory must explain. A theory that is only a phenomenological description, however useful it may be in practice, is not enough. If one believes this, quantum mechanics is very difficult to understand. To show why that is so, I list some requirements which such a theory ought to satisfy.

First, it should agree with experiment. Quantum mechanics does that. It yields accurate numbers for the wavelengths of spectral lines, the physical properties of molecules, and so on. And there is another class of results that it predicts just as well. Certain atomic events are unpredictable. They occur at random but appear to be governed by definite statistical distributions, and these derive correctly from the theory as needed. There are calculations we do not know how to make, or make accurately, but in 60 years no experimental fact has been found which shows that the basic ideas are wrong.

Second, since large objects are made of atoms, the theory should also govern the behavior of large objects. A human being is an example of a large object.

Third, the theory should give us the satisfaction of feeling that we understand what is going on. This is a subjective matter, concerning which we expect to find individual differences of opinion; little has yet been written about it.

Fourth, the theory should apply to individual cases. We do not ordinarily feel that a theory of gases need concern itself individually with every atom; in this case statistical results satisfy us, but I have said that the theory should apply also to large objects. I do not wish to treat the planet Jupiter or my friend John only as a member of a statistical ensemble of Jupiters or Johns. Newtonian mechanics did not require that; for quantum mechanics to require it seems to me regressive, though not everybody agrees with my judgement.

Atomic phenomena have an element of randomness which is believed to be primitive and fundamental and not in any way traceable to our ignorance or lack of skill. One may say that the theory is statistical or that it is probabilistic; it will help the discussion if I distinguish between these terms. Statistics is an objective science, and statistical laws contain as much positive knowledge as any other laws. My insurance company gathers data and makes its decisions on statistical grounds. But when I buy insurance I am thinking about probability, and I don't care at all about statistics. I have only one house to burn and one life to lose. Probability pertains to the individual case, and it has also an element of subjectivity. Suppose I am standing with you on a street corner and we watch a man walking on the other side of the street. I ask you: »What is the probability that he goes into house #*211?*« You mention some small number. Then I whisper into your ear: »But he lives in #*211.*« Nothing about the house or the man has changed, but suddenly you assign a larger probability.

Niels Bohr and Albert Einstein had different opinions about quantum mechanics but they agreed that it is a statistical theory, free from subjectivity and dealing only with the average behavior of things. There are others, however, who believe it should be probabilistic. They believe that probabilities change when we find out a new fact, say by making a measurement, and that the theory ought to be able to deal with things individually and not merely *en masse*. These are the people who say that nobody understands quantum mechanics. They find that the statistical interpretation is too phenomenological, not explanatory enough. They are ready to pay heavily for the enlightenment they seek, but do not seem to find it available at any price.

In Search of the Individual Case

The difficulties in interpreting quantum mechanics arise from two mathematical facts, both of which contain surprises for people who like to adopt a no-nonsense approach to physical description. The first is that an atom or other physical system can exist in a composite state. If first a violinist and then a horn-player sounds an A, the pitches are the same but the notes are readily distinguishable. We understand that this is because they are not pure tones; rather, the characteristic sound of each instrument can be resolved into several pure tones sounding at once with appropriate strengths, or amplitudes. We may say that the tones are composite (or complex). This is a commonplace in acoustics, but in 1925 it was not a commonplace that mechanical systems could exist in states which are in exactly the same sense composite: just as a horn produces several tones at the same time, an atom can occupy several of its possible states, with various amplitudes, at the same time; it can even be in several of its possible places at the same time. Consider for example a radioactive atomic nucleus, which at an initial instant we know to be of a certain kind. We put it into a counting device which in a few minutes sends an electronic signal to tell us that the nucleus has emitted a particle and become a nucleus of a different kind. The exact time at which we receive the signal cannot be predicted in advance, but it is possible to characterize any radioactive species by its half-life: the probability that the signal will occur during that period of time is one-half.

The theory of such a process seems at first very strange. The nucleus, originally in a certain state, begins to make a transition into another state, in which it exists simultaneously with the first state. As the amplitude of the second state increases, that of the first one goes down. At the end of a period equal to the half-life the amplitudes are exactly equal. At any moment, the square of the amplitude of the second state represents the probability that the nucleus has decayed by that time, while that of the first state represents the probability that it has not decayed. The two probabilities, of course, add up to 1. The theory says nothing about the sudden signal from the counter; it cannot, since it asserts that the decay of any individual atom occurs unpredictably, at random. Since radioactive samples ordinarily consist of a very large number of nuclei, the probabilistic description translates readily into a statistical one. We do not feel cheated if we are told that when the half-life has elapsed about half the sample will have decayed, but if we think about an individual nucleus we prefer to think of it as being either in its initial or its final state, and not in both at once.

I mentioned above that the problem of interpreting quantum mechanics stems from two mathematical facts. The superposition of states is the first one. The second is as follows. Let us think about the state of the counter into which we put the single radioactive nucleus. Initially, the counter has not fired. As time progresses, the probability that it has fired increases. If we assume that quantum mechanics governs large systems as well as small ones, the state of the counter develops exactly the same composite character as the state of the nucleus the counter is watching. It is a mathematical theorem of quantum mechanics that instead of being simply a counter that either has fired or else has not, it becomes a counter existing in two states at once. This is a predicted consequence of the strictly causal nature of quantum theory, but it is a thing that none of us has ever seen.

Bohr, if I understand him correctly, takes a commonsense view of the situation. For him, the quantum process ends with the »irreversible act of amplification« of the instrument which detects it. Nobody ever disputes whether or not an electronic signal has been sent, whether a hole has been punched in a paper tape, or a developed photographic film has a mark on it. These are all amplifications of the atomic act which bring it out of the microworld into the world of ordinary experience. At that point, discussion ceases. *Les jeux sont faits. Rien ne va plus.*

Nevertheless, there is the situation of the counter which is in two states at once. If I sit looking at the counter, am I then also in two states at once? Experience says no. It is necessary to assume that the process stops somewhere, and in the conventional interpretation it is the moment when the amplifier has served its purpose. This is required because we have never seen a counter in two quantum states at the same time, nor have we ourselves ever had an experience which we could explain as the result of being in two states at the same time. The logician's Law of Contradiction rests ultimately on experience.

My friend sits motionless, watching the counter. I sit in another room where I can watch her but cannot see the counter. If quantum mechanics applies to large systems it applies to my friend, watching the counter which watches the radioactive nucleus. After a while I ask her whether the light has flashed. She says yes. Now I know. Until now, she has been in a composite state as observed by me, though she herself has known for some minutes whether or not the counter has flashed and from her own point of view she was not at any moment in a composite state. I said earlier that probabilities often have a subjective element. So, apparently, have quantum states.

There is of course an enormous literature about all this, going back 50 years, but it still confronts me with a choice. If I stay with Bohr I stay with the statistical explanation, which does not answer my desire to know about what happens in individual cases. If I go with Eugen Wigner and many others I find myself confronting an apparent contradiction between what quantum mechanics says about the atoms and molecules of my body and brain and what it says about my mental state, which I know better than I know anything else. I say it is an apparent contradiction because I do not know. There is a great conceptual gap between talking about atoms and molecules of the brain and talking about a mental state. I do not require or expect a physical theory of consciousness. It is a question of nonoverlapping categories of discourse, as if I were to ask for a chemical theory of music. But still, there might be some physico-chemical scheme of mental function not amounting to a theory, some pattern of connections between the different categories which would serve as a guide for physicists who feel the need to incorporate observers into their theories and perhaps also for those at the other side of the gap, who think about thought.

What characteristics are necessary for an observer which participates in the study of the world at the quantum level? I think it must experience the flash of the counter as a distinct event, not as a gradual building up of probability. It must have a sense of Now, and our task is to see how this sense can be understood from the standpoint of physical science.

The Missing Now

If we think about squares or triangles, we normally draw them or imagine them drawn on a plane surface, a twodimensional section of spacetime. In fact the concept of a plane is an abstraction which must have developed out of the activities of surveying and drawing. So also for the threedimensional continuum; so also for four-dimensional spacetime, in which a dimension has been added to enable us to represent change. The representation used is that of graphs that represent time-dependent phenomena: ordinarily we borrow a dimension of space and label it time, but that is only for convenience; occasionally it is useful to have a real-time representation of events as they occur. The simplest possible case already contains all the intellectual problems, so let us consider it for a moment. An object moves in a straight line at constant speed, and we represent this notion by a formula,

distance = velocity x time,

or by the corresponding straight-line graph. The graph and the formula both represent the motion: give me a time and I will tell you what the distance was; give me the distance and I will tell you what time it was.

But these representations, neat as they are, omit an aspect of motion that is important in our experience of it, its continuous progression from start to finish: into the formula one can insert values of the time t in any order. But experience is not like that: In the formula all values of t are represented in the same way; in experience there is one value that is of supreme importance, the one called Now, and it seems to us to get constantly later and later. Now is when we have our experiences; Now is when we observe what we observe and know what we know; any other moment is a mental construction. None of this is visible on the graph, nor can it be made visible by trying to improve the representation. An obvious device would be to say that t continually increases, so that the point representing the present moment moves across the diagram as time passes. Then how quickly does it move? Unwillingly the answer is dragged out of us, since we can see the dangers ahead: at a rate of one second per second. How much is a guilder worth? One guilder. Nothing has been said. The guilder derives its value from aspects of reality that cannot be expressed in terms of guilders. Similarly the sense of Now and the so-called ›passage of time‹ derive their meaning from aspects of reality that cannot be expressed in terms referring to spacetime, which are the terms that physicists ordinarily use. And consequently a neurophysiology that reduces its arguments to physical (or biophysical, or chemical) terms cannot even in principle analyse the function of an observer or, *a fortiori*, explain consciousness.

It would be easy at this point to pronounce the word ›reductionism‹ in a disapproving tone, as though a certain scientific approach were basically faulty and its limits had been reached. But in fact no science capable of growth has limits that are well defined. It seems that we need an enlargement of the conceptual scheme of physics and the theories that depend on it, and a corresponding change in the way in which things are explained.

Requirements for an Observer

It is in quantifying the description of the world as we perceive it that time becomes spread out along a scale and the sense of Now is lost. I do not think that basic science can do without numbers, but surely numbers can be used in more than one way. Numerical attributes are usually stated in sentences containing the word ›is‹: the velocity of some particular electron *is* 2×10^7 m/s. But this kind of statement is a linguistic shortcut. It can neither be proved or disproved by direct observation. If we really need to justify it, we must give a long discussion of the logic behind the idea of an electron's velocity and the experimental procedures by which, indirectly, one determines its value. Procedures: somebody does something, something happens. Procedures belong to the world of experience, of Now, and they exist not so far from the domain of abstract concepts as one might have feared. I surmise that every scientific statement can be replaced, perhaps at the cost of many words, by statements referring to the perception of events and the performing of muscular acts. The results of science are usually stated, not narrated, and I have tried to show that these statements, tenseless in form, lose track of the human sense of time. The purpose was clarity and brevity. These are great virtues, but is there perhaps another way of talking, equally clear and brief, that does not throw away the sense of advancing time? If as I think this sense must be mentioned in any analysis of what consciousness is or how it works, that would have to be the language of any future science of the mind.

An observer needs, I believe, a sense of Now. It also needs a memory, but that is not hard to arrange. Does it need free will? In quantum mechanics, as I indicated, one cannot measure or specify simultaneously the position and momentum of a particle. An observer may choose to measure either but not both, for the measurements interfere with each other. I think the ideal quantum observer of these phenomena must be free to choose. We piece together our knowledge of this level of nature from the results of various observations, each of which requires a choice of procedure which excludes other observations from being made. In Bohr's language, these observations are said to reveal complementary aspects of a situation, and we learn from the totality of them. In order to acquire this knowledge the observer, it seems to me, must be free at each moment to choose what measurement to make. Whatever choice is made, the theory predicts what will happen; that is one of its great merits. If the observer is causally constrained to choose in a certain way, it becomes meaningless to talk about what would have happened if it had chosen otherwise. We are left in a situation that is a little strange. Most physicists do not feel competent to comment on the partly philosophical, partly physiological question of free will. Yet it turns out that for explanatory, as opposed to computational purposes, if there were no free will it would have been necessary to invent it.

I wrote earlier that the task of science is to explain the world in terms which, as far as possible, refer to the world as we perceive it and to ourselves as we perceive ourselves. We perceive the world as a series of individual cases; we believe that people think and learn. In trying to create a science which performs its task, we face situations which seem to require that physics be extended to include some model of consciousness. It appears that a model of consciousness based on current physics is not possible, and indeed there was no reason to expect that it would be. The brain is however a material system and so is amenable ultimately to physical law. My suggestion is that for the beginnings of a physical explanation of mental function physics as we know it does not need to be extended by the addition of new principles (though this might possibly be so), but rather that it may be necessary to express its arguments in a language of perceptions and acts more in harmony with the world of perceptions and acts which it seeks to explain. In this view, mathematics remains the language of calculation, but the language in which arguments are formulated and their conclusions drawn is plain, ordinary speech.

References

Davies, P. The eleven dimensions of reality. *New Scientist*, 1984, *nr. 1396*, 31-33.
Einstein, A., & Besso, M. *Correspondence 1903-1955*. Paris: Hermann, 1972.
Forsee, A. *Albert Einstein, theoretical physicist*. New York: McMillan, 1963.
Freedman, D.Z., & van Nieuwenhuizen, P. The hidden dimensions of spacetime. *Scientific American*, 1985, *252* (3), 74-83.
Jammer, M. *The philosophy of quantum mechanics*. New York: Wiley, 1974.
Lorentz, H.A., Einstein, A., Minkowski, H., & Weyl, H. *Das Relativitätsprinzip*. Leipzig and Berlin: Fortschritte der mathematischen Wissenschaften, 1923. English transl. *The principle of relativity*. New York: Dover, undated.
Peterson, A. *Quantum physics and the philosophical tradition*. Cambridge: MIT Press, 1968.
Poincaré H. *Science et hypothèse*. Paris: Flammarion, 1903. English transl. *Science and hypothesis*. New York: Dover, 1952.
Poincaré, H. *La valeur de la science*. Paris: Flammarion, 1904. English transl. *The value of science*. New York: Dover, 1958.
Poincaré, H. *Dernières pensées*. Paris: Flammarion, 1913. English transl. *Mathematics and science: last essays*. New York: Dover, 1963.
Taylor, E.F., & Wheeler, J.A. *Spacetime physics*. San Francisco: Freeman, 1966.

Chapter 4. The Use of the Biological Clocks in Time Perception[1]

Gerard Groos and Serge Daan

Recognition of Local Time vs. Time Interval Estimation

With the evolution of life on a rotating planet, biological ›clocks‹ have evolved as adaptations to periodic fluctuations in the environment. The role of such clocks, as innate temporal substrates for behavioral and physiological programs forming an adaptive match to predictable variations in the world outside, is well established (Aschoff, 1981). There are numerous ways in which the proper timing of behavioral events with respect to time of day contributes to the fitness of the individual organism. The behavior timed may either be genetically attached in a rigid manner to a particular phase of the endogenous, ›circadian‹ cycle or it may be flexibly adjusted in response to prior experience (e.g. Daan, 1981). For events which occur only once in a lifetime such a pupal eclosion in insect metamorphosis, precise timing in the day is crucial for individual survival but can not be altered by experience. Adjustment of daily behavioral programs to experience is widespread in animals. It shows, for instance, in increased intestinal and adrenal activity in rats at times of day corresponding with their customary feeding hours (Krieger, 1974; Suda & Saito, 1979). At the behavioral level it is known in the return of honeybees to an abundant food source 24 hours after training (Figure 1, part A; Beling, 1929), and in the enhanced probability of hunting and returning to a particular area in birds of prey, 24 hours after having made a successful hunt (Rijnsdorp et al., 1981). Such adjustment on a periodic basis is likely to contribute to daily habitual routines characteristic of the individual rather than of the species. Both types of daily timing of events imply the recognition of the time of day by living organisms. There is now widespread conviction that this is accomplished by direct reference to the phase of endogenous oscillators which themselves are phase-locked in synchrony with the external cycle of light and darkness. This recognition of phase and, hence, of local time does not necessarily involve recognition of the lapse of time. However, biological clocks potentially provide a mechanism usable for time interval estimation. In the case of behavioral adjustment to experience, circadian phase recognition is indistinguishable from 24-hour interval estimation. It is not excluded that circadian clocks can also be employed to estimate intervals that are shorter or longer than 24 hours. If this is the case, however, their functional meaning would extend beyond the recognition of time of day, since such clocks might provide organisms with a general sense of the passage of time. In this paper we shall address

[1] The authors gratefully acknowledge the kind permission of Prof. Dr. Jürgen Aschoff to incorporate his data in this paper.

the evidence pertinent to the involvement of circadian rhythmicity in time estimation. Before addressing this question a brief summary of current knowledge on the structural organization of circadian clocks is appropriate. We shall concentrate on the situation in mammals.

Circadian Pacemakers in Mammals

Circadian rhythmicity is probably ubiquitously present in eukaryotic animals and plants, including unicellulars. It is characterized by its persistence in experimental conditions where all periodic cues that can possibly indicate the time of day are removed. In such an aperiodic environment the organism develops ›freerunning‹ rhythms with an intrinsic period deviating slightly but significantly from 24 hours (Aschoff, 1981). In animals with a highly developed nervous system there are numerous physiological and behavioral processes with synchronous circadian

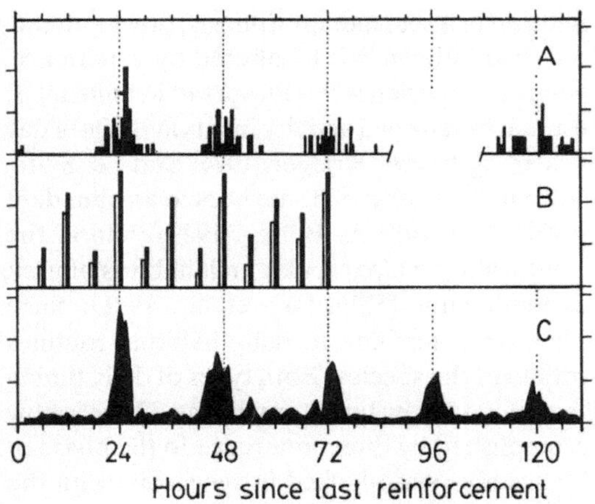

Figure 1. Circadian time interval estimation in honeybees and rats. If animals are exposed to the periodic availability of food at 24 hour intervals they will respond by concentrating their activity at times of day beginning just before the presentation of food. In active avoidance conditioning situations, rats learn the time of day at which to express the appropriate avoidance behavior. Both types of circadian interval estimation can be observed even in the absence of periodic reinforcement of the behavior. In part *A* this is shown for the frequency of appearance at a feeding site of honeybees which had been trained to feed at a fixed time of day (after Wahl, 1932). Circadian retention fluctuations after single-trial learning of an active avoidance response is shown in part *B* for the rat (after Holloway & Wansley, 1973a). The vertical bars indicate the percentage of animals meeting a fixed response criterion (open column) and the mean latency of response (filled column), measured at 6, 12, 18, 24, etc. hours after the conditioning trial. The persistence of anticipatory activity displayed by rats that have been exposed to a 24-hour restricted feeding schedule is presented in part *C* (after Boulos, Rosenwasser & Terman, 1980).

variation. This has led to the hypothesis of endogenous control by circadian pacemakers. This idea, in itself rather abstract, has prompted a search for anatomical substrates for such pacemakers.

This search has been successful in widely different groups of animals (mammals, birds, molluscs and insects) and it has shown that the task of temporal coordination is allotted to specific parts of the central nervous system. In various mammals, including primates, the suprachiasmatic nucleus has been identified as a major circadian pacemaker (Stephan & Zucker, 1972; Rusak & Zucker, 1979; Groos et al., 1983; Groos, 1984). This tiny group of nerve cells is located at the base of the brain just above the decussation of the optic nerves from which the suprachiasmatic nucleus receives a neural input (Moore, 1978). In man, a homologous nucleus has recently been identified.

The suprachiasmatic nucleus is capable of sustaining an endogenous circadian rhythm which is entrained to the environmental light-dark cycle via its retinal connection (Groos & Meijer, 1985). At the output side the entrained suprachiasmatic nucleus rhythm is transmitted through the neuro-endocrine system to drive various circadian rhythms (Rusak & Zucker, 1979). Two major functional characteristics of the suprachiasmatic nucleus pacemaker should be emphasized. First, the pacemaker rhythm is self-sustaining, that is, it persists independent of any extrinsic periodic drive. Second, since the intrinsic pacemaker period deviates somewhat from 24 hours, it has to be reset daily to retain synchrony with the environmental light-dark cycle. This process is called photic entrainment (Pittendrigh, 1981). Entrainment establishes an identical period between the pacemaker and the light-dark cycle in such a way that there is a stable, characteristic phase relation between pacemaker rhythm and entraining cycle. However, the pacemaker is only capable of achieving stable phase-locking to the external cycle if the latter's period falls within a limited range, the range of entrainment. Furthermore, due to the stability of the mechanisms involved in entrainment, re-entrainment after a phase-shift in the entraining cycle occurs gradually, usually requiring a number of cycles (Aschoff, 1981). Each of these pacemaker properties has also been observed in pacemakers other than the suprachiasmatic nucleus, for example in the pineal gland of birds. It seems justified to use these features of the freerunning and entrained rhythm as defining properties of an underlying pacemaker.

The Use of Biological Clocks in Circadian Time Interval Estimation

For quite some time students of animal behavior have known that, among other species, honeybees and birds use the highly reliable cue of the sun's celestial location for directional orientation during the day.

A major disadvantage of sun compass orientation, however, is that the sun continuously – though predictably – changes its position in the course of the day. Birds employ a circadian clock as a chronometer to correct for the passage of time in determining direction by observing the position of the sun (Hoffmann, 1960). This mechanism is known as time-compensated sun compass orientation (Von Frisch,

1950; Kramer, 1950). Time compensation is a clear example of time lapse measurement, since the response varies monotonically with the duration of time elapsed (Pittendrigh & Daan, 1976). The question should be raised whether mammals are similarly endowed with the ability to measure a lapse of time in the order of several hours by using their circadian pacemakers.

There is considerable evidence that the rat uses its circadian clock in the suprachiasmatic nucleus to estimate circadian time intervals (Holloway & Wansley, 1973a, 1973b; Tapp & Holloway, 1981; Holloway et al., 1980). Rats learning active or passive avoidance tasks at different times of day show optimal retention at intervals of 24, 48 and 72 hours after the initial learning trial (Figure 1, part B). Thus, the performance on these tasks peaks at a circadian interval and its multiples after the time of learning. This observation indicates that rats have a memory for the time of day at which they learned the avoidance reaction by measuring the lapse of consecutive 24-hour intervals from that time onward. They thereby periodically optimize their performance at corresponding clock times on subsequent days.

The ability to take the time at which a new response became useful into account is presumably of considerable adaptive significance for animals in general. Many events requiring appropriate novel behavior have an increased probability of occurrence at certain times of day. For instance, animals may reduce risk by avoiding a feeding site when predators or competitors are likely to be present, and enhance their fitness by visiting the same site at other times. Such responses should not be genetically coupled to a fixed phase of the circadian cycle since in many situations the critical phases for expressing or suppressing behavior cannot be predicted without specific temporal learning (Rusak & Zucker, 1975).

The case of learned avoidance responses in the rat with maximal retention at post-learning times that are multiples of 24 hours may be an instance of circadian time interval estimation based on a circadian pacemaker. Two observations provide compelling evidence for this view. If the rat suprachiasmatic nucleus is surgically lesioned many of the animal's circadian rhythms are eliminated because their important pacemaker is no longer available (Groos, 1984; Tobler et al., 1983; Rusak & Zucker, 1979). Similarly, rats sustaining suprachiasmatic lesions no longer show the circadian pattern in the retention of an avoidance response (Holloway et al., 1980). Moreover, when, in intact rats, the entrained pacemaker is phase-shifted by a shift in the timing of the light-dark cycle, the circadian retention rhythm is shifted accordingly (Tapp & Holloway, 1981). Thus, the ability to time this particular behavior on the basis of 24-hour interval measurements is critically dependent on the circadian pacemaker in the suprachiasmatic nucleus.

There are other ways in which animals express their ability to measure circadian time intervals. Rats, for instance, can readily be made to respond to a 24-hour interval between two short food presentations. When circadian feeding schedules are imposed on these animals within a few days they anticipate the meal time in terms of physiological and behavioral preparation for food consumption (Krieger, 1974; Suda & Saito, 1979). Feeding schedule anticipation, however, occurs independent of the circadian pacemaker in the suprachiasmatic nucleus. Lesions of this nucleus do not abolish the ability of rats to measure the inter-meal interval

(Boulos et al., 1980; Stephan et al., 1979a). In intact rats, the circadian pacemaker in the suprachiasmatic nucleus runs free, unentrained by the feeding schedule (Inouye, 1982; Gibbs, 1979; Aschoff et al., 1982). Yet, a circadian pacemaker is most probably involved in meal anticipation. Several arguments have been advanced to support this interpretation. First, if the period of the feeding schedule is varied between 17 and 34 hours, behavioral meal-anticipation occurs only within the limited range of periods between 23 and 29 hours (Boulos et al., 1980; Stephan et al., 1979b; Stephan, 1981). Thus, feeding schedules only entrain anticipatory activity rhythms within a narrow range of entrainment (Figure 2). Second, under different feeding cycle periods within this range a characteristic phase angle difference is established between the feeding schedule and the anticipatory activity (Figure 2; Stephan, 1981; Honma et al., 1983). Third, in response to an advance or delay shift in the entraining feeding cycle, rats gradually phase shift their anticipation rhythm to re-establish a stable phase relation with the schedule (Figure 2; Stephan, 1984). Fourth, after termination of the feeding schedule the anticipatory activity persists for a number of cycles while it gradually damps out (Figure 1, part C; Boulos et al., 1980). Thus, although the mechanisms underlying this type of circadian interval estimation in the rat are not located in the suprachiasmatic nucleus, they do involve a biological clock since the anticipation rhythm exhibits the defining characteristics of a process controlled by a circadian pacemaker.

The Use of Biological Clocks in Short Interval Estimation

While there is evidence that animals use biological clocks to estimate time intervals in the order of 24 hours, it remains to be asked whether interval durations considerably less than 24 hours are measured in a similar fashion. Apart from the specific case of time-compensated celestial orientation referred to above, the available evidence stems from human studies. Some information relevant to the question is available from early studies involving time estimation in freerunning human subjects (Macleod & Roff, 1936; Halberg et al., 1965; Lavie & Webb, 1975). The most systematic study of this kind, however, was undertaken recently by Aschoff (1984).

Aschoff asked 30 human subjects freerunning in complete temporal isolation, to ›produce‹ long (1 hour) and short (10-120 seconds) intervals. Since the freerunning periods of the sleep-wake rhythm developed by humans under constant conditions in isolation vary widely (Wever, 1979), it was possible to relate each of these estimates to the duration of wakefulness (α) during each cycle of the freerunning circadian rhythm. Because duration of wakefulness (α) varies in proportion to the total circadian cycle duration (τ), the time estimates could also be related to the circadian period of each subject. Aschoff found a strong, positive correlation in all subjects between 1-hour time estimates and duration of wakefulness (Figure 3). In contrast, the short interval estimates bear no systematic relation with the circadian period. These results suggest strongly that human subjects use a circadian mechanism for the estimation of 1-hour time intervals but not for estimating brief

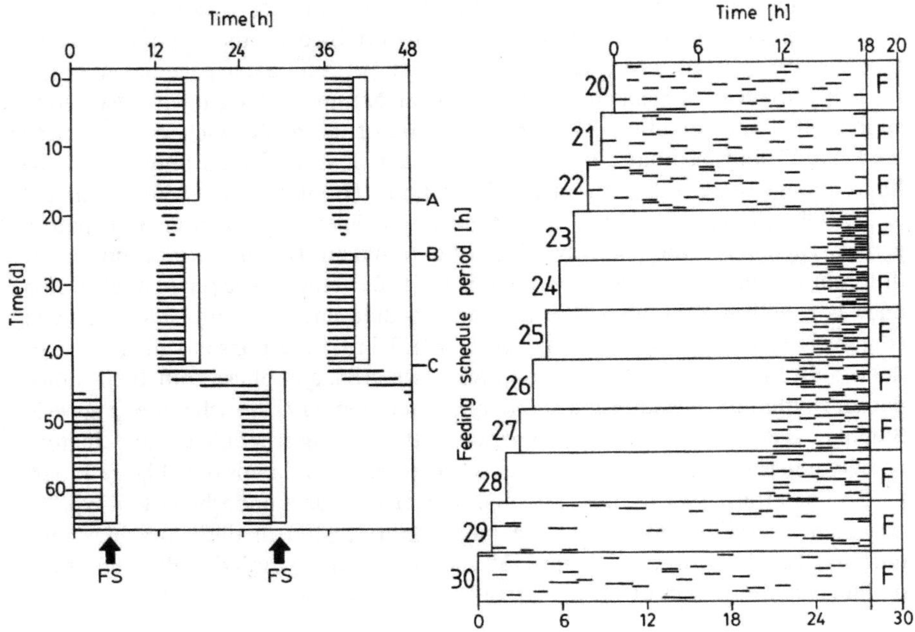

Figure 2. Schematic summary of feeding schedule anticipation characteristics (based on Stephan, 1981, 1984). The *left panel* shows the anticipatory activity pattern entrained to a 24-hour feeding schedule *(FS)*. Subsequent days of the record are plotted vertically, each day's record appearing underneath the previous one. The vertical rectangles indicate the daily 2-hour interval of food availability. The record is double plotted to facilitate visual inspection. Initially (day 0–18) the activity (short dark bars) anticipates the time of feeding. At *A* (day 19) the animal is starved for several days. Under these conditions the anticipatory activity persists for a few cycles and then damps out. At *B* the feeding cycle is reinstated and, as before, activity preceeds feeding. At *C* the feeding schedule is phase-shifted by 12 hours. This shift is followed by gradual re-entrainment of the anticipatory activity rhythm through a number of delay transients until a new stable phase relation is established by day 50. The *right panel* shows 18 day segments of activity recorded under feeding schedules (*F:* food availability) with periods from 20 to 30 hours. Within the entrainment range (23 to 28 hours) anticipatory activity is expressed. The phase of activity onset becomes more positive with respect to the phase of feeding with increasing feeding cycle length.

intervals. It has been demonstrated that the duration of wakefulness in humans may be determined by the circadian pacemaker in interaction with a process reflecting an increasing pressure to sleep (Daan et al., 1984). Hence, it remains open to future research to establish whether the pacemaker itself or ›sleep pressure‹ is responsible for human 1-hour interval estimation, as pointed out by Aschoff (1984). At any rate the available evidence supports the idea that the circadian system is somehow involved in the estimation of time intervals in the hours range, whereas entirely different mechanisms are apparently used for the perception of time lapses in the order of minutes.

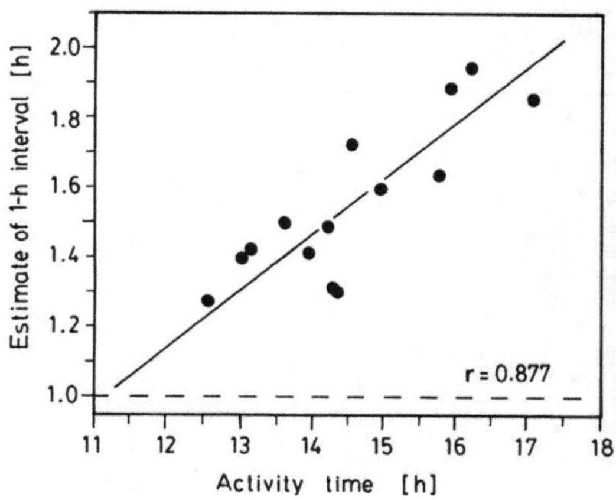

Figure 3. 'Daily' means of 1-hour estimates (method of production) for a single subject freerunning under conditions without time cues. The 1-hour estimates are plotted with respect to the duration of wakefulness ('activity time') in each circadian cycle. The subjective 1-hour estimate is positively correlated (r = 0.877) with activity time (data from Aschoff, 1984).

There is little comparative research on time interval estimation in animals other than man, at least for intervals in the hour range. Time perception in the minute range has been frequently studied (Richelle & Lejeune, 1980; Richelle et al., 1985, chapter 5 of the present volume; Gibbon & Allan, 1984). The production of time intervals in the hour range, comparable to the human experiments has been demonstrated in spontaneous animal behavior and by operant conditioning. Spontaneous production of short time intervals is exemplified by ultradian rhythms in animal behaviour, such as the remarkably precise 2-3 hour feeding rhythm in microtine rodents (Daan & Slopsema, 1978). It has been shown that this rhythm is under phase control of the circadian clock, both in natural and experimental conditions (Hoogenboom et al., 1984). Changes in the intrinsic circadian pacemaker's period, occurring either spontaneously or following experimental manipulations are usually paralleled by changes in the pattern of the ultradian rhythm (Figure 4). However, this does not represent sufficient evidence for a circadian pacemaker role in the production of the 2-3 hour interval *per se*. It might well be that the parallel ultradian changes result from the circadian pacemaker's phase control over the ultradian system.

Hourly interval estimation as studied by conditioning has consistently involved training sessions at different times of day. Considerable precision in time estimation has been demonstrated, most notably in honeybees, which are able to correctly discriminate between two odors alternatingly associated with a reward up to a frequency of 45 per minute (Koltermann, 1971). However, the problem with these and similar studies is that we can not discriminate between multiple recognition of local time and the perception of the lapse of time.

In summary, there are numerous examples for the involvement of circadian clocks in the recognition of local time. Circadian control over processes employed in the perception of time elapsed is documented for at least three cases: time compensated celestial orientation, ultradian behavioral rhythmicity and human time perception. In our view, the available evidence in all three cases is compatible

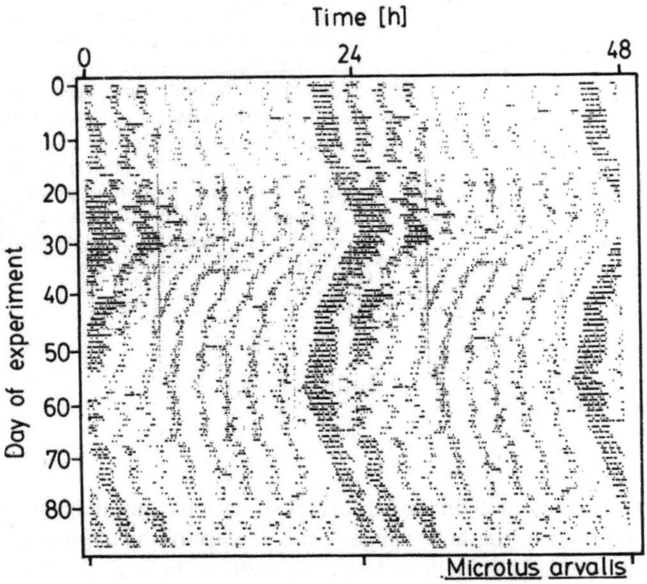

Figure 4. Circadian and ultradian wheelrunning activity and food intake in the common vole *(Microtus arvalis)*. The record shows the close phase correspondence of the ultradian (2–3 hour) rhythm with spontaneous fluctuations in the circadian period. Each day in the record (subsequent days plotted vertically) consists of a parallel activity and a food intake record. Activity is plotted above the food intake. The animal was exposed to constant darkness throughout this record (Daan, unpublished observations).

with two hypotheses: the time intervals (or orientational angles) produced are a direct reflection of the angular velocity of the circadian pacemaker, or they are dependent on interval timers the phase of which is set at least once per cycle by the circadian pacemaker. It should be rewarding to develop experimental designs discriminating between these hypotheses. While much of the basic physiology of time perception remains to be unraveled, it is clear that circadian pacemakers are involved in this faculty. In addition to allowing for the recognition of time of day and, via their role in photoperiodic time measurement, for the recognition of time of year, they apparently support at least in some documented instances the general perception of time elapsed.

The involvement of the circadian system in time perception, however, has only been documented for instances in which intervals of approximately one hour or longer were estimated. Whether biological clocks mediate the perception of brief intervals, that is, intervals which are relatively much shorter than one hour, remains to be established.

References

Aschoff, J. *Handbook of behavioral neurobiology.* Vol. 4. New York: Plenum Press, 1981.

Aschoff, J. Freerunning and entrained circadian rhythms. In: J. Aschoff (Ed.), *Handbook of behavioral neurobiology.* Vol. 4. New York: Plenum Press, 1981, pp. 81-94.

Aschoff, J. On the perception of time during prolonged temporal isolation. *Human Neurobiology,* 1984, in press.

Aschoff, J., Daan, S., & Honma, K.I. Zeitgebers, entrainment and masking. Some unsettled questions. In: J. Aschoff, S. Daan, & G. Groos (Eds.), *Vertebrate circadian systems.* Berlin: Springer, 1982, pp. 13-24.

Beling, I. Über das Zeitgedächtnis der Bienen. *Zeitschrift für vergleichende Physiologie,* 1929, *9,* 259-338.

Boulos, Z., Rosenwasser, A.M., & Terman, M. Feeding schedules and the circadian organization of behavior in the rat. *Behavioral Brain Research,* 1980, *1,* 39-65.

Daan, S. Adaptive daily strategies in behavior. In: J. Aschoff (Ed.), *Handbook of behavioral neurobiology.* Vol. 4. New York: Plenum Press, 1981, pp. 275-239.

Daan, S., & Slopsema, S. Short term rhythms in foraging behaviour of the common vole, *Microtus arvalis. Journal of Comparative Physiology,* 1978, *127,* 215-227.

Daan, S., Beersma, D.G.M., & Borbély, A.A. Timing of human sleep: Recovery process gated by a circadian pacemaker. *American Journal of Physiology,* 1984, *246,* 161-178.

Frisch, K. von. Die Sonne als Kompass im Leben der Bienen. *Experientia,* 1950, *6,* 210-221.

Gibbon, J., & Allan, L. (Eds.), *Timing and Time Perception.* Annals of the New York Academy of Sciences, Vol. 423, 1984.

Gibbs, F.P. Fixed interval feeding does not entrain the circadian pacemaker in blind rats. *American Journal of Physiology,* 1979, *236,* 249-253.

Groos, G. The physiological organization of the circadian sleep-wake cycle. *Experimental Brain Research Supplement,* 1984, *8,* 241-257.

Groos, G., Mason, R., & Meijer, J.H. Electrical and pharmacological properties of the suprachiasmatic nuclei. *Federation Proceedings,* 1983, *42,* 2790-2795.

Groos, G., & Meijer, J.H. The effects of illumination on suprachiasmatic nucleus electrical discharge. *Annals of the New York Academy of Sciences,* 1985, in press.

Halberg, F., Siffre, M., Engeli, M., Hillman, D., & Reinberg, A. Etude en libre-cours des rythmes circadiens du poulse, de l'alternance veille-sommeil et de l'estimation du temps pendant les deux mois de séjour sousterrain d'un homme adulte jeune. *Comptes Rendus de l'Academie des Sciences.* Paris, 1965, *260,* 1259-1262.

Hoffmann, K. Experimental manipulation of the orientational clock in birds. *Cold Spring Harbor Symposia on Quantitative Biology,* 1960, *25,* 379-387.

Holloway, F.A., & Wansley, R.A. Multiphasic retention deficits at periodic intervals after passive avoidance learning. *Science,* 1973a, *180,* 208-210.

Holloway, F.A., & Wansley, R.A. Multiphasic retention deficits at periodic intervals after active and passive avoidance learning. *Behavioral Biology,* 1973b, *9,* 1-14.

Holloway, F.A., Bird, D.C., Devenport, J., & Tapp, W.N. Effects of suprachiasmatic nucleus lesions and feed/starve entrainment conditions on activity and retention performance rhythms. *Society for Neuroscience Abstracts,* 1980, *6,* 832.

Honma, K., von Goetz, C., & Aschoff, J. Effects of restricted daily feeding on freerunning circadian rhythms in rats. *Physiology and Behaviour,* 1983, *30,* 905-913.

Hoogenboom, I., Daan, S., Dallinga, J.H., & Schoenmakers, M. Seasonal change in the daily timing of behaviour of the common vole, *Microtus arvalis. Oecologia,* 1984, *61,* 18-31.

Inouye, S.T. Restricted daily feeding does not entrain circadian rhythms of the suprachiasmatic nucleus in the rat. *Brain Research,* 1982, *232,* 194-199.

Koltermann, R. 24. Std. Periodik in der Langzeiterinnering an Duft- und Farbsignale bei der Honigbiene. *Zeitschrift für vergleichende Physiologie,* 1971, *75,* 49-68.

Kramer, G. Weitere Analyse der Faktoren, welche die Zugaktivität des gekäfigten Vogels orientieren. *Naturwissenschaften,* 1950, *37,* 377-378.

Krieger, D.T. Food and water restriction shifts corticosterone, temperature, activity and brain amine periodicity. *Endocrinology*, 1974, *95*, 1195-1201.

Lavie, P., & Webb, W.B. Time estimation in a long-term time-free environment. *American Journal of Physiology*, 1975, *88*, 177-186.

Macleod, R.B., & Roff, M.F. An experiment in temporal orientation. *Acta Psychologica*, 1936, *1*, 381-423.

Moore, R.Y. Central neural control of circadian rhythms. In: W.F. Ganong & L. Martini (Eds.), *Frontiers in neuroendocrinology*. New York: Raven Press, 1978, pp. 185-206.

Pittendrigh, C.S. Circadian systems: Entrainment. In: J. Aschoff (Ed.), *Handbook of behavioral neurobiology*, Vol. 4. New York: Plenum Press, 1981, pp. 95-124.

Pittendrigh, C.S., & Daan, S. A functional analysis of circadian pacemakers in nocturnal rodents. IV. Entrainment: Pacemaker as clock. *Journal of Comparative Physiology*, 1976, *106*, 291-331.

Richelle, M., & Lejeune, H. *Time in animal behaviour*. Oxford: Pergamon Press, 1980.

Richelle, M., Lejeune, H., Perikel, J., & Fery, P. From biotemporality to nootemporality: Toward an integrative and comparative view of time in behavior. In: J.A. Michon & J.L. Jackson (Eds.), *Time, mind and behavior*. Heidelberg: Springer Verlag, 1985. pp. 75-99.

Rijnsdorp, A., Daan, S., & Dijkstra, C. Hunting in the kestrel (*Falco tinnunculus*) and the adaptive significance of daily habits. *Oecologia*, 1981, *50*, 391-406.

Rusak, B., & Zucker, I. Biological rhythms and animal behavior. *Annual Review of Psychology*, 1975, *26*, 137-171.

Rusak, B., & Zucker, I. Neural regulation of circadian rhythms. *Physiological Review*, 1979, *59*, 449-526.

Stephan, F.K. Limits of entrainment to periodic feeding in rats with suprachiasmatic lesion. *Journal of Comparative Physiology*, 1981, *143*, 401-410.

Stephan, F.K. Phase shifts of circadian rhythms in activity entrained to food access. *Physiology and Behaviour*, 1984, *32*, 663-672.

Stephan, F.K., & Zucker, I. Circadian rhythms in drinking behavior and locomotor activity of rats are eliminated by hypothalamic lesions. *Proceeding s of the National Academy of Sciences USA*, 1972, *69*, 1583-1586.

Stephan, F.K., Swann, J.M., & Sisk, C.L. Anticipation of 24-hour schedules in rats with lesions of the suprachiasmatic nucleus. *Behavioral Neurological Biology*, 1979a, *25*, 346-363.

Stephan, F.K., Swann, J.M., & Sisk, C.L. Entrainment of circadian rhythms by feeding schedules in rats with suprachiasmatic lesions. *Behavioral Neurological Biology*, 1979b, *25*, 545-554.

Suda, M., & Saito, M. Coordinative regulation of feeding behavior and metabolism by circadian timing system. In: M. Suda, O. Hayashi, & H. Nakagawa (Eds.), *Biological rhythms and their central mechanisms*. Amsterdam: Elsevier/North Holland Biomedical Press, 1979, pp. 263-272.

Tapp, W.N., & Holloway, F.A. Phase shifting circadian rhythms produces retrograde amnesia. *Science*, 1981, *211*, 1056-1058.

Tobler, I., Borbély, A.A., & Groos, G. The effect of sleep deprivation on sleep in rats with suprachiasmatic lesions. *Neuroscience Letters*, 1983, *42*, 49-54.

Wahl, O. Neue Untersuchungen über das Zeitgedächtnis der Bienen. *Zeitschrift für vergleichende Physiologie*, 1932, *16*, 529-589.

Wever, R. *The circadian system of man*. New York: Springer Verlag, 1979.

Chapter 5. From Biotemporality to Nootemporality: Toward an Integrative and Comparative View of Time in Behavior[1]

Marc Richelle, Helga Lejeune, Jean-Jacques Perikel and Patrik Fery

The title of this paper refers to the levels of temporality as proposed by J.T. Fraser (1982) and, more precisely, to the two highest levels on a scale of cosmic evolution. Whatever the epistemological status of Fraser's levels concept (for a thoughtful discussion see Michon, 1985c, also chapter 20 of the present volume), it is obvious that time in living systems on the one hand and time in human and infrahuman behavior and mind on the other appear as two distinct fields of research. Roughly, the first is the field of chronobiology which deals with biological rhythms and their underlying mechanisms. The second is a branch of psychology. It addresses itself to problems such as time estimation, the emergence of the concept of time in the child, the ›time horizon‹ in normal and abnormal individuals and so on. If one looks upon these as purely cultural matters and if one favors a radical rupture between the biological and the cultural, the question of the relation between the two levels becomes irrelevant. If, on the contrary, one adheres to the evolutionary view, it is important to look for some continuity between successive levels of temporality, and to trace back to their biological roots the origin of behaviors and ideas related to time in humans. Human beings exhibit some pragmatic organization of time or time allocation in their daily life, they show capacities for elaborate cognitive treatment of temporal information, they build conceptual constructs about time, and they experience time with various affective connotations (*temps vécu* or subjective experience of time). Do all these emerge from more basic and more general forms of adjustments to time in living organisms, such as biological rhythms? And if so, how do they emerge? Students in the field of the psychology of time have, with a few exceptions (see Fraisse, 1967; Macar, 1980), neglected the issue. Even Piaget, in spite of his repeated emphasis on the essential continuity from the biological to the psychological and logical levels, did not pay much attention to concepts and data from chronobiology when he dealt with the ontogenesis of the time concept (Piaget, 1968). Some fifteen years ago, the senior author (Richelle, 1968) first advocated a synthetic approach that would bridge the gap between chronobiology and the psychology of time. The mutual ignorance in which these two fields held one another was all the more surprising because the study of time has otherwise been the locus of unusual multidisciplinary cross-fertilization. It was argued that an attempt at such a synthesis would inevitably imply a comparative approach, and would profitably gain from the experimental analysis of temporal regulations in animals.

---

1 Original experiments discussed in this paper were partly encouraged by a Twinning Grant from the European Science Foundation - ETPBBR to the two senior authors for the study ot time in young animals.

The term *temporal regulation*, as we have come to use it for several years, refers to all forms of behavioral adjustments to time, as exemplified by discriminations of duration of external stimuli, entrainment of motor responding in periodic schedules of reinforcement, temporal patterning of distribution of one's own responses, and the like (for a review of the available techniques and of the related problems, see Richelle & Lejeune, 1980). Since Pavlov's initial work on delayed, trace, and periodic conditioning, the study of time estimation and timing behavior in experimental animals has developed tremendously, due to the techniques of the operant laboratory designed by Skinner in the thirties, and constantly improved by his followers. However, important aspects of the problem have been neglected, and in spite of recent encouraging efforts to integrate various subfields in the psychology of time and to link them to chronobiology (see, for instance, Gibbon & Allan, 1984), a general theory of time in behavior and mind remains a far cry.

We do not intend to propose such a theory here. Instead, we shall focus on three questions which seem of special relevance to the general issue of the relations between *biotemporality* and *nootemporality*, and for which empirical answers, be it at an embryonic stage, are becoming available.

(a) *Temporal Regulations of Behavior and Biological Rhythms*. What is the relation, if any, between temporal regulations as studied in behavioral laboratories, which involve adjustments to arbitrary durations usually of the order of a few seconds or minutes, and biological rhythms rooted in ›natural‹, more fundamental periodicities? Do temporal regulations derive from biological rhythms? Do they obey the same underlying mechanisms or do they, on the contrary, correspond to quite different adaptation processes? Do temporal regulations of behavior possibly serve to make the organism free from chronobiological constraints?

(b) *Cross-species Differences and Similarities*. How do various species compare with respect to their timing capacities as assessed with behavioral techniques and ›arbitrary‹ durations (that is, not directly related to biological rhythmicities)? Are structurally simple and highly complex organisms equally endowed with such capacities as they seem to be with circadian or circatidal time keeping mechanisms? And, if this is not the case, is there some (bio)logical order to be found in the differences between species? Is there any ›evolutionary trend‹ detectable, that might lead us to account for the higher levels of nootemporality that characterize human behavior?

(c) *Ontogenesis*. Like any functional capacity of living systems, the question arises as to whether temporal regulations of behavior are merely the actualization of some innate basic competence, or are the result of a shaping of behavior by environmental conditions. A clear answer to this question can only be given by combining behavioral genetics and the ontogenetic approach. The latter also recommends itself if we want to understand the performances observed in mature subjects. Applied to humans, it should not only help us in following the deployment of cognitive capacities with respect to time, but also, by comparing with infrahuman organisms, in accounting for the crucial differences, if any, that make human time specifically human. Ontogenetic studies in animals are, of course, in that perspective a necessary complement to human studies.

We shall review some of the available facts and comment on selected examples of research in each of these three areas.

Temporal Regulation of Behavior and Biological Rhythms

A few years ago, we reviewed the literature dealing with the relations between temporal regulations and biological rhythms (Richelle & Lejeune, 1980). The number of relevant studies turned out to be extremely small. Only seven or eight references could be located, which reflects the general neglect of chronobiological variables by behavior analysts. Hobbs (1981) also denounced this neglect and pointed out that most experimenters do not care to give details about the maintenance and experimental lighting conditions, nor do they adapt the light-dark cycle to the species-specific habits of their animal subjects. Rats, though nocturnal animals, are usually tested during the day and are kept in the dark by night. Quoting Hobbs: »not only are we developing a psychology based extensively upon the rat, but one based upon ›sleepy‹ rats as well«. Curiously enough, even the behavior analysts who had engaged in the study of temporal regulations had, for years, shown no interest in the contribution of chronobiology.

A more recent review by Terman (1983) produced evidence of a somewhat increasing systematic interest in circadian periodicities as related to operant behavior: the list of papers referred to comes close to 20, most of them deriving from Terman's own pioneering research group. Terman is concerned with a more general issue than the one to which we are addressing ourselves here, namely the relation between biological rhythms and behavioral measures in general, not specifically temporal regulations. Circadian rhythmicity has been shown, for example, for responding reinforced by electrical brain stimulation in rats (Terman & Terman, 1970, 1975, 1976), for key-pecking reinforced by light in pigeons (Ghiselli & Thor, 1973), for visual detection thresholds (Rosenwasser et al., 1979). That various behavioral performances obey similar laws as physiological parameters traditionally measured in chronobiological studies should cause no surprise. Such data provide the required baseline against which experimenters can test the effects of manipulating crucial variables in behavioral control (schedule parameters, reinforcement magnitude or frequency, and the like). Especially suggestive in this respect are results obtained on a light-dark 12 : 12 hour cycle, by Zimmerman (Note 8, reported by Terman, 1983). Lever pressing for electrical brain stimulation was recorded throughout, as well as drinking. Inversion of the light-dark cycle resulted in gradual phase adjustment. Then the electrical brain stimulation was set at a higher, more rewarding, intensity during light periods. This produced an immediate shift of operant activity from dark periods to light periods, while drinking remained synchronized with darkness. An environmental condition known to exert important control of behavior, that is, reinforcement magnitude, can invert the rhythmicity of schedule controlled behavior and make it independent of other activities, such as drinking, that will continue to show the usual pattern. Rats were, so to speak, forced to take their pleasure during day time, quite contrary to the very nature of these nocturnal creatures. This experiment can serve as an experimental paradigm for research on the relation between biological rhythms and behavior, with the following general question in mind: To what extent can environmental contingencies counteract the natural periodicities that affect behavior, as they affect many other aspects of an organism's functioning?

Elsmore & Hursh (1982) have adopted an econometric approach to this question and contributed several ingenious experiments in which total daily food ration, availability of alternative sources of food, frequency of reinforcement, response cost or effort, and proximity to reinforcement were manipulated. They make a distinction between quantitative and qualitative aspects of behavior. *Quantitative* refers here to rate or frequency, *qualitative* to aspects not necessarily related to rate or frequency, such as accuracy in a discrimination task. Both aspects exhibit rhythmicity, but while the former is sensitive to manipulation of variables such as those mentioned above, the latter is not. This contrast was shown, among other situations, in an experiment in which monkeys had to count their own motor responses up to 10 to 15 in order to be reinforced.

Some of the experiments reported by Terman and by Elsmore & Hursh are relevant to the more specific issue that is our concern here. Zimmerman, in the experiment discussed above, delivered the electrical brain stimulation according to a schedule of Differential Reinforcement of Low Rates (DRL) 15 s. In that schedule, a subject is required to space its motor operant responses by at least 15 s if it is to be reinforced. If a response is given too soon, the timer is reset. In the normal conditions (median intensity reinforcement), in which operant activity was much higher during the dark part of the light-dark cycle, the Inter-Response Times (IRT) distribution showed better adjustment to the temporal requirements during dark periods than during light periods. Elsmore & Hursh also trained rats to discriminate the duration of auditory stimuli, long or short, with three levels of difficulty. Animals were tested every 3 hours under a light-dark 12:12 hour cycle. Both accuracy and rate of activity (as assessed by the number of trials completed) exhibited rhythmic fluctuations.

Submitting animal subjects to tests involving temporal regulations at regular intervals around the 24 h cycle, both in light-dark and constant conditions, is obviously a straightforward procedure for gathering data concerning the interaction between the circadian clock(s) and the capacity to adjust to arbitrary durations. An experiment of that type carried out in our laboratory illustrates the experimental strategy and provides some preliminary results (Perikel et al., Note 7). Naive female wistar rats were housed in isolated compartments, consisting of a home cage and an adjacent conditioning chamber. Water was available permanently in the home cage. They were exposed to a light-dark 12:12 cycle (light from 0630 h to 1830 h). After one week of habituation in these quarters, the subjects were food-deprived and shaped to press the lever for food. They were then run for 15 consecutive days on a continuous alternation of 30 min conditioning sessions and 90 min ›rest‹ periods.

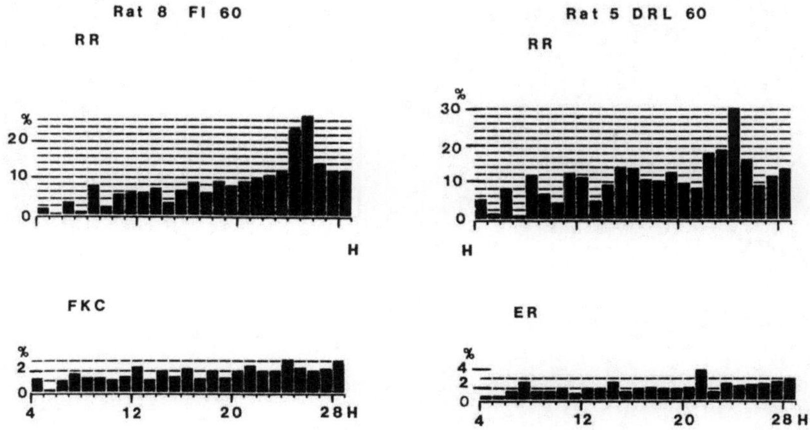

Figure 1. Periodograms of individual Wistar rats in Fixed Interval *(FI)* 60 s *(left)* and Differential Reinforcement of Low Rates *(DRL)* 60 s schedules *(right)*, computed after the root-mean-square-amplitude method (Enright, 1965 a, b). Periods tested range from 4 to 28 hours *(abscissa)*. *Ordinates* give, in percentage, an estimate of the importance of the rhythmic fluctuation for each parameter considered and at each period tested. The higher the block column at a given period, the more evident is the rhythmic fluctuation of the parameter. *Upper line:* Response rate *(RR)* for each schedule; *bottom line:* qualitative indices (*FKC:* Curvature Index of Fry et al., 1960; *ER:* Efficiency Ratio, i.e. the ratio between reinforced Inter-Response Time and total Inter-Response Times) (from Perikel et al., Note 7).

Some subjects were exposed during the conditioning sessions to a Fixed Interval (FI) 60 s schedule. In this schedule, the time that must elapse before a response can be reinforced remains constant from one interval to the next. The other subjects were exposed to a schedule of DRL of 60 or 30 s. Lever pressing responses were reinforced by food pellets, and conditioning sessions were signalled throughout by an auditory stimulus (white noise). Some individual subjects were submitted to additional manipulations: extension up to 60 days, reduction of amount of reinforcement, removal of auditory signals. Discrimination between conditioning and rest periods was learned within 24 h in FI, within three days or so in DRL. Periodogram analysis revealed a rhythmic fluctuation with periodicity close to 24 h in response rate (much more pronounced in subjects submitted to 60 days of experimental conditions, but already apparent after 15 days), but no such periodicity for qualitative indices of temporal regulations (curvature index in FI; efficiency index in DRL) (Figure 1). This is in line with Elsmore & Hursh's distinction between quantitative and qualitative aspects of performance. One remarkable byproduct of this procedure, in some subjects, has been the development of highly efficient low responding patterns under FI contingencies, their performance resembling a well adjusted DRL behavior, with a ratio of responses to reinforcement less than 2 (Figure 2). This induced ›quasi-DRL‹ performance under FI contrasts with the poor performances obtained from those

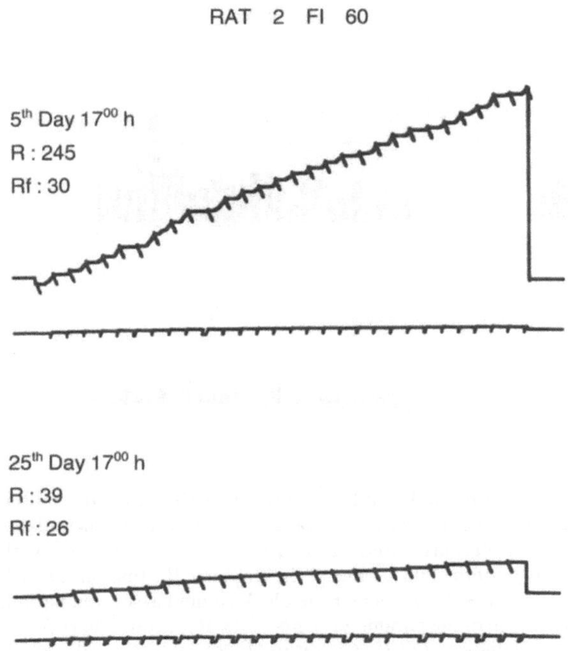

Figure 2. Cumulative records of responses for one individual rat under the Fixed Interval 60 seconds schedule, in experimental sessions starting at 1700 h on days 5 *(top)* and 25 *(bottom)*, that is the 50th and the 290th session, respectively. Number of responses emitted *(R)* and number of reinforcements obtained *(Rf)* are given in the figure. Downward pips indicate reinforcements on the cumulative graph, and reinforcement disponibility on the horizontal lines under each record. The abscissa corresponds to the duration of the session, that is, 30 minutes (from Perikel et al., Note 7).

rats run under DRL 60 s contingencies. These results illustrate once more, in a particularly clear way, the long-noticed paradox of spontaneous regulation observed under FI contingencies as opposed to the required spacing of responses for comparable delays under DRL.

Another potentially fruitful approach to elucidate the relation between biological rhythms and acquired temporal regulations of behavior, as suggested earlier (Richelle, 1968; Richelle & Lejeune, 1979, 1980), consists in testing the performance under FI contingencies, contrasting time intervals approximating the 24 h period with arbitrary shorter or longer intervals. Such studies give some insight in an animal's capacity to free itself from the basic biological periodicities. Bees visit with remarkable timing accuracy places where food has been available at a given hour each day. However, their sense of time, which is legendary in the chronobiological literature, is exhibited only if the interval between successive food presentations is close to the circadian period. Attempts to train them on FI 60 s have failed to reveal any spontaneous timing, so common among birds and mammals

(Grossman, 1973) (but see the following section for reservations about limited investigations of a given species in comparative studies). Increased freedom from natural periodicities was suggested by Richter (1965; see also Michon, 1985a, chapter 2 of the present volume) as an evolutionary trend leading to an increased flexibility and independence in the various time keeping mechanisms at work in an organism.

Studies on food availability and anticipation of food throw some light on this issue (Boulos & Terman, 1980; Aschoff, 1984; Rosenwasser et al., 1984; Fery, Note 1). When rats are fed *ad libitum* under light-dark 24 h cycles or under constant light conditions, they show eating periods fairly well synchronized with activity phases.

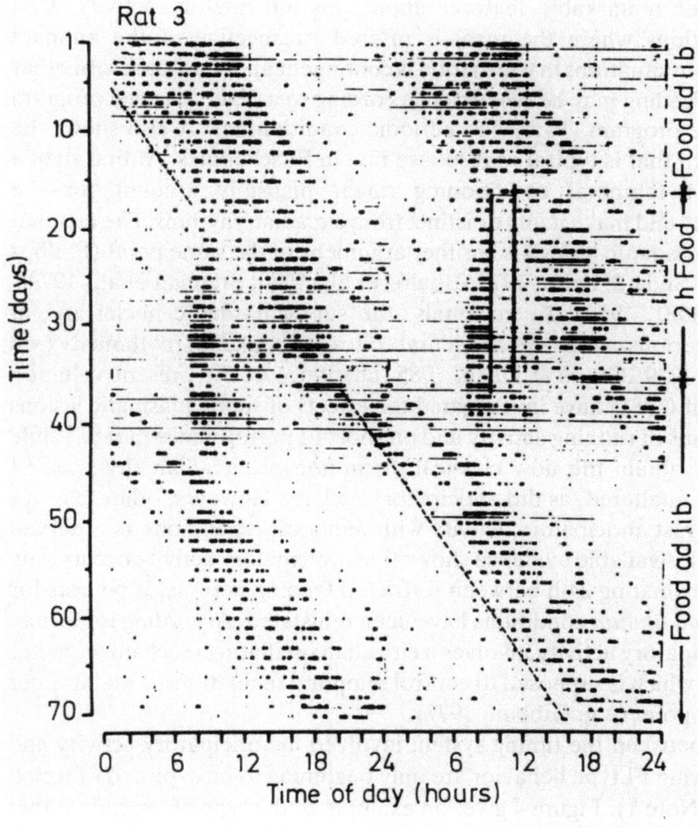

Figure 3. Wheel running activity of an individual rat kept in constant dim illumination (0.2 lux) and alternatively fed *ad libitum* or for two hours per day (between 0900 and 1100 h – see vertical lines to the right in the diagram). Experimental days are indicated on the left ordinate, time of day on the abscissa of the duplicate charts. Feeding conditions can be read on the right ordinate. During the restricted feeding period, a dissociation can be seen between wheel running anticipatory to food and wheel running that freeruns as in the preceding and following *ad libitum* feeding periods. (Adapted from Honma et al., 1983).

Eating is entrained, as is motor activity, by light synchronizers, and exhibits circadian rhythmicity when the synchronizer is removed. If feeding is restricted to limited periods during the 24 h cycle, anticipatory activity will occur prior to feeding time. This anticipatory activity persists under constant conditions, showing a dissociation from circadian activity. Rats can also anticipate meals spaced by intervals close to 24 h, viz. 23 to 27 h, but they seem unable to anticipate if the interval is outside this range. Though the anticipatory mechanism adjusts to the periodicity of environmental events that can be out of phase with circadian endogenous rhythms, it seems to exhibit circadian properties itself since it does not work for intervals differing markedly from the 24 h cycle. Figure 3, reprinted from a study by Honma et al. (1983), illustrates the basic effect of restricted feeding procedure under constant light conditions in rats.

There are other remarkable features about this anticipatory activity. It is observed in situations where the meal is offered irrespective of the animal's behavior as well as in situations in which food is contingent upon responses. In other words, restrictive feeding may be scheduled according to a Fixed Interval program or to a Fixed Time program (Pavlovian periodic conditioning). It also shows the acceleration pattern that is typical of response rate in FI schedules. At first sight a simple process of temporal conditioning might plausibly account for the anticipatory activity and make it quite distinct from circadian rhythms. The fact that it survives suprachiasmatic lesions is another argument to the same point (Phillips & Milkulka, 1979; Suda & Saito, 1979; Boulos et al., 1980; Stephan et al., 1979a, 1979b; Stephan, 1981, 1984). In mammals, the suprachiasmatic nuclei are, at present, the most serious candidates for neural control of circadian rhythmicity (see Rusak & Zucker, 1979; Groos & Daan, 1985, chapter 4 of the present volume). Innis & Vanderwolf (1981) have investigated the effects of suprachiasmatic lesions on circadian rhythms of running activity and on short FI performance in rats. While they were able to obtain the now classical disruption of circadian rhythms, FI patterns remained unaltered, as did anticipatory activity. However, other features paradoxically contrast anticipatory activity with temporal regulations as observed under FI schedules. Available evidence shows that anticipatory activity occurs only for intervals approximating 24 h between restricted feeding periods. It persists for several days after *ad libitum* conditions have been reinstated. According to Terman et al. (1984), anticipatory activity involves a circadian oscillating mechanism rather than a *scalar timer* which is supposed to control temporal regulations; a scalar timer can be reset at any point (e.g. Gibbon, 1977).

The relations between the timing system involved in anticipatory activity and the system underlying FI type behavior are only beginning to be explored (Terman et al., 1984; Fery, Note 1). Figure 4 gives an example of the type of procedure that can be used. The strategy essentially consists of making food available on a restricted feeding program and contingent upon the subject's activity. This amounts to programming the meals according to a FI schedule, the interval being chosen within the 22 to 27 h range that is known to be critical for anticipatory activity to occur. The interval is then progressively shifted to shorter values, to see if the circadian constraints on anticipatory activity can be removed.

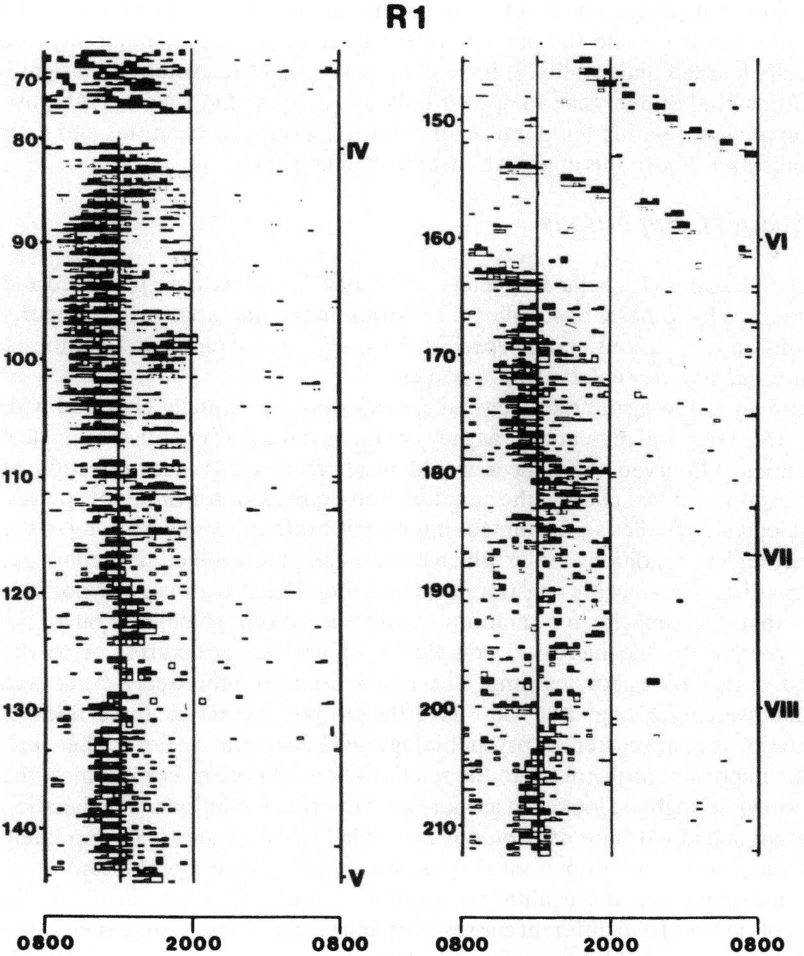

Figure 4. Quartile plots of lever pressing activity (*filled symbols*) and wheel running activity (*empty symbols*) under a conjunctive FI 24 h FR 50 schedule in one individual rat *(R1)* over successive experimental phases, from the top left to the bottom right. Prior to *phase IV*, the rat was exposed to a light-dark 12 : 12 h cycle under an *ad libitum* feeding schedule with no access to the lever; only the wheel running activity was recorded. In *phase IV*, the reinforcement was available every 2 h and for a 18 h availability period; it was delivered only after 50 responses had been emitted. In *phase V* the subject was maintained in constant dim light and under the same reinforcement schedule. *Phase VI* was like *phase IV*. In *phase VII ad libitum* free access of food was provided. *Phase VIII* was like *IV* and *VI*. The height of each symbol is proportional to the activity displayed per 30 min period. Successive experimental days are indicated at the left side of each chart. The thin vertical line halfway between the 0800-2000 h period indicates the occurence of reinforcement disponibility in *phases IV, V, VI and VIII* where the reinforcement schedule was in effect, or the moment at which it should have occurred during the *ad libitum* feeding phase *(VIII)*. Note the anticipatory pattern, in running wheel activity as well as in operant lever pressing activity, induced by the schedule of reinforcement in *phases IV, V and VIII*, and simultaneous free running of both activities in *phase V* (after Fery, Note 1).

This line of investigation is certainly promising for our understanding of the relative independence and the possible interactions between biotemporality and individually learned temporality. It is clear, however, that the evolutionary significance of this kind of analysis can appear only if it is framed in a consistent cross-species approach, relating temporal regulations to learning mechanisms and their neural substrate. It is to this approach that we shall now turn.

Cross-species Comparisons

On earlier occasions (Richelle & Lejeune, 1979, 1980, 1984) we have proposed and discussed three hypotheses concerning the distribution of the capacity for temporal regulations among various animal species: the *egalitarian-reductionist* hypothesis, the *ethological* and the *evolutionary* hypothesis.

According to the egalitarian view, all species would be equally endowed with capacities for temporal regulations, as they are for circadian or/and other biological rhythmicities. Observed differences would result from inadequate selection of experimental variables, making the situation non-equivalent for different species. The ethological hypothesis accounts for interspecific differences by pointing to the typical ecological conditions under which each species has evolved; it emphasizes the notion of species-specific constraints on learning. The evolutionary hypothesis assumes that the capacity for temporal regulations obeys some general trend, possibly parallel to the increased complexity of nervous structures, or to the increased potentialities for learning. The labels used are admittedly crude and somewhat oversimplifying, but they nevertheless correspond to three lines of interpretation that are current in psychobiology and behavioral research. There are, of course, important problems as to the relations between them. For instance, the ethological view might be looked at as more akin to a true evolutionary explanation than the view defined here as *evolutionary*, which could be more appropriately labelled *anagenetic*. For our present purpose we may forget these problems.

It would seem that the egalitarian hypothesis could be ruled out if the dissociation of (at least) two different classes of timing mechanisms, as suggested in the preceding section, could be confirmed. However, until such confirmation is obtained, it is heuristically advisable to question any interspecific difference by suspecting ›unfair‹ experimental conditions with respect to the species-specific behavioral repertoire. The concern for designing ›species-fair‹ tests is shared, though for quite opposite reasons, by advocates of the ethological view. Since temporal regulations as we define them are aspects of learned behavior, they can be looked at in the general framework of learning mechanisms and evolution. Current theorizing goes to the extreme view that there is no general universal learning mechanism, but that there are, in the limit, as many learning processes as there are species that learn different things under different conditions (Plotkin & Odling-Smee, 1979, 1982).

As a research strategy, the evolutionary view recommends itself by being clearly open to refutation. It expects, in temporal regulations as in other things, interspecific differences that can be described as an ordered trend in the same way as, in a given phylum, species can be ordered without any teleological implications, according to the number of neurons or to the encephalization quotient. Experiments can then be designed, based on the egalitarian or on the ethological hypothesis, for demonstrating either that the alleged differences are not real (viz. they fade out when another procedure is used or when another behavior is looked at), or that they show no trend but merely reflect the irreducible species-specific styles of adaptation. Adopting that research strategy does not, of course, relieve the experimenter from the task of asking an animal the sort of question it can understand, that is, selecting appropriate responses, reinforcers, stimuli, housing conditions, and so forth. It is also important to note that an evolutionary trend in temporal regulations may not show up in simple measures of behavior, such as the waiting time in a FI or DRL schedule, or the accuracy of time discrimination. It

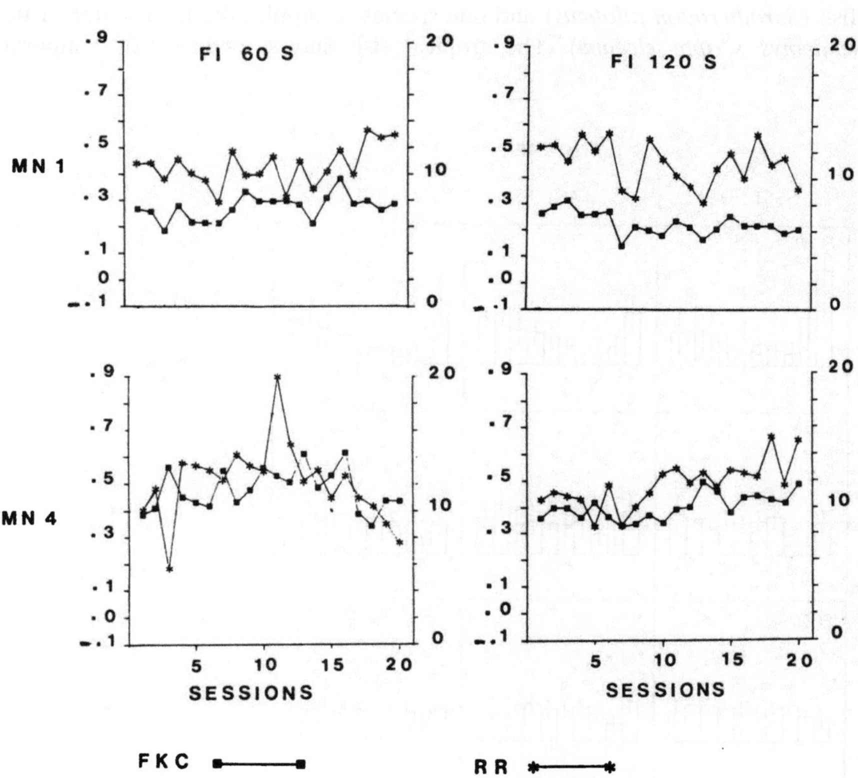

Figure 5. Evolution of performance of two individual fish *(MNI* and *MN4)* of the species *Sarotherodon niloticus* under Fixed-Interval contingencies *(FI 60 s* and *FI 120 s)*. Curvature index *(FKC)* and running rate *(PR)* – on the left ordinate scale and right ordinate scale respectively – are plotted as a function of session *(abscissa)*. The curvature index has been computed on distributions of responses in ten successive segments of the interval; it can take values from −0.9 to +0.9 (usually only positive values are obtained). The running rat is the average rate of responses per minute in the ›active‹ phase following post reinforcement pause (from Grailet, Note 2).

might be observable only when more complex dimensions of behavior are considered, such as flexibility in adjusting to various, rapidly changing temporal requirements or capacity to exhibit concurrent temporal regulations, and the like.

Whatever one's theoretical inclination, data are needed on a reasonably large sample of species if the issue is to be tested. The data available at present are far from sufficient for any coherent picture to emerge. From what we know, there is some hint that mammals rank highest (monkeys, cats, rats, and mice being the main species studied), followed by birds (mainly pigeons), fishes and reptiles in that order (see Richelle & Lejeune, 1984). Taking the phyletic scale in a simplistic way, one would expect to find reptiles ranking higher than fishes. But the number of species studied in these two groups, as well as the number of studies on any given species, is insufficient.

Figures 5 and 6 show samples of typical individual performances in one species of fish (*Sarotherodon niloticus*) and one species of reptile, the fresh water turtle (*Pseudemys scripta elegans*). The tropical fish shows evidence of temporal

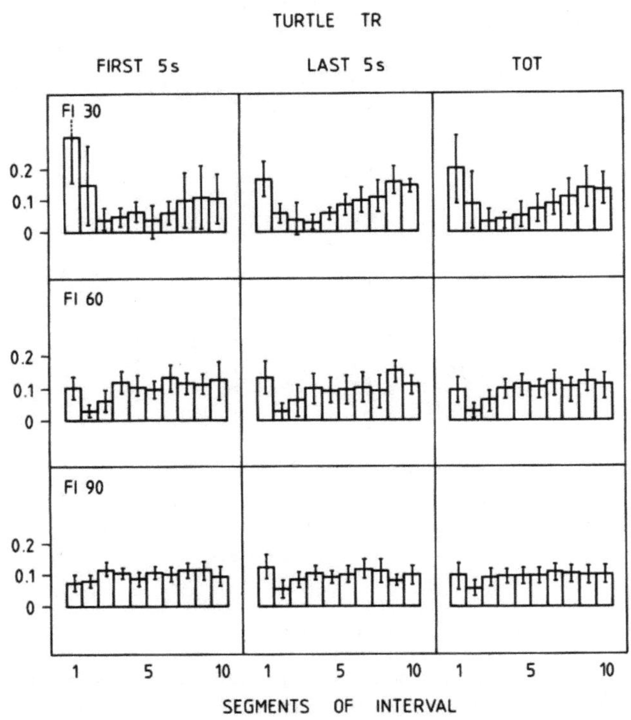

Figure 6. Relative frequencies distributions of responses in successive segments of the interval for FI values of 30, 60 and 90 s in an individual subject of the species *Pseudemys scripta elegans.* Average values are shown for the first five sessions (*left*) the last five sessions *(center)* and all sessions (*right,* 15 or 20 sessions) (from Laurent, Note 3).

regulation typical for the FI schedule (60 and 120 s), though qualitatively poorer than tends to be observed for the usual laboratory animals. The best subjects had a score between 0.3 and 0.5 on the so called curvature index (Figure 5). While the fish performed moderately well, the water turtles did not exhibit any evidence of temporal patterning of their responses even after having been exposed to the FI schedule (30, 60 and 90 s) for more than 100 sessions. Their curvature index remained between 0.003 and 0.18 (Figure 6).

One may wonder whether the experimental conditions were appropriate to the characteristics of these species. Provisions were made, of course, to adapt the situation to the animals. The fish were kept in individual watertanks, where they were trained to push with their head on an immersed lever. The turtles were housed in an aquaterrarium, and isolated for the period of a session in the conditioning chamber in which they were trained to push a specially designed lever with their head or paw for food reinforcers. At first sight, there is no reason to suspect the selected response to be more artificial or inappropriate than the familiar lever pressing of laboratory rats: subjects did learn it easily. The fish have also been trained without any difficulty to put the response to use under other schedules of reinforcement not involving temporal regulation. Other important variables, however, may have been responsible for their poor performance in this case. The feeding schedule involved in FI contingencies may not be compatible with feeding habits of *Pseudemys scripta* and thus account for its unexpectedly poor temporal patterning. It remains to be demonstrated, therefore, that digestive physiology has not obscured timing competence in this particular case.

That a change of response or of some other variable suffices to produce drastically different performances in conditioned behavior has been widely documented in the 10 years since the investigators' attention was first drawn to the so called biological constraints on learning.

Key pecking in pigeons has been subjected to close scrutiny, with the result that the notion of arbitrariness of that particular operant (and possibly of any response selected by the experimenter) had to be abandoned definitively. Pecking, because of its status in the natural repertoire of preparatory and consummatory eating behavior, was initially expected to be amenable to control by contingencies that require temporal regulations. The observed limitation on the pigeon's capacity to space pecking by more than 10 to 15 s in a DRL schedule has become a classical illustration of a species-specific constraint, and explains why experimenters began to look for some other responses that would more reliably reveal true timing competence of pigeons. Somewhat better performances have indeed been obtained with treadle pressing (Hemmes, 1975; Mantanus et al., 1977) but this is a rather unnatural motor response for birds. We have explored another operant, namely sitting on a perch. Can pigeons, obviously unable to space their pecks by a given delay, keep perching for that duration? In a study reported earlier (Lejeune & Richelle, 1982) pigeons were trained to perch for durations up to 50 s in order to get food. The reinforcer was presented only if the bird jumped off the perch after the critical duration had elapsed. This schedule of Differential Reinforcement of Response Duration (DRRD) produced excellent timing as evidenced by IRT

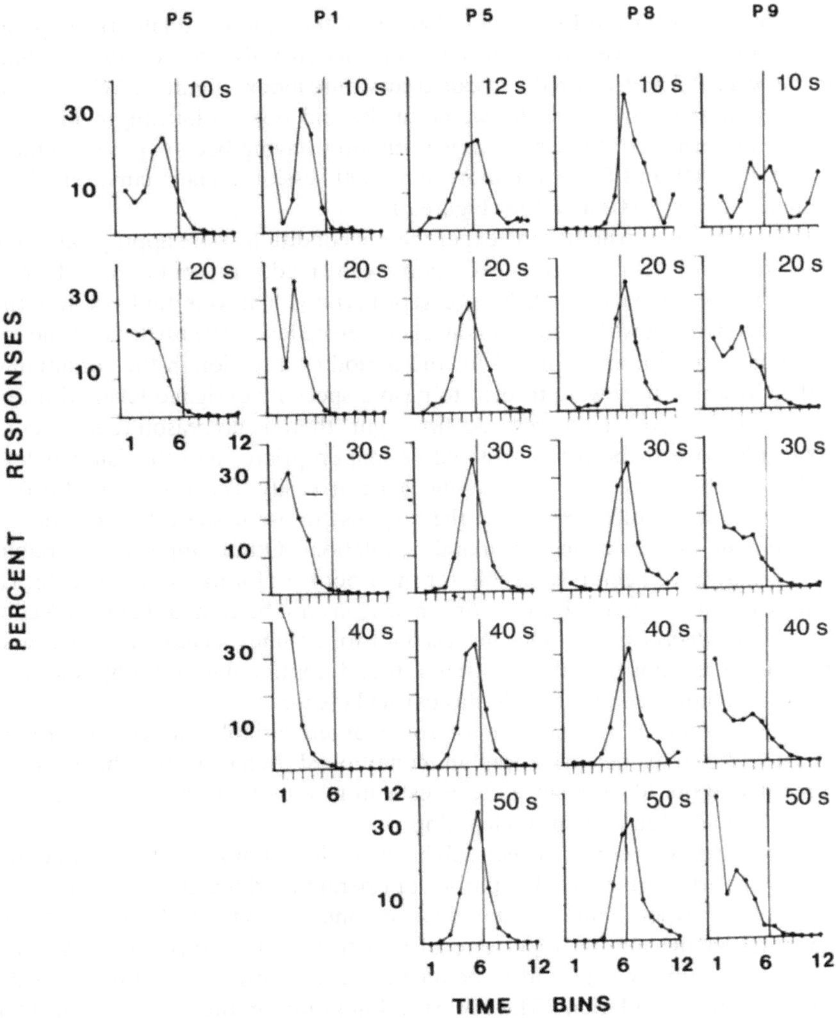

Figure 7. Temporal regulation of operant motor behavior as a function of the response topography and the schedule characteristics in pigeons. *From left to right:* bird P5 under differential reinforcement of low rates *(DRL)* of key pecking; bird *P1* under *DRL* of treadle pressing; bird *P5* under differential reinforcement of response duration *(DRRD)* for perching; bird *P8* under DRL of brief perching responses; bird *P9* under the same conditions – to which it was shaped and maintained for the 10s delay – switched to ›neck stretching‹ response from 20 s on. Critical value of the temporal parameter is indicated on each graph in which relative frequency is plotted as a function of the interresponse time or response duration class. Each step on the abscissa corresponds to one sixth of the critical values, marked by the vertical line. Interresponse time (or response durations for central column) were recorded over the 3 last sessions out of 20 at each schedule values for birds P1, P8 and P9, over 3 sessions at 10 and 20 s and over the 3 last sessions out of 15 at 30, 40 and 50 s for bird P5 (see Lejeune & Richelle, 1982).

distributions, illustrated by bird P 5 in the third column of Figure 7. This contrasts sharply with key pecking performances, in which spacing of responses hardly matches the critical value beyond 10 s, as illustrated for the same individual in the first column of Figure 7. The same procedure was applied to turtle doves, with the same outcome (Richelle & Lejeune, 1984). Pecking under DRL is not, therefore, a valid procedure to uncover the timing capacities of these two species. One might object, however, that the schedule factor (DRRD versus DRL) was more important than the type of response as a determinant of the difference, as was pointed out by Fantino (1984).

A recent experiment (Lejeune et al., Note 4) provides a solution. It compared DRL performance in pigeons using two topographically distinct operants, treadle pressing and discrete, brief perching. Reinforcement, in both cases, was contingent upon minimal spacing of responses. The critical delay was progressively increased from 35 to 40 s (for treadle pressing) or 50 s (for brief perching) in 15 steps. Eight adult homing pigeons served as subjects in this experiment. Interresponse time distributions for one bird in each group (Pl and P8 respectively) are presented in Figure 7. Other subjects showed similar results with one exception that will be discussed below. Treadle pressing performance, supposedly slightly better than keypecking, is far from matching the DRL requirement. Brief perching, in sharp contrast, produces symmetric distributions, with a mode coinciding with or slightly longer than the critical delay. Bursts of responses with short interresponse times are clearly eliminated by the particular topography of perching, but such bursts alone cannot explain the poor timing obtained with treadle pressing or pecking. Figure 8 shows the median interresponse times or response durations plotted against the critical values for typical individuals in each condition. Up to 50 s, perching birds exhibit excellent performances, compared with treadle pressing. Efficiency ratios remained high (above 0.40) for perching subjects and for all delay values, while they fell from between 0.20 and 0.30 to less than 0.10 for the other birds.

An interesting case is offered by bird P9 in the perching group, whose results are also shown in Figure 7 and 8. Its performance is similar to that of the treadle pressing birds. In fact, after initial shaping of the common perching response, this bird shifted to a different strategy consisting in pushing the perch down with its neck, thereby achieving the same mechanical effect as it would by jumping on it. This shift took place between training on DRL 10 and 20 s, so that results shown are for ›neck stretching‹, except at 10 s. The various parameters of its performance shifted accordingly.

These results clearly demonstrate that response topography, not the schedule, is the crucial factor in the differences observed between perching and key pecking, or treadle pressing. That brief perching responses of birds can be spaced in time just as well as lever presses in rats is an intriguing observation, especially if we compare it with the relatively poor treadle pressing performance, a response that seems no less ›natural‹ for birds than key pecking. Intriguing is also the idiosyncratic behavior developed by bird P9. Possibly, neck stretching and pushing are topographically quite close to key pecking and equally resistant to temporal control. Watching the birds in the conditioning chamber revealed the peculiar preparatory activity

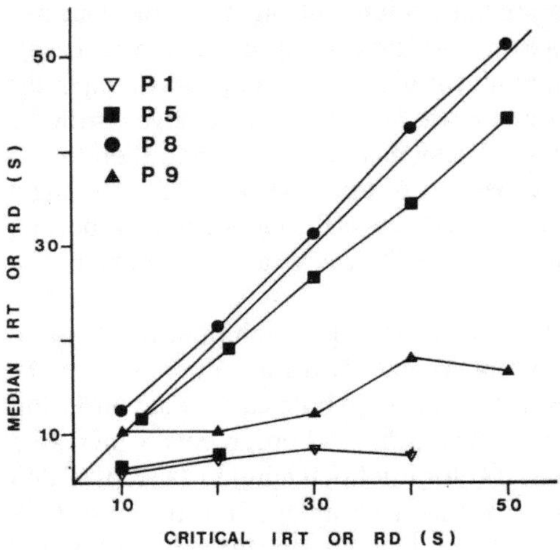

Figure 8. Median Inter Response Times *(IRT)* or response durations *(RD)* as a function of the critical IRT or RD value and response topography in pigeons. *From top to bottom,* birds P8 under DRL of brief perching (●), P5 under differential reinforcement of perching duration (■ from 10 to 50 s), P9 (▲) under DRL of brief perching (at 10 s) followed by DRL of ›neck stretching‹ (from 20 to 50 s), P5 under DRL keypecking (■ at 10 and 20 s) and P1 under DRL of treadle pressing (▽ from 10 to 40 s). Data are computed as in Figure 7. Points would have fit with the diagonal line in the case of perfect matching between median values and schedule requirements.

exhibited by the perching birds: movements of the paws and of the neck (stretching upward) preceded jumping onto the perch. This anticipatory activity seemed to build up until the ›decision to jump‹ was made. These movements were not observed prior to jumping in the home cage, where a similar perch was also available.

This example of within species comparison in temporal regulations as exhibited with different motor behaviors illustrates the complexity of the cross-species approach. What performance or performances should we retain as revealing the underlying timing competence in a given species? Should we systematically explore the essentially unlimited range of responses, reinforcers, stimuli, etc., and characterize the animal's capacity on the basis of its best performance? Or should we, perhaps more appropriately, evaluate an animal's ability to apply its clock(s) to a variety of responses and situations with equal efficiency? Does it make sense to dissociate a hypothetical underlying clock and the temporally regulated behavior, to make a distinction between *competence* and *performance*? Why not admit that there are as many clocks as there are behaviors exhibiting timing properties and that every behavior is inherently timed? We suggest that those and similar questions remain open to further inquiry. They should not be dodged in the current attempts, perhaps premature, to build conceptual unifying mechanisms.

Ontogenetic Studies

The study of ontogenetic development has proved a fruitful way to gain insight on the organization and the function of behavior. In spite of its long standing and widely recognized merits the ontogenetic approach has been surprisingly neglected in the field of time. A few data are available on the ontogeny of circadian rhythms (Davis, 1981). Though circadian rhythms are admittedly controlled by some basic mechanism of genetic origin, their emergence requires some time in early development. Thus it takes 3 to 4 months in the human infant before the sleep-wake rhythmicity gets organized. In rats, adrenal rhythmicity develops after 3 to 4 weeks; pineal rhythmicity, however, appears much earlier, within 3 to 4 days. Animals need not experience 24 h cycles in their early stages to exhibit circadian rhythmicity as adults. Young mice kept under 20 h or 28 h light-dark cycles show persistence of the induced rhythm when transferred to constant conditions, but only for 15 days or so, after which the circadian rhythm takes over (Davis & Menacker, cited in Davis, 1981).

Not much is known about the development of biological rhythms with aging. In humans, the main trend in old age seems to be a decrease in amplitude that eventually results in the fading out of rhythms. There may also be a change in phase relationships within the circadian system, and a decreased capacity to adjust to time schedule modifications. Maintaining good temporal organization, in the chronobiological sense, would be a warrant for healthy aging. Humans might profitably ponder over some experimental data obtained in insects. The life of flies is shortened when they are exposed to unusual periodicities, that is, to schedules differing from the 24 h cycle, or constant light, or to phase shifts simulating jet lags (Aschoff et al., 1971; Saint-Paul & Aschoff, 1978).

At another level of the study of time, the ontogeny of time estimation and of the time concept has been extensively studied since Piaget's pioneering contribution (Piaget, 1946). Most of this research, however, starts after the *infant* stage, when the human subject begins to master symbolic and linguistic tools. There are good reasons to assume that language is causing changes no less significant here than in the other areas of behavior and cognition. What, precisely, makes the difference cannot be assessed unless we know something of the infant stage on the one hand, and of development in non-speaking animals on the other. Temporal regulations of behavior in the human baby are being currently explored by Pouthas, whose approach and data are illustrated in chapter 6 of the present volume (Pouthas, 1985). We shall concentrate here on studies of animal development.

To our knowledge, no study has been devoted specifically and explicitly to the exploration of temporal regulations of behavior in animals as a function of age, except for Goodrick (1969) who compared performances of young (that is, 9 to 10 months old) and senescent (28 months) male rats under a FI schedule but who, unfortunately, provided no other measure than overall response rate. Incidental relevant data can be found in a few studies not directly addressing the present question. It was found that newly hatched chickens and ducklings exhibit the typical pattern of behavior generated by FI contingencies (Marley & Morse, 1966; De Paulo & Hoffman, 1981). In squirrel monkeys this pattern is more marked in young (2 to 3 years) than in old (15 years and more) subjects (Harrison & Isaac, 1984).

Systematic developmental studies on temporal regulations have been in progress in our laboratory for a couple of years. In one of these studies (Lejeune et al., Note 5), young quails (*Coturnix coturnix japonica*) are used as subjects. This species is rarely used in operant laboratories (see, however, Reese & Reese, 1962; Cloar & Melvin, 1968), but it recommends itself for our purpose; it is easily produced in incubators, it is a nidifuge species, it has been extensively described by ethologists and is frequently used in psychophysiological (and more specifically neuroendocrinological) research with important chronobiological implications (Balthazart, 1983; Balthazart et al., 1979). A preliminary study on adult subjects, using key pecking as a response, was run to test general conditionability and behavior under FI schedules up to 2 minutes. Then, a comparative study was undertaken in young and adult subjects. Ontogenetic studies raise technical and methodological problems quite similar to those raised by cross-species studies. Just as one must look for species-fair tests, one must design age-fair situations for young animals. The reinforcer must be adequate (flour was used rather than grain); so must the response (treadle pressing was preferred to key pecking in this exploratory study, because of the difficulty to calibrate the key to the excessively frail beak of freshly hatched quails). The amount of reinforcer as the response determining force must be adjusted as a function of growth.

Any experimenter working on development of acquired behavior is confronted with another universal problem: learning takes time and young organisms grow older every day. Reaching asymptotic stabilized performances in temporal regulations in adult subjects usually requires 30 to 50 daily sessions, including shaping, training to the particular contingencies, bringing to the final value of the critical temporal parameter, and exposing the subject to it long enough to extract valid data. However, quails are mature at about 40 days. Therefore, subjects were run in the conditioning chamber for 30 minutes four or five times per day in this case, so that a reasonable number of sessions could be obtained within twelve days.

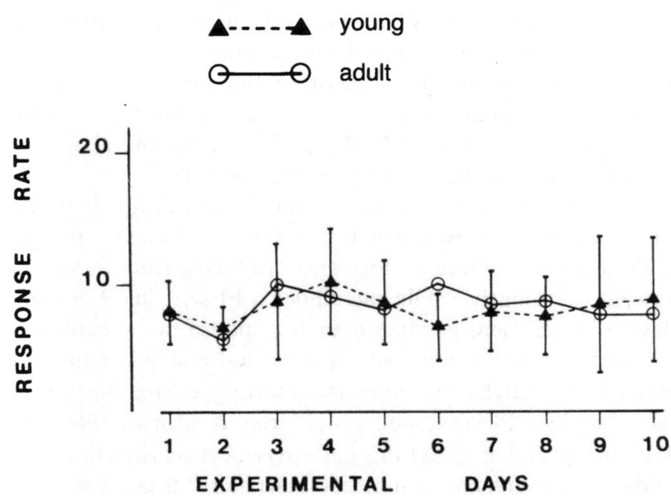

Figure 9. Evolution of response rates over 10 experimental days in two groups of quails of the species *Coturnix coturnix japonica* (young and adult) under a FI 60 s schedule of treadle pressing. Each point presents the daily average of response rates computed for each group over all subjects and over all daily sessions. Standard deviations are given only for the young subjects (7 days old at experimental day 1).

This procedure reduces, of course, the number of subjects that can be run in one experimental unit, so that the results presented here are as yet from a very small number of subjects (N=10 for the young group; N=3 for adults).

All young subjects were successfully conditioned by the fourth day of life and exposed to FI contingencies from the fifth day on, with the interval increased to 60 s within 8 sessions. Although the response rate of young subjects did not differ from that of adults (Figure 9) the quality of their temporal regulation, assessed by the curvature index was clearly lower, as shown in Figure 10. The young quails improved from day to day and showed evidence of emerging temporal regulation. However, their performance did not match the performance of adults. This could mean that young animals have not yet completely developed their capacity for temporal regulations. More data and controls for a number of variables are needed, however, before such a conclusion can be drawn. Control for the effects of food deprivation on the physical growth of the organism is essential in all ontogenetic studies. By running the subjects several times a day, we may have cancelled these effects to some extent. However, the young animals submitted to conditioning ended with a 15 to 25 percent loss of weight compared to control subjects maintained on an *ad libitum* feeding schedule. A precise description of spontaneous eating habits should provide criteria for choosing intersession intervals and number of sessions per day.

The momentary isolation required by the conditioning sessions may be a source of emotional disturbance interfering with temporal regulations. Although subjects were kept in a group in a compartment adjacent to the conditioning chamber and remained in auditory contact with their fellows when isolated during experimental sessions, this may not have completely eliminated emotional reactions.

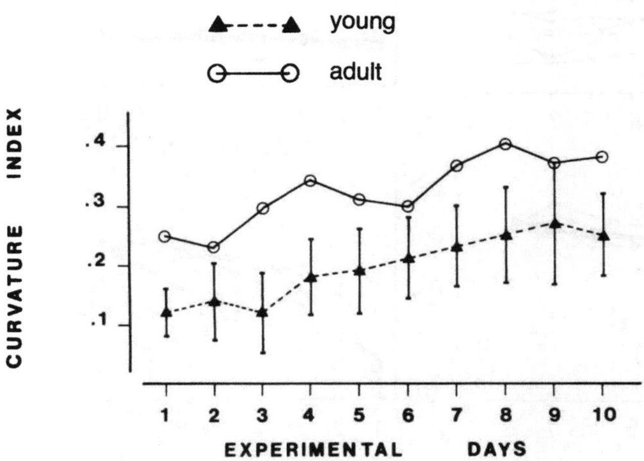

Figure 10. Evolution of the curvature index value over 10 experimental days in two groups of quails (young and adults) under a FI 60 s schedule of treadle pressing. Each point presents the curvature index value computed for each group over all subjects and all daily sessions. See the legend of Figure 9 for other details.

In another study (Lejeune et al., Note 6) young rats tested at time of normal weaning (20 days) were compared for performance under FI 60 s, with senile (approximately 26 months old) rats. The young animals were shaped to press a lever for food by day 21 and then put on FI schedule for five 30 minute sessions per day for 8 consecutive days. Old subjects were naive animals that had been housed individually for about two years. The main results are summarized in Figure 11, which shows that response rate was lower and much less variable, both intraindividually and interindividually, in senile rats than in young rats. In contrast, temporal regulation (as assessed by the curvature index) was less variable in younger subjects. lt was clearly better early in training and remained somewhat better throughout the experiment. Young subjects reached values which seniles did not ever attain. Given their initial poor performances, old animals showed considerable improvement but they showed a marked interindividual variability, sharply in contrast with the reduced interindividual variability of their response rate.

The young rats in this study performed much better than young quails but, of course, one must ask the question: How young is *young*? Data for rats were from day 23; those for quails from day 5. What does this difference mean develop-

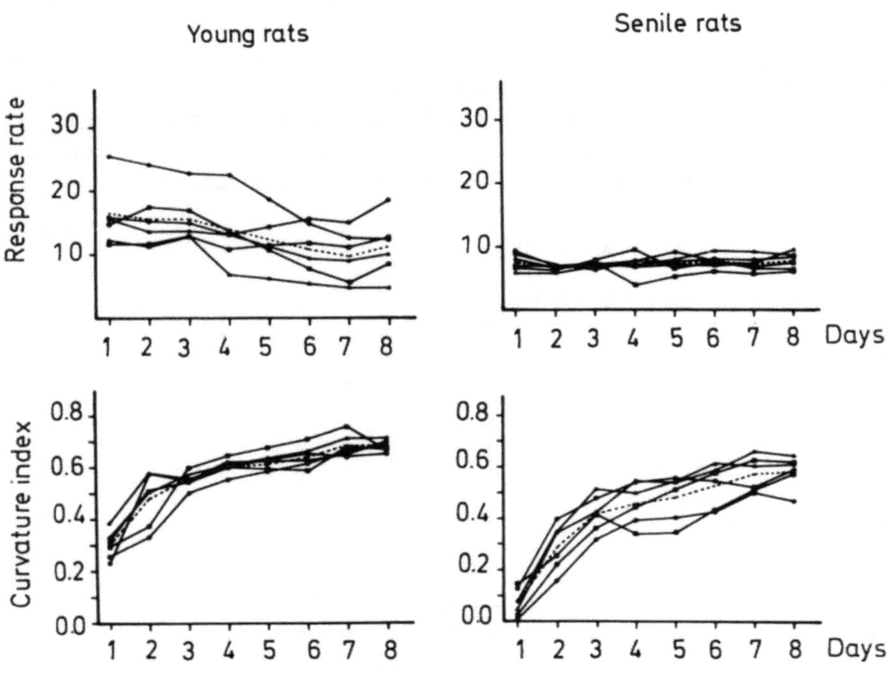

Figure 11. Evolution of individual and group average *(broken lines)* response rates *(upper half)* and curvature index values *(lower half)* over 8 experimental days *(abscissa)* in 6 young and 6 senile male Wistar rats under a FI 60 s schedule. Each point takes into account the complete daily performance (5 sessions), be it for individual subjects or group average.

mentally? It might be appropriate to derive embryogenic and epigenetic criteria of a morphological nature and to relate behavioral measures to some index of neural development.

However, in the absence of a more sizeable body of empirical data, it seems better to consider some prospects for future developmental research. Restricting ourselves to the behavioral level, we must be aware that the kind of temporal regulation explored in the reported studies is only one among many to reveal developmental properties. If, as we have suggested (Richelle & Lejeune, 1980), temporal regulations involve inhibition depending to various degrees upon the characteristics of the situation, more demanding contingencies should be explored concurrently with FI schedules. If the FI schedule has been preferred as a starting point, it is because it is more feasible and also because it has some relevance to the problem of the relation between periodic conditioning and circadian rhythmicity.

How prenatal and early exposure to a selection of ›arbitrary‹ periodicities (of food presentation, for instance) influence further adjustment to similar periodicities in conditioning situations, is another as yet unexplored line of research that should merge with similar studies on the circadian system of food anticipation. Can ›temporal sets‹ be induced by early exposure? Is there a sensitive period for that induction? To what extent can it go against the ›natural‹ rhythm? Are there, in this respect, notable cross-species differences? We cannot predict the outcome of investigations addressing such questions, but we can foresee their potential relevance to applied problems in humans.

As suggested with respect to species differences, age and development may not be critical with respect to accuracy of temporal regulagions, but to the flexibility of timing systems. Research should therefore focus on the capacity of an organism to adjust to rapidly varying delays or to master various delays concurrently. Neither do we know as yet anything about the tolerance as a function of age to temporal contingencies that remain unchanged for a long period of time; but there are a few hints on the possibly aversive properties of constant temporal schedules that should be further explored in young and old subjects. Can young organisms be maintained as long on an arbitrary temporal regulation task as adults or seniles? Speculating a bit more, one might also ask about the advantages or the disadvantages of periodic schedules or regular allocation as a function of age. Is the regular time pattern of classroom activities a source of efficiency and motivation for the child or the adolescent, or does it generate, by its inherent predictability, boredom and loss of interest? Ontogenetic animal studies of temporal regulations may perhaps help us some day in analyzing school situations.

Final Comments

Much of what has been said in the preceeding pages is about *Time* in the *Behavior* of animals. What about their *Mind*? In a way, the sort of experiments we have been discussing or suggesting as a necessary path toward a synthetic view integrating chronobiological, comparative, and ontogenetic aspects of temporal regulations might look alien to the dominant trend in psychological research on time today. Intensive attempts are currently made to characterize the properties of the internal clocks that time behavior in animals and in humans. There is no question about the fruitfulness of such theorizing of which the present volume offers many and convincing examples. Models proposed to account for animal timing capacities are usually based on conceptual and mathematical tools developed in the field of psychophysics of time in humans (see chapters 2, 7 and 13 of the present volume). In that respect, they contribute to the integration of domains unduly kept apart and, in the long run, to a comprehensive theory of psychological time (see Church & Deluty, 1977; Gibbon, 1977; Gibbon & Church, 1981; Gibbon et al., 1984). Such models, however, derive their validity from their applicability to a wide array of empirical facts. But, as we have seen, whole areas of fundamental importance are still *terrae incognitae*. Elaborating highly sophisticated models runs the risk of being pointless if there are not enough data to apply such models to. Thus it is impossible at the present stage to claim wide validity for a model based on the assumption that temporal regulations for short interval are strictly distinct from biological periodicities such as circadian rhythms. We do need a more systematic inquiry on the relations between the two sets of phenomena. Similarly, we cannot contend that »the accuracy of (time) discrimination does not vary substantially among vertebrates« (Roberts, 1983, p. 347) when available comparative data cover no more than a dozen species and hardly support that kind of generalization anyway.

In conclusion, what we want to say is the following. A balance should be maintained between the construction of models for inferred mechanisms and the collection of facts guided by a sense of diversity in nature. Students of time in animals should not abandon simple, straightforward curiosity to guide them. We must explore many more species, many more procedures, many more responses, reinforcers, stimuli, ages, etc. And finally, sticking to the observable may well be for some time to come, an efficient way to reach correct inferences about the clocks inside the animals that might indeed be part of their minds.

References

Aschoff, J. Circadian timing. In: J. Gibbon & L. G. Allan (Eds.), *Timing and time perception*, Annals of the New York Academy of Sciences, 1984, Vol. 423, pp. 443-468.

Aschoff, J., Saint-Paul, V. von, & Wever, R. Die Lebensdauer von Fliegen unter dem Einfluss von Zeitverschiebungen. *Naturwissenschaften*, 1971, *58*, 574.

Balthazart, J. Hormonal correlates of behavior. In: D.F. Farner, J.R. King, & K.C. Parkes (Eds.), *Avian biology*. New York: Academic Press, 1983, pp. 221-365.

Balthazart, J., Massa, R., & Negri-Cesi, P. Photoperiodic control of testosterone metabolism, plasma gonadotrophins, cloacal gland growths and reproductive behavior in the Japanese quail. *General and Comparative Endocrinology*, 1979, *39*, 222-235.

Boulos, Z., & Terman, M. Food availability and daily biological rhythms. *Neuroscience and Behavioral Review*, 1980, *4*, 119-131.

Boulos, Z., Rosenwasser, A.M., & Terman, M. Feeding schedules and the circadian organization of behavior in the rat. *Behavioral and Brain Research*, 1980, *1*, 39-65.

Church, R.M., & Deluty, M.Z. Bisection of temporal intervals. *Journal of Experimental Psychology: Animal Behavior Processes*, 1977, *3*, 216-228.
Cloar, T., & Melvin, K.B. Performance of two species of quail on basic reinforcement schedules. *Journal of the Experimental Analysis of Behavior*, 1968, *11*, 187-190.
Davis, F.C. Ontogeny of circadian rhythms. In: J. Aschoff (Ed.), *Handbook of behavioral neurobiology*, Vol. 4. New York: Plenum Press, 1981, pp. 257-274.
De Paulo, P., & Hoffman, H.S. Reinforcement by imprinting stimulus versus water on simple schedules in ducklings. *Journal of the Experimental Analysis of Behavior*, 1981, *36*, 151-169.
Elsmore, T.F., & Hursh, S.R. Circadian rhythms in operant behavior of animals under laboratory conditions. In: F.M. Brown & R.C. Graeber (Eds.), *Rhythmic aspects of behavior*. Hillsdale, NJ: Lawrence Erlbaum Associates, 1982, pp. 273-310.
Enright, J.T. Accurate geophysical rhythms and frequency analysis. In: J. Aschoff (Ed.), *Circadian clocks*. Amsterdam: North Holland Publishing Company, 1965a, pp. 31-42.
Enright, J.T. The search for rhythmicity in biological time series. *Journal of Theoretical Biology*, 1965b, *8*, 426468.
Fantino, E. Timing in animal learning and behavior: Discussion paper. In: J. Gibbon & L. G. Allan (Eds.), *Timing and time perception*. Annals of the New York Academy of Sciences, Vol. 423, 1984, pp. 292-295.
Fraisse, P. *Psychologie du temps* (2nd edition). Paris: Presses Universitaires de France, 1967.
Fraser, J.T. *The genesis and evolution of time*. Brighton: Harvester Press, 1982.
Fry, W., Kelleher, R.T., & Cook, L. A mathematical index of performance on fixed interval schedule of reinforcement. *Journal of the Experimental Analysis of Behavior*, 1960, *3*, 193-199.
Ghiselli, W.B., & Thor, D.H. Light reinforcement of operant responding in the pigeons. *Proceedings of the 81th Annual Convention of the American Psychological Association*, 1973, pp. 881-882.
Gibbon, J. Scalar expectancy theory and Weber's Law in animal timing. *Psychological Review*, 1977, *84*, 279-325.
Gibbon, J., & Allan, L. (Eds.), *Timing and time perception*. Annals of the New York Academy of Sciences, Vol. 423, 1984.
Gibbon. J., & Church, R.M. Time-left: Linear *vs* logarithmic subjective time. *Journal of Experimental Psychology: Animal Behavior Processes*, 1981, *7*, 87-108.
Gibbon, J., Church, R.M., & Meck, W. Scalar timing in memory. In: J. Gibbon & L.G. Allan (Eds.), *Timing* and *time perception*. Annals of the New York Academy of Sciences, Vol. 423, 1984, pp. 52-77.
Goodrick, C.L. Operant responding of nondeprived young and senescent male albino rats. *Journal of Genetic Psychology*, 1969, *114*, 29-40.
Groos, G., & Daan, S. The use of biological clocks in time perception. In: J.A. Michon & J.L. Jackson (Eds.), *Time, mind, and behavior*. Heidelberg: Springer Verlag, 1985, pp. 65-74.
Grossman, K.E. Continuous fixed-ratio and fixed-interval reinforcement in honey bees. *Journal of the Experimental Analysis of Behavior*, 1973, *20*, 105-109.
Harrison, D.W., & Isaac, W. Disruption and habituation of stable fixed-interval behavior in younger and older monkeys. *Physiology and Behavior*, 1984, *32*, 341-344.
Hemmes, N.S. Pigeons performances under differential reinforcement of low rate schedule depends upon the operant. *Learning and Motivation*, 1975, *6*, 344-357.
Hobbs, S.H. Circadian rhythms and animal behavior research. *Animal Learning and Behavior*, 1981, *9*, 604-605.
Honma, K., Goetz, C. von, & Aschoff, J. Effects of restricted daily feeding on freerunning circadian rhythms in rats. *Physiology and Behavior*, 1983, *30*, 905-913.
Innis, N.K., & Vanderwolf, C.H. Neural control of temporally organized behavior in rats: The suprachiasmatic nucleus. *Behaviour Analysis Letters*, 1981, *1*, 53-62.
Lejeune, H., & Richelle, M. Differential reinforcement of perching duration in the pigeon: A comparison with differential reinforcement of low rate keypecking. *Behaviour Analysis Letters*, 1982, *2*, 49-57.
Macar, F. *Le temps: perspectives psychophysiologiques*. Bruxelles: Mardaga, 1980.

Mantanus, H., Fayt, C., & Ferette, R. Nature de l'opérant et renforcement de débits lents chez le pigeon. *Psychologica Belgica*, 1977, *17*, 135-142.

Marley, E., & Morse, W.H. Operant conditioning in the newly hatched chicken. *Journal of the Experimental Analysis of Behavior*, 1966, *9*, 95-103.

Michon, J.A. The compleat time experiencer. In: J.A. Michon & J.L. Jackson (Eds.), *Time, mind and behavior*. Heidelberg: Springer Verlag, 1985a, pp. 20-52.

Michon, J.A. Temporality and metaphor. In: J.A. Michon & J.L. Jackson (Eds.), *Time, mind and behavior*. Heidelberg: Springer Verlag, 1985b, pp. 288-296.

Michon, J.A. J.T. Fraser's ›Levels of temporality‹ as cognitive representations. To be published in J.T. Fraser et al. (Eds.), *The study of time V*. Amherst, MA.: University of Massachussets Press, 1985c.

Phillips, J.L.M., & Milkulka, P.J. The effects of restricted food access upon locomotor activity in rats with suprachiasmatic lesions. *Physiology and Behavior*, 1979, *23*, 257-262.

Piaget, J. *Le développement de la notion de temps chez l'enfant*. Paris: Presses Universitaires de France, 1946.

Piaget, J. Problèmes psychologiques et épistémologiques du temps. In: J. de Ajuriaguerra (Ed.), *Cycles biologiques et psychiatrie*. Genève: Georg et Cie, Paris: Presses Universitaires de France, 1968, pp. 267-279.

Plotkin, H.C., & Odling-Smee, F.J. Brain-behaviour studies and evolutionary biology. In: D.A. Oakley & H.C. Plotkin (Eds.), *Brain, behaviour and evolution*. London: Methuen, 1979, pp. 52-77.

Plotkin, H.C., & Odling-Smee, F.J. Learning in the context of a hierarchy of knowledge gaining processes. In: H.C. Plotkin (Ed.), *Learning, development and culture. Essays in evolutionary epistemology*. Chichester and New York: Wiley, 1982, pp. 443-471.

Pouthas, V. Timing behavior in young children: A developmental approach to conditioned spaced responding. In: J.A. Michon & J.L. Jackson (Eds.), *Time, mind and behavior*. Heidelberg: Springer Verlag, 1985, pp. 100-109.

Reese, E.P., & Reese, T.W. The quail *Coturnix coturnix* as a laboratory animal. *Journal of the Experimental Analysis of Behavior*, 1962, *5*, 265-270.

Richelle, M. Notions modernes de rythmes biologigues et régulations temporelles acquises. In: J. de Ajuriaguerra (Ed.), *Cycles biologiques et psychiatrie*. Genève: Georg et Cie, Paris: Masson, 1968, pp. 233-255.

Richelle, M., & Lejeune, H. L'Animal et le temps. In: P. Fraisse (Ed.), *Du temps biologique au temps psychologique*. Paris: Presses Universitaires de France, 1979, PP, 73-128.

Richelle, M., & Lejeune, H. *Time in animal behaviour*. Oxford: Pergamon Press, 1980.

Richelle, M., & Lejeune, H. Timing competence and timing performance: A cross-species approach. In: J. Gibbon & L. Allan (Eds.), *Timing and time perception*. Annals of the New York Academy of Sciences, Vol 423, 1984, pp. 255-277.

Richter, C.P. *Biological clocks in medecine and psychiatry*. Springfield, IL: Charles C. Thomas, 1965.

Roberts, S. Properties and function of an internal clock. In: R.I. Mellgren (Ed.), *Animal cognition and behavior*. Amsterdam: North Holland Publishing Company, 1983, pp. 345-397.

Rosenwasser, A.M., Pechat, R.J., & Adler, N.T. Memory for feeding time: Possible dependence in coupled circadian oscillations. *Physiology and behaviour*, 1984, *32*, pp. 2530.

Rosenwasser, A.M., Raibert, M., Terman, J.S., & Terman, M. Circadian rhythms of luminance detectability in the rat. *Physiology and Behaviour*, 1979, *23*, 17-21.

Rusak, B., & Zucker, I. Neural regulation of circadian rhythms. *Physiological Reviews*, 1979, *59*, 449-526.

Saint Paul, V. von, & Aschoff, J. Longevity among blowflies *Phormia terranovae R.D.* kept in non-24 hours light-dark cycles. *Journal of Comparative Physiology*, 1978, *127*, 191-195.

Stephan, F.K. Limits of entrainment to periodic feeding in rats with suprachiasmatic lesions. *Journal of Comparative Psychology*, 1981, *143*, 401-410.

Stephan, F.K. Phase shifts of circadian rhythms in activity entrained to food access. *Physiology and Behaviour*, 1984, *32*, 663-671.

Stephan, F.K., Swann, J.M., & Sisk, C.L. Anticipation of 24 hours feeding schedules in rats with lesions of the suprachiasmatic nucleus. *Behavioral and Neurological Biology,* 1979a, *25*, 346-363.

Stephan, F.K., Swann, J.M., & Sisk, C.L. Entrainment of circadian rhythms by feeding schedules in rats with suprachiasmatic lesions. *Behavioral and Neurological Biology,* 1979b, *25*, 545-554.

Suda, M., & Saito, M. Coordinative regulation of feeding behavior and metabolism by a circadian timing system. In: M. Suda, O. Hayaishi & H. Nakagawa (Eds.), *Biological rhythms and their central mechanism*. Amsterdam: North Holland Elsevier, 1979, pp. 263-271.

Terman, M. Behavioral analysis and circadian rhythms. In: M.D. Zeiler & P. Harzem (Eds.), *Advances in analysis of behaviour,* Vol. 3. Chichester: Wiley, 1983, pp. 103-141.

Terman, M., Gibbon, J., Fairhurst, S., & Waring, A. Daily meal anticipation: Interaction of circadian and interval timing. In: J. Gibbon & L. Allan (Eds.), *Timing and time perception*. Annals of the New York Academy of Sciences, Vol. 423, 1984, pp. 470-487.

Terman, M., & Terman, J.S. Circadian rhythm of brain selfstimulation behavior. *Science*, 1970, *168*, 1242-1244.

Terman, M., & Terman, J.S. Control of the rat's circadian self-stimulation rhythm by light-dark cycles. *Physiology and Behaviour,* 1975, *14*, 781-789.

Terman, M., & Terman, J.S. Circadian functions of reinforcing brain stimulation. In: A. Wauquier & E.J. Rolls (Eds.), *Brain stimulation reward*. Amsterdam: North Holland Publishing Company, 1976, pp. 419-422.

Reference Notes

1. Fery, P. *Chronobiologie et conditionnement au temps: Contribution à l'étude de l'implication des oscillateurs circadiens dans les régulations temporelles acquises*. Unpublished Thesis for the Licenciate in Psychology, University of Liège, 1984.
2. Grailet, J.M. *Regulation temporelle acquise chez un poisson cichlidé africain Sarotherodon niloticus* (L.). Unpublished MasterThesis. Laboratory of Experimental Psychology, University of Liège, 1983.
3. Laurent, E. *Regulation temporelle acquise chez une tortue d'eau douce: Pseudemys scripta elegans* (Wied). Unpublished Master Thesis, Laboratory of Experimental Psychology, University of Liège, 1983.
4. Lejeune, H., Jasselette, P., Lachaussée, S., Lambotte, A., Piette, V., Salah, D., & Servais, M. *Etude chez le pigeon de deux types de réponse: Coups de patte et perchage dans un programme DRL*. Unpublished research report. Laboratory of Experimental Psychology, University of Liège, 1984a.
5. Lejeune, H., Nagy, J., Gougnard, I., Janssen, A.F., Leclercq, J., & Schyns, Ph. *Ontogenèse de la régulation temporelle acquise chez la caille*. Unpublished research report. Laboratory of Experimental Psychology, University of Liège, 1984b.
6. Lejeune, H. Nagy, J., & Jasselette, P. *Regulation temporelle acquise en programme FI chez le rat de 21 jours: Comparaison avec le rat sénile*. Unpublished research report. Laboratory of Experimental Psychology, University of Liège, 1984c.
7. Perikel, J.J., Fery, P., Lorea, Ph., Malaise, N., Maréchal, Ph., & Pirlet, C. *Etude de la régulation temporelle chez le rat dans une perspective chronobiologique*. Unpublished research report. Laboratory of Experimental Psychology, University of Liège, 1984.
8. Zimmerman, J.C. *Circadian functions of operant behavior under interresponse time contingencies*. Ph. D. Dissertation, Northeastern University, Univ. Microfilm No 7724430, 1977, (quoted by Terman, M., 1983).

Chapter 6. Timing Behavior in Young Children: A Developmental Approach to Conditioned Spaced Responding

Viviane Pouthas

Introduction

The early development of timing and the estimation of temporal duration have been under-researched for theoretical reasons and perhaps to a greater extent for methodological ones. One might be led to think that humans cannot perceive and estimate duration until they master general concepts of time and measurement. Results of numerous studies (Fraisse, 1948; Fraisse & Orsini, 1958; Goldstone & Goldfarb, 1966) seem to be consistent with this assumption: they demonstrate that 6 to 8 years of age is a transition period in several respects. When asked to reproduce or estimate intervals, 8-year olds respond accurately while younger children perform erratically. This dramatic change in the capacity of 6- to 8-year olds to estimate duration correctly suggests that this development may be connected to the acquisition of logical and conventional time.

However, there are »modes of adaptation to time« (Fraisse, 1982), present prior to the mastery of the notion of time, which do not presuppose conceptual thinking and knowledge. Using a habituation technique, Demany et al. (1977) have shown that 3- to 5-month old infants can discriminate between two different sound sequences. This demonstrates that infants are capable of perceiving brief durations and rhythmic structures. Conditioning to a periodic duration has also been evidenced in the infant during the first few days following birth. Thus, Marquis (1941) found that two groups of infants, fed either on a 3-hour or a 4-hour schedule adapted to each rhythm although their natural rhythm was 3.2 hours. The measure of adaptation was the amount of agitation prior to meal times recorded via actographic cribs. And Brackbill et al. (1967) demonstrated that time as conditional stimulus (CS) – that is, a sequence of 20 s interval, 4 s UCS (light on, light off), 20 s interval, 4 s UCS – was effective for establishing both conditioned pupillary constriction and dilatation. Previous research using conditioning paradigms (classical and operant) has consistently shown very precise adjustments to time in animals even though they lack any advanced conceptual apparatus.

Methodological reasons may also have precluded an adequate description of developmental trends in the capacity for timing behavior and estimating duration. Traditional psychophysical methods, largely dependent upon verbal techniques are difficult to apply at early ages: moreover some of these methods, like estimation and production, require knowledge of conventional units of time. A few studies (Crowder & Hohle, 1970; Friedman, 1977) using reproduction tasks, however, clearly demonstrate that, given sufficient practice and feedback, time intervals can indeed be reproduced by children as young as 4 years.

To explore capacities to estimate time in very young children, a temporal operant conditioning paradigm may be more appropriate. As Richelle & Lejeune have suggested (1980, p. 6), »the description of operant techniques clearly shows that part of the study of behavioral adjustments to time in animals is very close to the study of time estimation in humans using psychophysical methods.«

The operant technique used in the experiments I shall describe is the Differential Reinforcement of Low Rates (DRL) schedule. In this schedule, reinforcement is contingent upon a response which terminates an Interresponse time (IRT) of a specified minimum duration. Responses emitted before the critical delay are not reinforced and, furthermore, reset the timer. Thus, to obtain optimal reinforcement under the DRL schedule the child *must learn both waiting behavior and temporal discrimination*. This ›DRL situation‹ is not too far removed from situations in which the very young child experiences duration. As Janet (1928) remarked, our first encounter with duration occurred every time a delay was imposed between a desire and its realization, and he suggested that the most complex forms of adaptation to time involving conceptual and symbolic processes would derive from the basic properties of waiting behavior (a question also raised by Richelle et al., 1985, in chapter 5 of the present volume).

The few developmental studies which have investigated temporally conditioned responses in young children with DRL schedules have not obtained consistent results. Weisberg & Tragackis (1967) and Weisberg (1970) showed that DRL 10 s and DRL 18 s schedules generated and maintained low rates of responding in children ranging in age from 15 to 41 months. At the end of conditioning, more than 40 percent of all responses resulted in a reinforcement. Yet Stein & Landis (1978) found that children do not develop precise temporal discriminations before age 7 when reinforced for keypressing with a DRL 5 s schedule. These findings are consistent with their hypothesis that 7-year old children have a greater capacity than younger ones to inhibit motor behavior. However, my own previous results (1981a) differ from those of Stein & Landis. Half of the 4½-year old children, who were deliberately not informed of the waiting contingency, were found capable of precise temporal discrimination when trained with DRL schedules. The subjects' spontaneous verbalizations (Pouthas, 1981b) suggest that the necessary conditions for precise adaptation to DRL type tasks are the ability to represent the DRL contingencies to oneself and to instruct oneself (for example by saying: »I must wait«).

The main objective of this research paper is to present the most salient results I have obtained in experiments designed to investigate the temporal performance of 10-months to 11-year old children on DRL schedules of reinforcement. These results may serve as a basis for further consideration of the way in which this ontogenetic line of research using operant conditioning paradigms can furnish information on how new forms of temporal behavior are superimposed on those observed prior to the appearance of the symbolic function and to cognitive representations of time (see, Richelle et al., 1985, chapter 5 of the present volume for further discussion of this point; in particular on the ontogenetic approach pp. 76-90).

Method

Apparatus (Figure 1)

One of the main problems in using operant conditioning paradigms in young children is to find an adequate operant and a motivating reinforcer. The operant response selected consisted of pressing a large red button mounted on a telegraph key. The reinforcement consisted of a colored slide projected for 4 seconds on the screen of a toy theater or a big white box.

For the infant group a large red plastic plate was mounted on the telegraph key. The reinforcer consisted of a teddy bear hand puppet manipulated for 4 seconds by an experimenter on the stage of the toy theater. Trial experiments had shown that it was easier to maintain the attention of the 10-month and 2-year olds, if an auditory stimulus was coupled with the visual stimulus. A tape recorder therefore provided 4 seconds of music.

Procedure

Critical delays (5 or 8 s) of the DRL schedule, number of sessions, and criteria for terminating the sessions depend on the aim of each experiment and more precisely on the age range which the experiment assessed. They will be described briefly before presenting results for the youngest, intermediate and oldest children to account for procedural differences across age groups and to facilitate comparisons with previous research data.

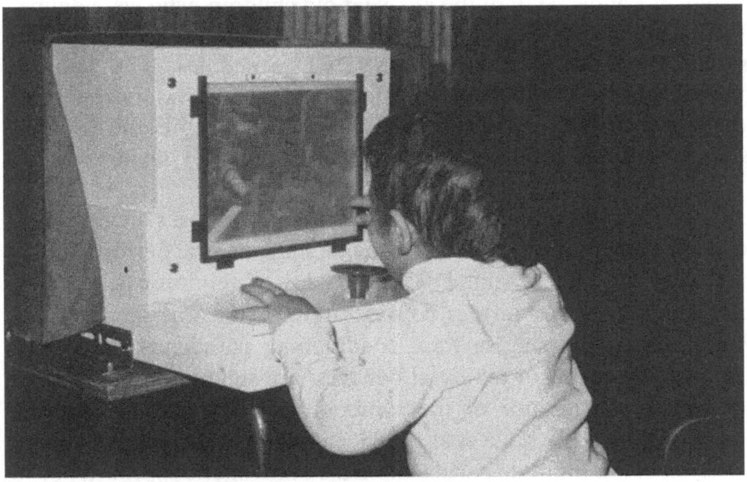

Figure 1. A 18-month old child in front of the large picture box and attached response box.

Results

Infant Group and 2-Year-Old Group

The subjects were recruited from Parisian and suburban daycare centers: 8 to 10 months (2 boys); 2 years (1 boy, 2 girls). At the beginning of training, the infants were shaped to press the plastic plate and the 2-year old children were simply told: »You are going to play at making pictures come on the screen«. The experimenter let the children discover the operant response alone. They were then trained with a DRL 5 s. For the 8 to 10-month olds the session ended when the subject became uncomfortable (nappy wetting, crying...). The end of the session occurred after the subject had received between 10 and 25 reinforcers. A criterion of 16 reinforcers was established for the 2-year olds.

Table 1 presents the number of responses per minute, the percentage of IRTs < 1 s (that is, brief IRTs) and the percentage of IRTs ≥ 5 s (measure of performance efficiency) for the 2 ten-month olds and the 3 two-year olds during the first and last training session on the DRL 5 s schedule. The number of sessions is indicated for each subject.

The data show that DRL training reduces the response rate and the percentage of brief IRTs. These findings are consistent with those of Weisberg & Tragackis (1967) and Weisberg (1970). However in terms of the percentage of reinforced IRTs, these authors report that at least 40 percent of all responses resulted in a reinforcement. Our own data show poorer efficiency: between 20 and 30 percent, except for Subject 5 who attained 44 percent.

A more informative way of looking at temporal performance is to inspect interresponse time distributions. These distributions are obtained after grouping IRTs into classes according to duration (Figure 2). The absolute frequencies are then converted into relative frequencies. If the mode of the distribution coincides with the critical delay period (in Figure 2 it is, for example, 5 s), it can be assumed that precise temporal regulation is present.

In order to compare Weisberg's data (1970) with ours (Pouthas & Jacquet, 1983; Pouthas et al., Note 1) both data sets are presented in the same figure (Figure 3). The distributions of IRTs are shown for the sixth DRL 18 s session. The data are

Table 1.

Subject (age in months)	Number of sessions	Response per min.		% IRTs < 1 sec.		% IRTs ≥ 5 sec.	
		First	Last	First	Last	First	Last
S1 (8)	12	25	13	56	47	21	25
S2 (10)	18	19	9	35	34	23	28
S3 (18)	6	26	20	66	46	4	19
S4 (24)	6	32	14	66	55	4	19
S5 (24)	10	38	8	75	21	5	44

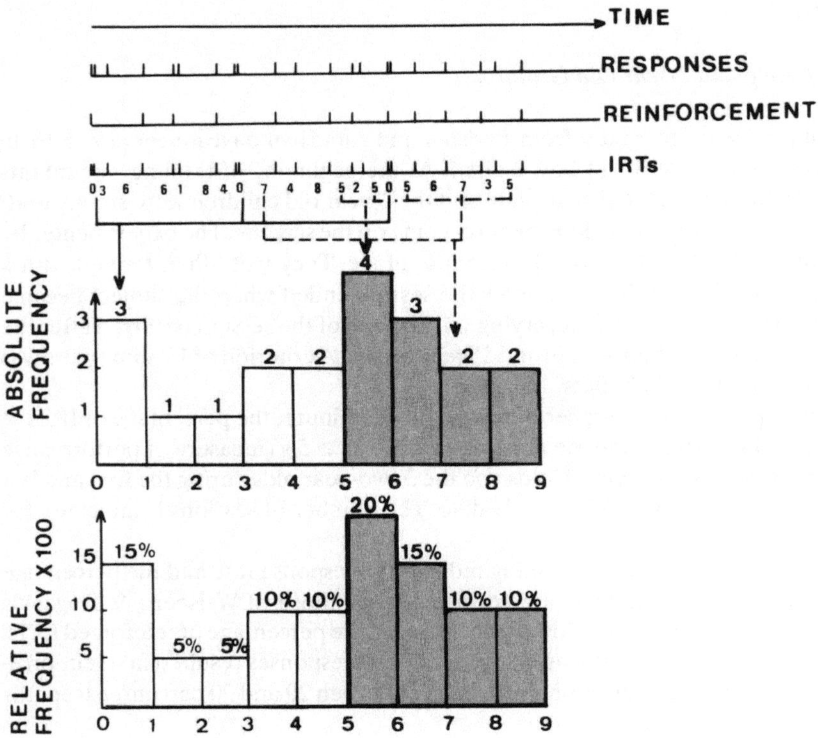

Figure 2. Theoretical histogram illustrating steps in calculation of IRT distributions (Adapted from Richelle & Lejeune, 1980).

those of three children who received limited experience in two DRL 2 s training sessions and three children who were trained on a DRL 10 s schedule for six sessions before the DRL 18 s acquisition, from Weisberg's (1970) study.

On the right are shown the distributions of IRTs for the last DRL 5 s session obtained from our subjects. Figure 3 shows that subjects in Weisberg's study were more efficient than those in our study. However, in neither group do the modes of the distributions of IRTs coincide with the value of the critical delay as is generally the case in highly conditioned animal subjects (Richelle & Lejeune, 1980). For most children, the proportion of IRTs corresponding to the longest waits is relatively high.

Taken together, the results of Weisberg and of my studies show that very young children, who are usually considered to be unable to wait, are in fact capable of refraining from responding, as evidenced by a decrease in response rates. They are able to withhold their behavior for a certain amount of time, but they cannot

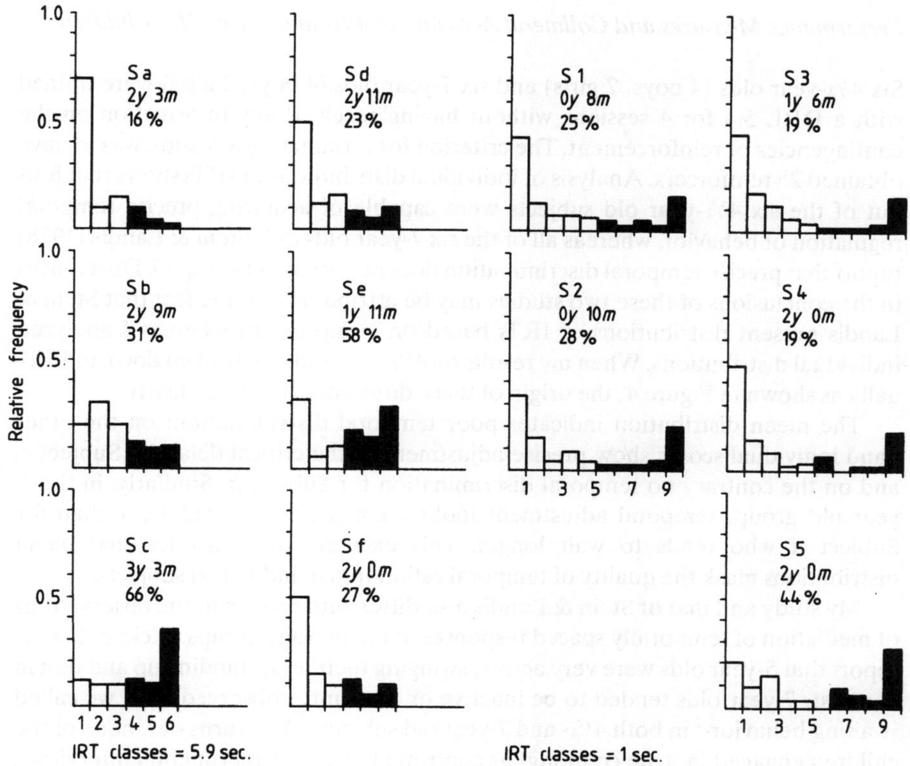

Figure 3. Distribution of IRTs for the sixth DRL 18 s session (Weisberg, 1970) and for the last DRL 5 s session (Pouthas & Jacquet, 1983; Pouthas et al., 1984). Age and percentage of reinforced IRTs appear under subject designation. Shaded columns indicate reinforced IRTs.

precisely adjust their responses to the required delay. Their waits are often too long, perhaps because they are engaged in gross motor activities, as shown by a systematic observation of collateral behavior, that is, behavior other than the explicitly reinforced operant response. Collateral activities have been associated with superior temporal performance in a number of studies on animal subjects (for a review, see Richelle & Lejeune, 1980, p. 188-199). The three 2-year olds developed collateral behavior which became systematic throughout the training: S3 followed a systematic ›route‹ around the experimental room; S4 turned over the building block used as a seat; S5 leaned to one side of the big white box then to the other side. This suggests that very young children who establish a particular repetitive action scheme to ›wait‹ are estimating a duration by relying upon a motor rhythm.

Performance Measures and Collateral Activities of 4½- and 7-Year Old Children

Six 4½-year olds (4 boys, 2 girls) and six 7-year olds (4 boys, 2 girls) were trained with a DRL 5 s for 4 sessions without having received any information on the contingencies of reinforcement. The criterion for terminating a session was to have obtained 25 reinforcers. Analysis of individual distributions of IRTs shows that four out of the six 4½-year old subjects were capable of acquiring precise temporal regulation of behavior, whereas all of the six 7-year olds did. Stein & Landis (1978) report that precise temporal discrimination does not occur before age 7. Differences in the conclusions of these two studies may be attributable to the fact that Stein & Landis present distributions of IRTs based on group means whereas I analyzed individual distributions. When my results on 4½-year olds are broken down individually as shown in Figure 4, the origin of these differences appears clearly.

The mean distribution indicates poor temporal discrimination: on the other hand individual scores show precise adjustment to the critical delay for Subject *a*, and on the contrary no temporal discrimination for Subject *b*. Similarly, in the 7-year old group, temporal adjustment looks more precise for Subject *c* than for Subject *d* who tends to wait longer. This example demonstrates that mean distributions mask the quality of temporal estimation in individual subjects.

My study and that of Stein & Landis also differ with respect to the observations of mediation of temporally spaced responses in the two age groups. Stein & Landis report that 5-year olds were very active, swinging their legs, standing up and that in contrast, 7-year olds tended to be inactive or to count. I observed what we called ›waiting behavior‹[1] in both 4½- and 7-year old subjects. As it turns out, none of the children engaged in ›true counting‹ as confirmed by post-experimental interviews. Some of the subjects developed fine motor repetitive activities. For example, a 4½-year old subject plucked his thumb nail with the forefinger of his other hand during each IRT; three times for IRTs lasting on the average 5.5 s. Such ›counting activities‹ may have provided the subject with temporal cues. Precise temporal regulation was evidenced in 4½- and 7-year old subjects: either they showed waiting behavior or repetitive behavior.

Effects of Counting on DRL Performance of 9- and 11-Year Old Children

Around 8 to 9 years of age, children begin to use and understand chronometric time concepts, such as the concept of a clock second (Smythe & Goldstone, 1957; Goldstone & Goldfarb, 1966; Friedman, 1978). An experiment was conducted to investigate the effects of counting on the efficiency of DRL performance among 9- and 11-year olds in comparison with DRL performance of 7-year olds. Three groups of 7 subjects aged 7 (3 boys, 4 girls), 9 (3 boys, 4 girls) and 11 (4 boys, 3 girls)

1 Waiting behavior: the subject, while attentive, remained inactive in terms of motor activity except for approaching his/her hand towards the response box to give the operant response.

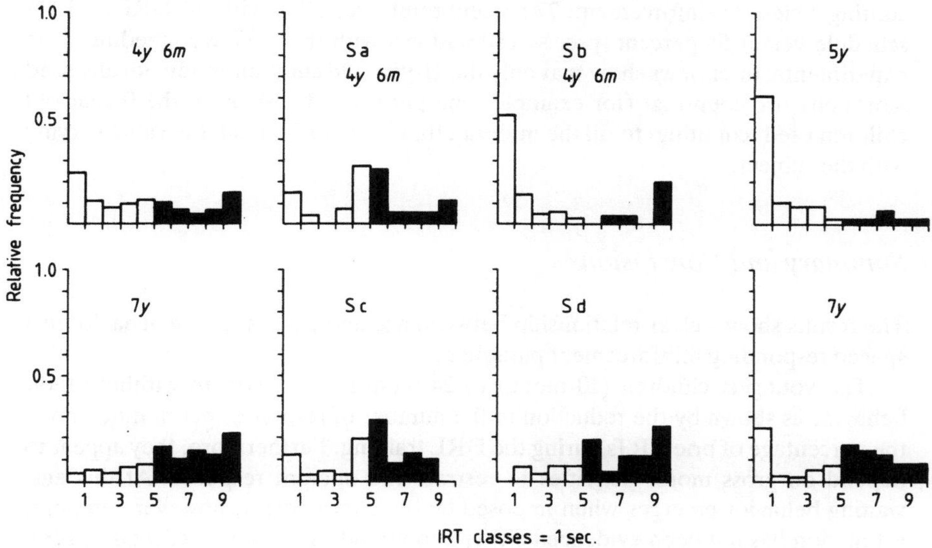

Figure 4. Distributions of IRTs during the last DRL 5 s session. *From the left:* mean distributions of 4 $^1/_2$- and 7-year old groups and examples of individual distributions, from Pouthas & Jacquet (1983). *Rightmost diagrams:* after Stein & Landis (1978).

recruited from the same elementary school, first received two training DRL 5 s sessions, then were exposed to a DRL 8 s schedule for two sessions, and finally were shifted to a DRL 8 s LH4 schedule for two additional sessions. In DRL with a limited hold (LH), there is an upper limit for reinforced IRTs. Thus, in a DRL 8 s LH4 responses must be spaced by at least 8 s and at most 12 s, in order to be reinforced. The criterion for terminating a session was to have received 25 reinforcers.

Table 2 illustrates that 7-year olds are less efficient than the 9- and 11-year olds. At 9 years of age, temporal regulation is more precise when imposed by the

Table 2.

Groups	DRL 8 s		DRL 8 s LH 4
	a % of reinforced IRTs	b % of IRTs between 8 s and 12 s	% of reinforced IRTs
7-year olds	66	42	40
9-year olds	84	55	77
11-year olds	85	75	86

b: measure indicating precision of adjustment to critical delay contingencies of reinforcement: 77 percent reinforced IRTs with the DRL 8 s LH4 schedule versus 55 percent ›precise estimations‹ with the DRL 8 s schedule. Post experimental interviews show that only the 11-year old children systematically used ›chronometric counting‹ (for example, one per second). Many of the 9-year old children used ›counting‹ to fill the interval, that is, up to 20 or 30, the value varying with the subject.

Summary and Conclusions

The results show a clear relationship between age and type of timing behavior in a spaced responding reinforcement paradigm.

The youngest children (10 month to 24 month olds) learn to withhold their behavior as shown by the reduction in the number of responses per minute and in the percentage of brief IRTs during the DRL training. Furthermore, they appear to depend on gross motor activities to restrain the operant response. At this age, waiting behavior emerges when imposed by the environment; however, temporal estimation has not been evidenced either in our studies (Pouthas & Jacquet, 1983; Pouthas et al., Note 1) or in the literature. Capacity of acquiring precise temporal regulation of behavior is present in some 4½-year olds and well established in 7-year olds, even when they have received no information on the temporal constraints of reinforcement. Knowledge that in certain circumstances a temporal interval must elapse between two events appears to be acquired in children of this age. At the beginning of training the subject probably compares this prior knowledge with the characteristics of the situation at hand, and this leads to the self instruction »I have to wait«. Spontaneous verbalizations support this argument (Pouthas, 1981b). However this ›cognitivist explanation‹ is not sufficient to explain the formation of precise temporal discrimination in children in the 4½- to 7-year age range. The notion of control by a ›biological clock‹ remains an unresolved issue. At 9 and 11 years of age, children use ›chronometric counting‹ (for example, 1 per second) but in this case can we define DRL performance as conditioned responses to duration?

References

Brackbill, Y., Fitzgerald, H.E., & Lintz, L.M. A developmental study of classical conditioning. *Monographs of the Society for Research in Child Development*, 1967, *32*, no. 8.

Crowder, A.M., & Hohle, R.H. Time estimation by young children with and without informational feedback. *Journal of Experimental Child Psychology*, 1970, *10*, 353-365.

Demany, L., McKenzie, B., & Vurpillot, E. Rhythm perception in early infancy. *Nature*, 1977, *266*, 718-719.

Fraisse, P. Etude comparée de la perception et de l'estimation de la durée chez les enfants et les adultes. *Enfance*, 1948, *1*, 199-211.

Fraisse, P. The adaptation of the child to time. In: W.J. Friedman (Ed.), *The developmental psychology of time*. New York: Academic Press, 1982, pp. 113-139.

Fraisse, P., & Orsini, F. Etude expérimentale des conduites temporelles. III: Etude génétique de l'estimation de la durée. *l'Année Psychologique*, 1958, *58*, 1-6.

Friedman, E.R. Judgments of time intervals by young children. *Perceptual and Motor Skills*, 1977, *45*, 715-720.
Friedman, W.J. Development of time concepts in children. In: H.W. Reese & L.P. Lipsitt (Eds.), *Advances in child development and behavior.* London: Academic Press, 1978, pp. 267-298.
Goldstone, S., & Goldfarb, J.L. The perception of time by children. In: A.H. Kidd & J.L. Rivoire (Eds.), *Perceptual development in children.* New York: International Universities Press, 1966, pp. 445-487.
Janet, P. *L'évolution de la mémoire et de la notion de temps.* Paris: Chahine, 1928.
Marquis, D.P. Learning in the neonate. *Journal* of *Experimental Psychology,* 1941, *29*, 263-282.
Pouthas, V. Adaptation à la durée chez l'enfant de 2 à 5 ans. *l'Année Psychologique*, 1981a, *81*, 33-50.
Pouthas, V. Adaptations comportementales à la durée chez le très jeune enfant. In: *Memoire, conditionnement, evolution.* Publications de la Sorbonne, série »Etudes«, 1981b, *18*, pp. 59-69.
Pouthas, V., & Jacquet, A.Y. Attente et adaptation à la durée chez l'enfant, dans la Psychogénèse du Temps: Cinq approches. *Cahiers de Psychologie Cognitive*, 1983, *3*, 397-407.
Richelle, M., & Lejeune, H. *Time in animal behaviour.* Oxford: Pergamon Press, 1980.
Richelle, M., Lejeune, H., Perikel, J., & Fery, P. From biotemporality to nootemporality: Toward an integrative and comparative view of time in behavior. In: J.A. Michon & J.L. Jackson (Eds.), *Time, mind, and behavior.* Heidelberg: Springer Verlag, 1985, pp. 75-99.
Smythe, E.J., & Goldstone, S. The time sense: A normative genetic study of the development of time perception. *Perceptual and Motor Skills*, 1957, *7*, 49-59.
Stein, N., & Landis, R. Effects of age and collateral behavior on temporally discriminated performance of children. *Perceptual and Motor Skills*, 1978, *47*, 87-94.
Weisberg, P. Effects of reinforcement history on timing (DRL) performance in young children. *Journal of Experimental Child Psychology*, 1970, *9*, 348-362.
Weisberg, P. & Tragackis, C.J. Analysis of DRL behavior in young children. *Psychological Reports*, 1967, *21*, 709-715.

Reference Note

1. Pouthas, V., Provasi, J., & Jacquet, A.Y. *Régulation temporelle acquise en programme DRL chez le bébé.* Paris: Mémoire de maîtrise, Université de Paris V, 1984.

Part II. Processes: The Perception and Retention of Time

Chapter 7. Time Psychophysics and Related Models[1]

Françoise Macar

Introduction

This paper will deal with the major recent models of time (duration) perception and provide a brief summary of the psychophysical data on which these models are based. Three major issues which pertain to the question of temporal mechanisms will be considered successively: the thresholds for duration, the psychophysical function for time, and the interactions between temporal and nontemporal information.

Research in time psychophysics has produced multifarious results by means of a considerable variety of methods. These methods differ in several respects related either to external or to organismic variables such as number of stimulus durations involved, extent of difference between durations, reference to conventional chronometric units or to other types of measurement, and involvement of verbal or of other responses. Allan (1979) has provided a clear description of the major psychophysical methods. Her nomenclature distinguishes two main categories on the basis of the objective durations involved. *Duration scaling* tasks concern perceived duration when a set of clearly distinct intervals is presented, whereas *duration discrimination* tasks require discrimination of highly confusable intervals.

In duration scaling, six different methods are listed. (1) Verbal estimation: the verbal response reflects either a two-term judgment (›long‹ or ›short‹), or the reference to chronometric units such as seconds or minutes. (2) Magnitude estimation: the subject translates the magnitude of the perceived duration into a number, instead of referring to temporal units. A standard duration may be provided first. No categories are defined by the experimenter. (3) Category rating: ordered categories, specified by the experimenter, are used by the subject to classify his or her perceived duration. (4) Production: motor or verbal responses are required in order to delimit a particular interval. (5) Ratio-setting: the subject generates a specified proportion of the interval presented by the experimenter. This method is further subdivided into reproduction, if the proportion is equal to 1, fractionation, if it is inferior to 1, and multiplication if it is superior to 1. (6) Synchronization: the subject responds in synchrony with the end of a fixed standard duration, or makes a sequence of responses at a rate identical to that of a set stimulus presented by the experimenter.

Duration discrimination comprises altogether four methods. (1) Comparison: two intervals are presented (the ›standard‹ and the ›test‹), and the subject judges

1 This work was supported by a grant from the Ministry of Industry and Research no. 83C0913.

their relative duration. In two subcategories of this method, the forced-choice with either a fixed or a roving standard (that is, varying from trial to trial), the standard interval is presented either first or second in a trial, and the subject judges its position. (2) Single stimulus: one of two possible durations occurs on each trial, and the subject decides whether the longer or the shorter was presented. (3) Many-to-few: the response is the same as defined above, but more than two durations are provided. (4) Identification: the number of possible responses is increased as well as the number of possible durations.

It has long been known that the data yielded by different methods are usually weakly correlated. It is difficult to determine which method produces greatest accuracy and least discrepancy within or between subjects (Carlson & Feinberg, 1970; Doob, 1971; Fraisse, 1957; Fraisse et al., 1962; Goldfarb & Goldstone, 1963; Hornstein & Rotter, 1969; McConchie & Rutschmann, 1971). Variability is a stumbling block in time psychophysics. It seems particularly marked in the verbal estimation method, probably because recourse to conventional chronometric units complicates the task and involves memory and cognitive processes. Problems of comparison between methods add to difficulties in confronting data obtained in different modalities and with different durations. Thus, it is worth querying whether common mechanisms underlying time perception do exist. Some arguments favoring commonality of mechanism can be extracted here and there from the vast time literature. Specific training for a particular duration can ameliorate the reproduction of a longer one (Bakan et al., 1959). Transfer effects between the auditory and visual modalities have been noticed in the reproduction of intervals (Warm et al., 1975). High correlations have been found between these modalities in data concerning the detection of duration lengthening (Eijkman & Vendrik, 1965).

Absolute Threshold for Perceived Duration, Successiveness, and Temporal Order

Values reported for the absolute duration threshold range from less than 10 to about 150 milliseconds, depending on the modality and physical characteristics of the stimulus and on the level of practice of the subject (Durup & Fessard, 1930; Servière et al., 1977). The shortest minimum duration is found in the auditory modality. For perception of successiveness, the threshold for auditory stimuli approaches 10 ms after sufficient practice. It lies around 20 ms for perception of order. Hirsh & Sherrick (1961) found that this particular value was independent of differences in modality, intensity, or spatial localization. This conclusion was contested later by several authors and attenuated by Hirsh (1974) himself. Various other values of the threshold were reported by Rutschmann (1973) depending on the site of stimulation and intensity of brief flashes, and by Warren (1974) with complex sequences of auditory stimuli (see Jones, 1985, chapter 13 of the present volume). Nevertheless, the largely hypothetical constancy of the temporal order threshold has suggested a role of central mechanisms rather than specific sensory ones. Even some kind of ›simultaneity center‹ common to all sensory modalities has been postulated (Efron, 1963a, 1963b; Corwin & Boynton, 1968).

The Hypothesis of Channel Independence

The mechanisms leading to perception of succession and of temporal order between two stimuli can be described in the terms of the independent channels theory (e.g. Sternberg & Knoll, 1973). According to this theory, a signal is detected after a random arrival latency. This latency essentially depends on the signal's specific characteristics, and perhaps on certain detection criteria. Information is transmitted through a channel, and reaches the central processor at a certain arrival time. An order judgment is determined by the difference between the arrival times of the information yielded by two stimuli presented through distinct channels. No interference is possible at the transmission level but a decision rule may be applied to the output of the central mechanism: each particular difference between arrival times would correspond to a certain probability of correct temporal order judgment.

The difference between arrival times can be affected by several distinct limiting factors. For instance, in Kristofferson's (1970) attention switching model, attention is focused upon one channel at a time. Therefore, information transmitted through different channels cannot enter the central processor at the same time. Thus, all information conveyed by non-privileged channels is neglected until attention is switched to them, after an irreducible refractory period. Stimuli are judged as simultaneous or as successive depending on whether or not attention has switched fast enough to each of the channels containing relevant information. This kind of model has also been applied to discrimination of very brief durations (Allan et al., 1971). As an alternative to this central limitation assumption, the limits of order perception have also been thought to depend on different rates of information transmission between the channels (Rutschmann, 1973).

The independent channels theory has inspired several related models. Nevertheless, the theory has not been supported consistently by empirical studies: interaction has been observed between different inputs as early as the transmission stage (Pastore, 1983; Patterson & Green, 1970).

The Status of Internal Periodicities

Another hypothesis which seeks to account for the limitations of the central mechanism is Stroud's (1956) perceptual moment. In this conception, time is measured on the basis of periodic sampling of the environmental input. Two different inputs can be ordered only if they occur in different samples. The extent of a sample is supposed to be of the order of 100 ms (see also White, 1963). As such it could be the basis for an ›internal clock‹ mechanism, the ticking of which would refer to the electroencephalographic alpha waves (8 to 12 Hz in frequency) elicited in unfocused waking state (Wiener, 1958). A similar entity, the ›time quantum‹ postulated by Kristofferson (1967) is set at approximately 50 ms. According to this author, equallyspaced points in time would be generated by an internal clock mechanism. These points would determine the minimum possible delay of attention switching, as well as the transfer of information from one to another stage of central processing. More recently, Pöppel (1976) opted for a period of 20 to 30 ms as the perceptual moment or time quantum. This value has been corroborated by studies in various fields: time psychophysics, reaction times and visual information processing. Several other periods have also been postulated including a lenghty 700 ms, known as the ›indifference interval‹ and evidenced in temporal discrimination studies. In contrast with these various discrete perceptual moment hypotheses, Allport's (1968) ›traveling moment‹ supposed that the sensory input is continuously sampled. Accordingly, the ›moment‹ would correspond to a running segment of the input function.

One may reconcile the various values that have been proposed as perceptual moments by the argument that interactions between them and with all kinds of organic periodicities are the very basis of temporal processing. According to Gooddy (1958), the nervous system acts as a ›clock form‹ by integrating multiple time bases which derive from organic periodicities, such as glandular and autonomous cycles, or periodic neuronal discharges. However, the electroencephalogram is assumed to be a tangible sign of such integration processes. The infinite flexibility of the clock form system makes it both appealing and difficult to verify experimentally, since the proper combination of time bases for a particular response can be extracted from a considerable array. It would evidently be essential to define the mechanisms which regulate the choice and adequate combination of such time bases (see also Michon, 1985, chapter 2 of the present volume).

Electrophysiological Correlates

Research on event-related potentials has produced interesting data that are relevant to absolute duration thresholds. ›On‹ and ›off‹ responses evoked by onset and offset of brief visual stimuli have been recorded directly from the visual pathways and visual cortex in anaesthetized cats (Efron, 1973). Off responses were found to be elicited by stimulus offset only when stimulus duration was longer than 50 ms. For durations between 5 and 50 ms off responses always appeared after an almost fixed delay following stimulus onset. Apparently they were linked to the onset rather than to the offset of the stimulus (Figure 1). In human subjects provided with scalp electrodes, an irreducible duration of 30 ms has been obtained with similar stimulus parameters (Servière et al., 1977). It should be remembered, however, that additional delays, among others in response preparation, evidently occur before the overt response stage.

Other correlates of duration thresholds have been found at the cellular level. In the cat's optic pathways, a certain number of the neurons which responded to presentation of two successive flashes showed stable activity patterns. Neuronal discharges evoked by each flash were distinct when interstimulus interval exceeded 100 ms. At 80 ms and below they partially overlapped. The interstimulus interval was perceived in all trials beyond 100 ms, and in 50 percent of the trials longer than 40 ms (Peck & Lindsley, 1972, 1973). In addition, Levick & Zacks (1970) showing

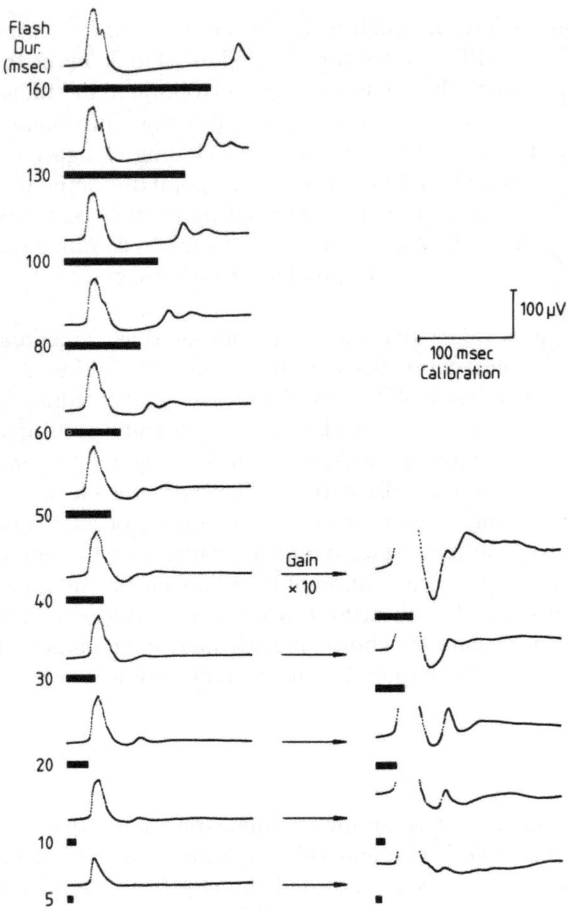

Figure 1. ›On‹ and ›off‹ responses from the cat optic nerve for flashes of different duration. Each tracing represents a computer average of 175 identical stimuli. The tracings on the right are identical with those on the left but at higher vertical gain. (After Efron, 1973).

that comparable unitary activity patterns exist in the cat's retina, demonstrated that discharges evoked by weak flashes in tonic cells persisted from 50 to 70 ms.

These data fit the time quantum hypothesis; the irreducibility of certain central responses would be responsible for the perceptual data obtained. Stimuli would be judged as instantaneous or simultaneous whenever the central responses are not modulated by their temporal characteristics. According to Efron (1967), a stimulus is perceived after a delay during which information is integrated. The duration of this processing period approximates 60 ms. It has been demonstrated that flashes induce similar reaction times whether their duration is 20 or 40 ms. This suggests that a processing period of identical length elapses before either flash is perceived. If the stimulus is too brief, it will be over before the processing stage is reached. In this case, its offset cannot be distinguished from its onset, and consequently no duration will be perceived. Similarly, two stimuli will be judged as successive only if the processing period of the first has been completed when the second occurs.

Differential Threshold and Psychophysical Function

Another major issue in time psychophysics concerns the differential threshold between two durations. According to Woodrow (1930), this threshold is about 10 percent of the shortest duration with auditory stimuli ranging from 0.2 to 2 s. It is reduced to 8 percent with 0.6 to 0.8 s stimuli but increases up to 16 percent with 4 to 30 s ones. Similar or smaller values were later reported, depending on physical characteristics of the stimulus, experimental procedures, or psychological variables (Fraisse, 1967; Rossi, 1972; Small & Campbell, 1962).

Weber's Law

If Weber's law holds in time psychophysics, the differential threshold (or the ›just noticeable difference‹, JND) between two durations should be a constant proportion of the shorter duration. This has been supported by a number of studies. Confirming earlier results, Thompson et al. (1976) reports a stable Weber fraction of about 0.055 for duration values ranging around 1 and 2 s. Temporal discrimination research in animals usually yield differential thresholds around 25 percent (Catania, 1970; Church et al., 1976; Stubbs, 1968; Perikel et al., 1974).

In the framework of animal conditioning, support for Weber's law was provided by Gibbon (1977, 1981). Other relevant data come from the paradigms allowing computation of a temporal bisection function (see Church & Deluty, 1977)[2].

However, Weber's law has been questioned in studies by Allan (1979), Allan & Kristofferson (1974) and Woodrow (1951), which covered a stimulus range between 0.05 and 30 s. Allan, for instance stated that although the JND is an increasing function of the shorter duration in discrimination studies, generally no linear relationship is found. In scaling tasks, the standard deviation of the response

2 The temporal bisection function relates the discriminability of a sample interval (T) to two other durations (a ›short‹ one and a ›long‹ one) with which the animals are first trained. Reinforcement is delivered for a response on one manipulandum (e.g. response key) after ›short‹ and on another manipulandum after ›long‹. Following adequate training, intermediate values of T are presented, without reinforcement. Response probabilities on each manipulandum for different values of T are calculated. The value at which equal probability of ›short‹ and ›long‹ reports are observed, that is, the bisection point, reflects the subjective middle between the two extremes. Bisection is found at the geometric mean. Gibbon demonstrated that the psychometric bisection function confirms Weber's law, which assumes equal probability for either ›short‹ or ›long‹ report at T and at kT, when the short and long values are kS and kT (where k is a constant).

bution increases in a monotonic fashion with the mean response value but it is by no means directly proportional to it. Allan & Kristofferson (1974) even reported constant discriminability over a considerable range of duration values, in highly experienced subjects.

Getty (1975, 1976) elaborated a generalized form of Weber's law, including a variance component independent of stimulus magnitude. He argued that many duration discrimination studies have revealed an increase in the Weber fraction for decreasing duration below 0.2 s (Abel, 1972a, 1972b; Allan et al., 1971; Woodrow, 1951). This has been found to be the case for various sensory dimensions and has generally been interpreted as a consequence of a sensory noise component, which would be added to the magnitude dependent variability. Getty reported data from a forced choice discrimination procedure, which were in good agreement with his model, for a stimulus range of 0.05 to 2 s. The Weber fraction was around 0.05 in the region of constancy, from approximately 0.2 to 2 s. It rose for shorter and longer durations, as had already found by Blakely (Note 1) and others.

The Psychophysical Function

The form of the psychophysical function, which expresses the relationship between objective and subjective time, has always been much debated. Weber's law generates logarithmic scale if it is assumed that constant subjective differences correspond to constant objective time ratios. A stable coefficient of variation has been found over various duration ranges in magnitude estimation data (Divenyi & Danner, 1977; Stevens, 1971). However, quite a number of duration scaling studies do not favor a logarithmic law at all, especially when intra-individual data are examined (Allan, 1983; Jesteadt et al., 1977; Luce & Mo, 1965).

Eisler (1976) reviewed more than one hundred experiments which he found in agreement with a power function. He proposed a ›parallel clock model‹ (Eisler, 1975) based on a power function as the psychophysical law (Figure 2). In his conception, there are two sensory registers (›clocks‹) which may run simultaneously. A comparison between two durations is made by counting the total subjective duration (from the beginning of the first duration to the end of the second) in one sensory register and counting the second subjective duration (from the beginning to the end of the second duration) in the other register. A ›comparator‹ then compares the difference between the contents of the two registers with the content of the second one. The two quantities are continuously processed until the correct ratio is reached after which the response is executed. For instance, in a reproduction task the subject will end the second duration when the two quantities are judged equal. In a single stimulus task, the comparison is between the current trial and the preceding one. This raises the question of what happens if an interstimulus interval separates two durations. Eisler proposes that the first register will continue running during this ›empty‹ interval, but it will accumulate ›zeros‹ rather than duration ›pulses‹ (Eisler, 1981). This model has been

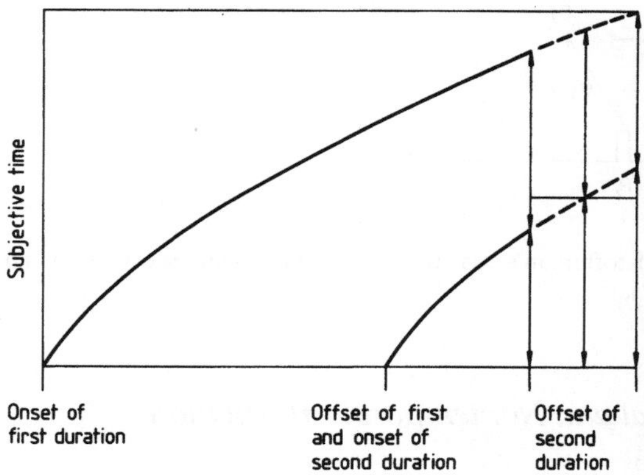

Figure 2. Schematic time course of Eisler's parallel-clock model. The upper curve indicates the accumulation of the *total* duration (first plus second interval). Note that it does not terminate at the offset of the first duration. The lower curve indicates the accumulation of the second duration. The two-pointed arrows show the compared quantities, the difference between total and second subjective duration, and second subjective duration, respectively. The middle pair of arrows marks the point where the first and second duration are experienced as equal; for the left pair, the first duration is experienced as longer and for the right pair of arrows, the second duration. (After Eisler, 1981).

recently questioned by Ellerman & Michon (1983), who found that some derivations from Eisler's formulae are not borne out by his data.

Allan et al. (1971) presented a model based on the assumption of a linear psychophysical law. This model was later refined by Kristofferson (1977, 1980). In his real time criterion theory, the variability of subjective duration as a function of objective time depends entirely on an internal criterion mechanism. Duration discrimination tasks are assumed to be solved by simple discrimination of the temporal order of perceptual offset latencies (Figure 3). Suppose that the subject has to decide whether a duration delimited by two auditory pulses is long or short relative to the whole range of stimulus durations presented during the experiment. The first pulse (P_1) triggers an internal time point event (B_1), following a constant afferent latency (α_1). B_1 triggers an internal criterion duration (I), the end of which is an internal time point event (C). The second pulse (P_2) triggers another time point event (B_2), after a constant perceptual offset latency (α_2). The judged temporal order of C and B_2 determines response choice. If B_2 occurs first, the response will be »short«. If C occurs first, the response will be »long«. This implies that »short« responses are time-locked to stimulus onset and do not depend on stimulus duration. These responses can even be made correctly *before* stimulus offset, which rules out the eventuality that the subject's decision is based on a measure of the objective duration. Kristofferson provided data showing that the latency distribution of the »short« and the »long« responses do indeed differ. The »short« responses show a distribution similar to a reaction time distribution and strictly linked to stimulus offset, while the »long« responses, which occur after a certain delay following stimulus onset, are distributed as time estimation responses. The model further assumes a perfect mapping of objective duration onto subjective duration: all durations occur in real time. Most discrimination data seem in agreement with this conclusion (Allan & Kristofferson, 1974).

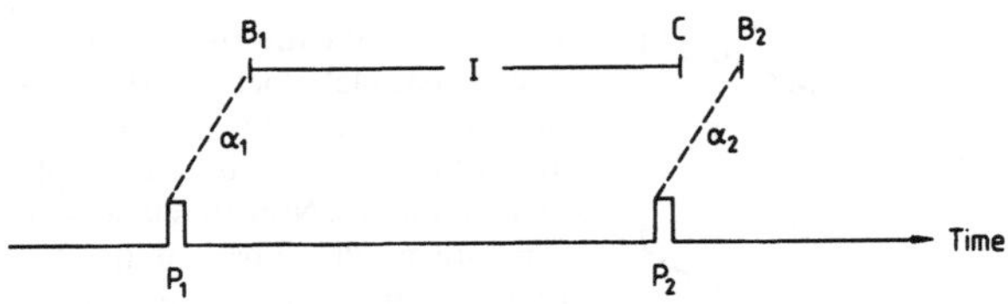

Figure 3. Schematic illustration of Kristofferson's real-time criterion model. See text for an explanation. (After Kristofferson, 1977).

Interactions between Temporal and Nontemporal Information Processing

One of the implications of Kristofferson's real time criterion theory is that certain duration judgments may be independent of the measurement of stimulus duration itself. Although this conclusion only concerns a limited number of data, it appears that duration measurement is always in part dependent on nontemporal parameters, for instance on the modality and the energy level of the stimulus (for reviews, see Allan, 1979; Allan & Kristofferson, 1974; Doob, 1971; Fraisse, 1957; Macar, 1980). This holds mostly for durations shorter than 1 s (Divenyi & Danner, 1977; Oléron, 1952). Accordingly, durations ›filled‹ with an auditory or a visual stimulus are generally overestimated in comparison to ›empty‹ durations (Thomas & Brown, 1974). Particularly important influences on subjective duration stem from stimulus frequency and the organization of the stimulus set. An interval is judged longer when the frequency of the intermittent stimulations of which it is composed increases (Adams, 1977; Fraisse, 1957, 1965; Schiffman & Bobko, 1977). Enhanced organization has an opposite effect (Schiffman & Bobko, 1974; Avant et al., 1975; see also Block, 1985, chapter 11 of the present volume).

Such results are basic to several theoretical conceptualizations. Most models proposed around the early sixties assumed that subjective time depends on accumulation of discrete elements: density of perceived changes for Fraisse (1957), for whom such changes essentially consist in objective indices yielded by the task performed; amount of ›mental content‹ per unit of duration, that is the mean number of events perceived during the relevant interval, for Frankenhaeuser (1959); and ›internal pulses‹ for Creelman (1962) and Treisman (1963). In his counting model inspired by signal detection theory, Creelman assumed that pulses accumulating in an internal counter are emitted by several independent stochastic sources. Treisman postulated a single pacemaker, but submitted it to the influence of the arousal level. These are two different ways of accounting for the well-known fact that there are numerous distinct factors that may influence subjective duration, such as drugs, boredom, or sensory deprivation. Most models of this kind involve five basic components, as argued by Michon (1967): a time base produces pulses and transmits them to a counter provided with a gating mechanism; from the counter, the pulses are transferred to a comparator, eventually after a transitory

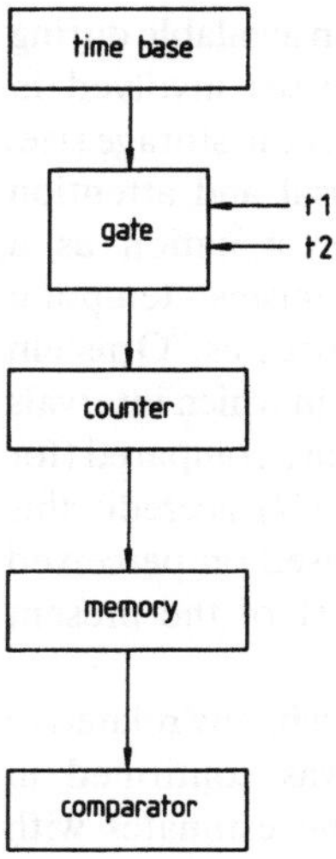

Figure 4. The five basic components in models of time perception. See text for an explanation. (After Michon, 1967).

stage in a memory (Figure 4). Important in these models is the emphasis that is placed on arousal factors or, in certain varieties, on attention and motivation as the modulator of the speed of the pacemaker. Killeen's (1984) recent incentive theory for animal conditioning for instance, assumes the existence of adaptive clocks whose speed varies with rates of reinforcement.

Attention processes were also emphasized by Green & Luce (1974), who attempted to explain intraindividual variability through the existence of attention ›windows‹ which would be differentially directed at the relevant stimulus dimensions in the course of an experiment.

Memory processes were given a prominent place in an information processing model proposed by Gibbon et al. (1984), and designed in the context of animal conditioning. In this model, pulses produced by a pacemaker are fed into an accumulator in working memory. They are stored in a more permanent reference memory when a trial is reinforced, in order to be compared to the new contents of the working memory in all subsequent trials. When the contents of the two memories are similar, a response identical to that which previously produced the reinforcement is generated.

These models based on chronometric units still inspire research. Fraisse's change model was recently re-emphasized by Block & Reed (1978), and by Poynter & Homa (1983), who proposed that attention shifts induced by external, or by physiological and chemical changes constitute the superordinate mechanism of time perception (See also Block, 1985, chapter 11 of the present volume). Revised versions of Creelman's model have also been advanced, with modifications essentially concerning the frequency of pulses emitted (Divenyi & Danner, 1977; Kinchla, 1972). However, a definitive step towards the rejection of specific chronometric mechanisms and the adoption of information procession notions was made by Ornstein (1969) when he introduced the cognitive approach to time. His

›storage size metaphor‹ designates both the amount of information available during the relevant interval, and the perceptual and memory processes involved in information encoding. The more efficient the encoding, the smaller the storage size, and the shorter the subjective duration. Higher levels of arousal and attention increase the amount of information stored, with time overestimation as a consequence. Thus, there is no need for specific time mechanisms: temporal processing depends on perception, memory, and attention processes. Ornstein applied his model to several types of temporal data. Most studies in which intervals of equal duration, but containing different amounts of elements, are compared (for instance, Adams, 1977; Fraisse, 1957; Schiffman & Bobko, 1977) accredit this thesis, though they argue just as well in favor of the models based on perceived changes or on internal pulses (see also Block, 1985, chapter 11 of the present volume).

One specific assumption, namely that subjective duration is positively related to the amount of information stored rather than perceived, was confirmed in Ornstein's study. Other authors have also reported longer time estimates with increased recall scores in memory experiments (Block, 1974; Poynter, Note 4). On the other hand, a major problem resides in the direct relationship supposed by Ornstein between stimulus complexity and subjective duration. Although supporting data do exist (Mulligan & Schiffman, 1979; Schiffman & Bobko, 1974), the effect of complexity appears ambiguous and probably induces distinct individual strategies of responding. Thus for instance, data from subjects required to estimate the duration of a certain time period during which mental arithmetic operations had to be solved contradicted Ornstein's predictions (Wilsoncroft et al., 1978). Hogan (1978) suggested that time perception is a U-shaped function of both personality and stimulus complexity dimensions.

Michon (1972; see also Michon, 1985, chapter 2 of the present volume) also anchored his thesis in information processing theory, but he postulated that information can be of a temporal as well as a nontemporal nature. Temporality, that is the temporal structure of the multiple interactions which relate the organism to the temporally organized environment, is processed either automatically, or under subject's control. Examples of automatic processing are rhythm perception, and the adaptation to sustained external periodicities. Through the mechanism of ›tuning‹, the organism tries to adapt itself as accurately as possible to the temporal structure of current events. This suggests that temporal ›schemata‹ (that is, internal representations of reality), either innate or precociously acquired, are available. Controlled processing is necessary when these schemata are insufficient. Time estimation, for instance, is a deliberate process which involves selective attention and memory (Michon & Jackson, 1984; Jackson, 1985, chapter 12 of the present volume). Information is transitorily stored in short-term memory, and then analyzed and synthesized by the central nervous system. Thus, according to Michon, temporal information is processed in the same fashion as nontemporal information. Since processing capacity is limited, the two types of information may be competitive. Subjective time is reduced when the processing of temporal information is sacrified in favor of nontemporal information.

Thomas & Cantor (1978) extensively analyzed the outcome of this competition between temporal and nontemporal information. In their opinion, attention is generally devoted first to nontemporal information. For instance, when subjects have to evaluate either the duration, or the size, or both features of a stimulus, size is judged smaller in the third condition than in the second. In contrast, comparable judgments of duration are obtained in the first and third conditions. This suggests that the first condition (pure duration) involved divided attention as much as the third (mixed) condition. According to Thomas & Weaver (1975) attention is divided between a time processor which encodes stimulus duration, and an information processor which encodes nontemporal features as well as the time necessary for their processing. This processing time adds to the output of the timer in their model. When the required processing time of nontemporal information is increased, or when more attention is (retrospectively) directed to the information processor, the experimental results reflect an increase in subjective duration. The conception of a timer as proposed by Thomas & Cantor (1978) is very close to Kristofferson's (1977) real time criterion model. The role of processing time has also been thoroughly investigated by Massaro & Idson (1976, 1978; Idson & Massaro, 1977), whose time perception model primarily concerned particular discrimination tasks involving the masking method applied to auditory stimuli.

It should be mentioned in passing that Thomas & Weaver's model was conceived for stimuli which do not exceed 100 ms. It may probably be applied to longer durations (see Michon, 1985, chapter 2 of the present volume), although Poynter & Homa (1983) suggested an inverse relationship between subjective duration and processing time for stimuli longer than 10 s.

Conclusion: Some Biopsychological Speculations

The integration of time perception into the framework of information processing theory, and the appeal for the functional identity of mechanisms for the processing of temporal as well as nontemporal information, appear as central trends in current theorizing. The question how information is processed is more often investigated than the nature of the time bases, even though the existence of specific time bases is frequently assumed. There seems to be consensus about the necessity of admitting the involvement of several distinct mechanisms, according to duration range, particular task requirements, or individual strategies (see Gibbon & Allan, 1984, and several chapters in the present volume). The hiatus emphasized by Fraisse (1957) between the perception and the estimation of time, around 2 seconds, is partial evidence for such multiplicity of processes. This hiatus has been confirmed in ontogenetic (Fraisse, 1948), psychopathological (Ajuriaguerra et al., 1967) and pharmacological (Mitrani et al., 1977) studies. Other evidence for multiple mechanisms in quite divergent duration ranges does exist. In rhythmic tapping, Michon (1967) measured temporal quanta that were stable within a given experimental condition but ranged between 25 to 125 ms across conditions. He suggested the existence of a processing time unit that would adapt specific to each particular situation. In sum, the available data seem to favor eclectic hypotheses, much like the flexible ›clock form system‹ suggested by Gooddy (1958). It would appear as though multiple potentialities but no permanent time basis exist in the organism; the time base convenient for each particular situation would be constructed from this vast array of possibilities and would be extracted from the background of ›noisy‹ experiences according to certain decision criteria. It would be somehow encoded with practice, in view of sustained use, but would be deleted when no longer functional. Physiological research, though not very successful so far in identifying specific time bases (for a review see Macar, 1980), has yielded data which suggest that these time bases do at least vary intraindividually, as formerly assumed by Piéron (1923). For instance, it has been postulated that sensorimotor activity is modulated by cardiac and respiratory cycles. Although large individual variations are the rule, it is surprising to observe that systematic trends may affect temporal judgments quite noticeably, though this is in general true for only one or two subjects per experimental group (Granjon & Requin, 1970). In studies concerned with potential proprioceptive time bases, data collected in animals support a similar hypothesis. When trained to press a lever without interruption for a constant minimum duration, a relationship between the level of performance and certain criteria of pressure force was found in one animal per group (Fowler et al., 1972; Greenwood, Note 2). Clear hypothesizing and clever experimental designs may uncover more of these idiosyncratic strategies (see for instance Jackson, 1985, chapter 12 of the present volume).

Time bases may also rely on neuronal interactions. Analogies can be found in spatial perception. The visual pathways and brain areas contain specialized neurons which fire only when the stimulus has a definite spatial orientation or moves in a specific direction (for instance Hübel & Wiesel, 1962; Hyvärinen & Poranen, 1974; Mountcastle et al., 1975). Also cellular networks which group units with common or perhaps complementary properties have been identified (MacKay, 1980). At the neuronal level, the processing of visual information undoubtedly relies on an extremely complex computation of all spatio-temporal relations between units (Marr, 1982). In the same fashion the proper time basis for a particular situation may consist in the combination of distinct patterns of discharges yielded by neurons of different origin (Figure 5). Certain neurons would fire only in response to definite configurations of parameters in various modalities: they would be feature detector neurons, comparable to the specialized units involved in spatial perception. Others, the periodic neurons, would be time conditioned to one or another organismic periodicity. Neurons with very stable firing rate and not dependent on external stimulation have been found in the midbrain reticular system of cats and rats (Schlag et al., 1971; Skvarill et al., 1959). Moreover, it has been demonstrated that neuronal discharges are sensitive to temporal conditioning (Mednikova, 1975). A third category of neurons, which would be aspecific with respect to duration but specific to modality, could also be postulated. Their pattern and rate of discharge would yield information about the physical characteristics of the stimulus set. These neurons would be responsible for the effects of nontemporal information on subjective time.

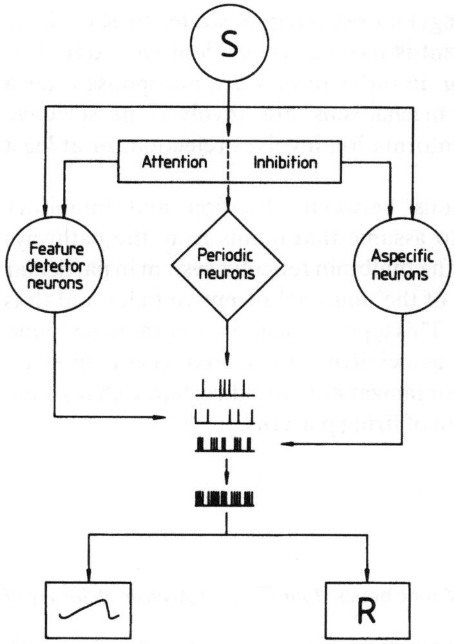

Figure 5. A time-netwerk model. The relative weight of all three constituent sources of neural activity, and also of the attention and inhibition mechanisms, depend on each particular stimulus configuration and on individual parameters.

All discharges would be encoded on the basis of both parallel and serial processing. The degree of synchronization between the firing of such specialized units, or between a few salient elements of such discharges, might be crucial for the constitution of the behavioral time bases. Synchronization in a particular time basis would become enhanced with practice. Perhaps the salience of the system would also increase, because a larger number of neuronal discharges would progressively be extracted from the background constituted by all discharges in the central nervous system. This would necessarily involve active filtering of the discharges under the control of selective attention mechanisms. Increase in synchronization and in the amount of units engaged would induce more accurate temporal judgments. Certain event-related potentials, such as the contingent negative variation (CNV) which develops in situations where two salient events are temporally related, do perhaps constitute a global index of such synchronization phenomena (Macar, 1977; Macar & Besson, Note 3; Macar & Vitton, 1979).

To account for all data which indicate the role of attention and arousal on temporal judgments, it is necessary to assume that attention, in addition to its role in the filtering of discharges, modulates the neural ›time networks‹. It is currently understood that attention processes include two components, the one specific and stimulus directed (selective attention), the other nonspecific (corresponding to activation or arousal in certain terminologies). Limitations of the time base system could depend on attention shifts rather than on the time network proper, in agreement with Kristofferson's conceptions. In addition, there must be interactions between timers and inhibitory temporal regulations, as demonstrated by Richelle & Lejeune (1980). The inhibitory component is particularly evident each time it is necessary to regulate one's own behavior, in order to produce a response after a specified delay. Moreover, inhibitory mechanisms are involved in selective attention, since attending to privileged information involves rejection, or at least weakening, of concurrent inputs.

One way to explain the interactions between attention and inhibitory mechanisms and the ›time network‹ is to assume that inputs from the pathways which mediate attention and inhibition – the midbrain reticular system in particular – can affect the form and the dimension of the neuronal receptive fields and thus lead to reorganization within the network. This type of phenomenon has been found in visual neurons after stimulation of the associative cortical areas (Dewson et al., 1966; Spinelli & Pribram, 1967). Such reorganizations would certainly change the coding of the spatio-temporal combination of firing patterns.

References

Abel, S.M. Duration discrimination of noise and tone bursts. *Journal of the Acoustical Society of America*, 1972a, *51*, 1219-1223.

Abel, S.M. Discrimination of temporal gaps. *Journal of the Acoustical Society of America*, 1972b, *52*, 519-524.

Adams, R.D. Intervening stimulus effects on category judgments of duration. *Perception and Psychophysics*, 1977, *21*, 527-534.

Ajuriaguerra, J. de, Boehme, M., Richard, J., Sinclair, M., & Tissot, R. Désintégration des notions de temps dans les démences dégénératrices du grand âge. *Encéphale*, 1967, *5*, 385-438.

Allan, L.G. The perception of time. *Perception and Psychophysics*, 1979, *26*, 340-354.

Allan, L.G. Magnitude estimation of temporal intervals. *Perception and Psychophysics*, 1983, *33*, 29-42.

Allan, L.G., & Kristofferson, A.B. Psychophysical theories of duration discrimination. *Perception and Psychophysics*, 1974, *16*, 26-34.

Allan, L.G., Kristofferson, A.B., & Wiens, E.W. Duration discrimination of brief light flashes. *Perception and Psychophysics*, 1971, *9*, 327-334.

Allport, D.A. Phenomenal simultaneity and the perceptual moment hypothesis. *British Journal of Psychology*, 1968, *59*, 395-406.

Avant, L.L., Lyman, P.J., & Antes, J.R. Effects of stimulus familiarity upon judged visual duration. *Perception and Psychophysics*, 1975, *17*, 253-262.

Bakan, P., Nagle, L.G., & Denny, M.R. Learning, transfer and retention in judgment of time intervals. *Papers of the Michigan Academy of Science, Arts and Letters*, 1959, *44*, 219-266.

Block, R.A. Memory and the experience of duration in retrospect. *Memory and Cognition*, 1974, *2*, 153-160.

Block, R.A. Contextual coding in memory: Studies of remembered duration. In: J.A. Michon & J.L. Jackson (Eds.), *Time, mind, and behavior.* Heidelberg: SpringerVerlag, 1985, pp. 169-178.

Block, R.A., & Reed, M.A. Remembered duration: Evidence for a contextual change hypothesis. *Journal of Experimental Psychology: Human Learning and Memory*, 1978, *4*, 656-665.

Carlson, V.R., & Feinberg, I. Time judgment as a function of method, practice and sex. *Journal of Experimental Psychology*, 1970, *85*, 171-180.

Catania, A.C. Reinforcement schedules and psychophysical judgment: A study of some temporal properties of behavior. In: W.N. Schoenfield (Ed.), *The theory of reinforcement schedules*. New York: Appleton, 1970, 1-42.

Church, R.M., & Deluty, M.Z. Bisection of temporal intervals. *Journal of Experimental Psychology: Animal Behavior Processes*, 1977, *3*, 216-228.

Church, R.M., Getty, D.J., & Lerner, N.D. Duration discrimination by rats. *Journal of Experimental Psychology: Animal Behavior Processes*, 1976, *2*, 303-312.

Corwin, T.R., & Boynton, R.M. Transitivity of visual judgments of simultaneity. *Journal of Experimental Psychology*, 1968, *78*, 560-568.

Creelman, C.D. Human discrimination of auditory duration. *Journal of the Acoustical Society of America*, 1962, *34*, 582-593.

Dewson, J.H. III, Noble, K.W., & Pribram, K.H. Corticofugal influence at cochlear nucleus of the cat: Some effects of ablation of insular-temporal cortex. *Brain Research*, 1966, *2*, 151-159.

Divenyi, P.L., & Danner, W.F. Discrimination of time intervals marked by brief acoustic pulses of various intensities and spectra. *Perception and Psychophysics*, 1977, *21*, 125-142.

Doob, W. *Patterning of time*. New Haven: Yale University Press, 1971.

Durup, G., & Fessard, A. Le seuil de perception de durée dans l'excitation visuelle. *Année Psychologique*, 1930, *31*, 52-62.

Efron, R. The effect of handedness on the perception of simultaneity and temporal order. *Brain*, 1963a, *86*, 261-284.

Efron, R. The effect of stimulus intensity on the perception of simultaneity in right and left-handed subjects. *Brain*, 1963b, *86*, 285-294.

Efron, R. The duration of the present. In: R. Fischer (Ed.), *Interdisciplinary perspectives of time*. Annals of the New York Academy of Sciences, vol. 138, 1967, pp. 713-729.

Efron, R. An invariant characteristic of perceptual systems in the time domain. In: S. Kornblum (Ed.), *Attention and performance IV*. New York: Academic Press, 1973, pp. 713-736.

Eijkman, E., & Vendrik, A.J.H. Can a sensory system be specified by its internal noise? *Journal of the Acoustical Society of America*, 1965, *37*, 1102-1109.

Eisler, H. Subjective duration and psychophysics. *Psychological Review*, 1975, *82*, 429-450.

Eisler, H. Experiments on subjective duration 1868-1975: A collection of power function experiments. *Psychological Bulletin*, 1976, *83*, 1154-1171.

Eisler, H. The parallel-clock model: Replies to critics and criticisms. *Perception and Psychophysics*, 1981, *29*, 516520.

Ellerman, H., & Michon, J.A. The psychophysics of time reproduction. In: C. Micko & U. Schulz (Eds.), *Formalization of psychological theories*. Bielefeld: University of Bielefeld, 1983, pp. 38-50.

Fowler, S.C., Morgenstein, C., & Notterman, J.M. Spectral analysis of variations in force during barpressing time discrimination. *Science*, 1972, *176*, 1126-1127.

Fraisse, P. Etude comparée de la perception et de l'estimation de la durée chez les enfants et chez les adultes. *Enfance*, 1948, *1*, 199-211.

Fraisse, P. *Psychologie du temps*. Paris: Presses Universitaires de France, 1957.

Fraisse, P. L'Oppel-Kundt temporel ou l'influence de la fréquence des stimulations sur la perception du temps. *Revue de Psychologie Française*, 1965, *4*, 352-358.

Fraisse, P. Le seuil différentiel de durée dans une suite régulière d'intervalles. *Anné Psychologique*, 1967, *67*, 43-49.

Fraisse, P., Bonnet, C., Gelly, N., & Michaut, G. Vergleich der Zeitschätzungsmethoden. *Zeitschrift für Psychologie*, 1962, *167*, 268-277.

Frankenhaeuser, M. *Estimation of time: An experimental study*. Stockholm: Almqvist & Wiksell, 1959.

Getty, D.J. Discrimination of short temporal intervals: A comparison of two models. *Perception and Psychophysics*, 1975, *18*, 1-8.

Getty, D.J. Counting processes in human timing. *Perception and Psychophysics*, 1976, *20*, 191-197.

Gibbon, J. Scalar expectancy theory and Weber's law in animal timing. *Psychological Review*, 1977, *84*, 279-325.

Gibbon, J. On the form and location of the psychometric bisection function of time. *Journal of Mathematical Psychology*, 1981, *24*, 58-87.

Gibbon, J., & Allan, L. (Eds.), *Timing and time perception*. Annals of the New York Academy of Sciences, vol. 423, 1984.

Gibbon, J., Church, R.M., & Meck, W.H. Scalar timing in memory. In: J. Gibbon & L. Allan (Eds.), *Timing and time perception*. Annals of the New York Academy of Sciences, vol. 423, 1984, pp. 52-77.

Goldfarb, J.L., & Goldstone, S. Time judgment: A comparison of filled and unfilled durations. *Perceptual and Motor Skills*, 1963, *16*, 376.

Gooddy, W. Time and the nervous system. *The Lancet*, 1958, *7031*, 1139-1141.

Granjon, M., & Requin, J. Un role de la périodicité cardiaque dans la régulation temporelle de l'activité sensorimotrice? *Psychologica Belgica*, 1970, *10-2*, 141-167.

Green, D.M., & Luce, R.D. Variability of magnitude estimates: A timing theory analysis. *Perception and Psychophysics*, 1974, *15*, 291-300.

Hirsh, I.J. Temporal order and auditory perception. In: H.R. Moskowitz, B. Scharf, & J.C. Stevens (Eds.), *Sensation and measurement*. Dordrecht: Reidel, 1974, pp. 251-258.

Hirsh, I.J., & Sherrick, C.E. Jr. Perceived order in different sense modalities. *Journal of Experimental Psychology*, 1961, *62*, 423-432.

Hogan, H.W. A theoretical reconciliation of competing views of time perception. *American Journal of Psychology*, 1978, 3, 117-128.

Hornstein, A.D., & Rotter, G.S. Research methodology in temporal perception. *Journal of Experimental Psychology*, 1969, *79*, 561-564.

Hubel, D.H., & Wiesel, T.N. Receptive fields, binocular interaction and functional architecture in the cat's visual cortex. *Journal of Physiology*, 1962, *160*, 106-144.

Hyvärinen, J., & Poranen, A. Function of the parietal associative area 7 as revealed from cellular discharges in alert monkeys. *Brain*, 1974, *97*, 673-692.

Idson, W.L., & Massaro, D.W. Perceptual processing, and experience of auditory duration. *Sensory Processes*, 1977, *1*, 316-337.

Jackson, J.L. Is the processing of temporal information automatic or controlled? In: J.A. Michon & J.L. Jackson (Eds.), *Time, mind, and behavior.* Heidelberg: Springer Verlag, 1985, pp. 179-190.

Jesteadt, W., Luce, R.D., & Green, D.M. Sequential effects in judgments of loudness. *Journal of Experimental Psychology: Human Perception and Performance*, 1977, *3*, 92-104.

Jones, M.R. Structural organization of events in time. In: J.A. Michon & J.L. Jackson (Eds.), *Time, mind, and behavior.* Heidelberg: Springer Verlag, 1985, pp. 192-214.

Killeen, P.R. Incentive Theory III: Adaptive clocks. In: J. Gibbon & L. Allan (Eds.), *Timing and time perception*. Annals of the New York Academy of Sciences, vol. 423, 1984, Pp. 515-527.

Kinchla, J. Duration discrimination of acoustically defined intervals in the 1 to 8 sec range. *Perception and Psychophysics*, 1972, *12*, 318-320.

Kristofferson, A.B. Attention and psychophysical time. *Acta Psychologica*, 1967, *27*, 93-100.

Kristofferson, A.B. Attention. In: R.M. Patton, T.A. Tanner, J.J. Markowitz, & J.A. Swets (Eds.), *Applications of research on human decision making*. Washington: WASA, 1970, pp. 49-64.

Kristofferson, A.B. A real-time criterion theory of duration discrimination. *Perception and Psychophysics*, 1977, *21*, 105-117.

Kristofferson, A.B. A quantal step function in duration discrimination. *Perception and Psychophysics*, 1980, *27*, 300-306.

Levick, W.R., & Zacks, J.L. Responses of cat retinal ganglion cells to brief flashes of light. *Journal of Physiology*, 1970, *206*, 677-700.

Luce, R.D., & Mo, S.S. Magnitude estimation of heaviness and loudness by individual subjects: A test of probabilistic response theory. *British Journal of Mathematical and Statistical Psychology*, 1965, *18*, 159-174.

Macar, F. Signification des variations contingents négatives dans la dimension temporelle du comportement. *Année Psychologique*, 1977, *77*, 439-474.

Macar, F. *Le temps: Perspectives psychophysiologiques*. Bruxelles: Mardaga, 1980.

Macar, F., & Vitton, M. Contingent negative variation and accuracy of time estimation: A study on cats. *EEG and Clinical Neurophysiology*, 1979, *47*, 213-228.

MacKay, D.M. Sensory processing in the brain and evoked potentials. In: H.H. Kornhuber & L. Deecke (Eds.), *Motivation, motor and sensory processes of the brain: Electrical potentials, behavior and clinical use. Progress in Brain Research*. Amsterdam: North Holland Elsevier, 1980, pp. 634-639.

Marr, D. *Vision: A computational investigation into the human representation and processing of visual information*. San Francisco: Freeman, 1982.
Massaro, D.W., & Idson, W.L. Temporal course of perceived auditory duration. *Perception and Psychophysics*, 1976, *20*, 331-352.
Massaro, D.W., & Idson, W.L. Target-mask similarity in backward recognition masking of perceived tone duration. *Perception and Psychophysics*, 1978, *24*, 235-236.
McConchie, R.D., & Rutschmann, J. Human time estimation: On differences between methods. *Perceptual and Motor Skills*, 1971, *32*, 319-336.
Mednikova, T.V.S. Conditioned reactions to time in hypothalamic neurons. The perifornical nucleus. *Zhurnal Vysshei Nervnoi Deyatel'nosti*, 1975, *25*, 1022-1030.
Michon, J.A. *Timing in temporal tracking*. Soesterberg: Institute for Perception - TNO, 1967.
Michon, J.A. Processing of temporal information and the cognitive theory of time experience. In: J.T. Fraser, F.C. Haber, & G.H. Mueller (Eds.), *The study* of time. Heidelberg: Springer Verlag, 1972, pp. 242-258.
Michon, J.A. The compleat time experiencer. In: J.A. Michon & J.L. Jackson (Eds.), *Time, mind, and behavior*. Heidelberg: Springer Verlag, 1985, pp. 20-52.
Michon, J.A., & Jackson, J.L. Attentional effort and cognitive strategies in the processing of temporal information. In: J. Gibbon & L. Allan (Eds.), *Timing* and time *perception*. Annals of the New York Academy of Sciences, 1984, vol. 423, pp. 298-321.
Mitrani, L., Shekerdjiiski, S., Gourevitch, A., & Yanev, S. Identification of short time intervals under LSD 25 and mescaline. *Activitas Nervosa Superior*, 1977, *19*, 103-104.
Mountcastle, V.B., Lynch, J.C., Georgopoulos, A., Sakata, H., & Acuna, C. Posterior parietal association cortex of the monkey: Command functions for operations within extrapersonal space. *Journal of Neurophysiology*, 1975, *38*, 871-908.
Mulligan, R.M., & Schiffman, H.R. Temporal experience as a function of organization in memory. *Bulletin of the Psychonomic Society*, 1979, *14*, 417-420.
Oléron, G. Influence de l'intensité d'un son sur l'estimation de la durée apparente. *Année Psychologique*, 1952, *52*, 383-392.
Ornstein, R.E. *On the experience of time*. Harmondsworth: Penguin Books, 1969.
Pastore, R.E. Temporal order judgment of auditory stimulus offset. *Perception and Psychophysics*, 1983, *33*, 54-62.
Patterson, J.H., & Green, D.M. Discrimination of transient signals having identical energy spectra. *Journal of the Acoustical Society of America*, 1970, *48*, 894-905.
Peck, C.K., & Lindsley, D.B. Average evoked potential correlates of two flash perceptual discrimination in cats. *Vision Research*, 1972, *12*, 641-652.
Peck, C.K., & Lindsley, D.B. Single unit and evoked responses in cat optic tract to paired lights flashes. *Experimental Brain Research*, 1973, *16*, 371-382.
Perikel, J.J., Richelle, M., & Maurissen, J. Control of key pecking by stimulus duration. *Journal of Experimental Analysis of Behavior*, 1974, *22*, 131-134.
Piéron, H. Les problèmes psychophysiologiques de la perception du temps. *Année Psychologique*, 1923, *24*, 1-25.
Pöppel, E. Time perception. In: H.L. Teuber, R. Held, & H. Leibowitz (Eds.), *Handbook of sensory physiology*, Vol. 8. New York: Springer, 1976, pp. 713-729.
Poynter, W.D., & Homa, D. Duration judgment and the experience of change. *Perception and Psychophysics*, 1983, *33*, 548-560.
Richelle, M., & Lejeune, H. (Eds.), *Time in animal behaviour*. Oxford: Pergamon Press, 1980.
Rossi, M. Le seuil différentiel de la durée. In: A. Valdman (Ed.), *Papers in linguistics and phonetics to the memory of Pierre Delattre*. The Hague: Mouton, 1972, pp. 435-450.
Rutschmann, R. Visual perception of temporal order. In: S. Kornblum (Ed.), *Attention and performance IV*. New York: Academic Press, 1973, pp. 687-701.
Schiffman, H.R., & Bobko, D.J. Effects of stimulus complexity on the perception of brief temporal intervals. *Journal of Experimental Psychology*, 1974, *103*, 156-159.
Schiffman, H.R., & Bobko, D.J. The role of number and familiarity of stimuli in the perception of brief temporal intervals. *American Journal of Psychology*, 1977, *90*, 85-93.
Schlag, J., Radil-Weiss, T., Bohdanecky, Z., & Syka, J. Spontaneous discharge patterns of mesencephalic neurons: Interval histogram and mean relationship. *Kybernetik*, 1971, *9*, 11-15.

Servière, J., Micelli, C., & Galifret, Y. A psychophysical study of the visual perception of ›instantaneous‹ and ›durable‹. *Vision Research*, 1977, *17*, 57-63.

Skvarill, J., Faidherbe, J., & Deliège, M. Note sur la possibilité de caractériser les unités réticulées mésencéphaliques par leur activité spontanée. *Archives Internationales de Physiologie et Biochimie*, 1959, *67*, 227-239.

Small, A.M. Jr., & Campbell, R.A. Temporal differential sensitivity for auditory stimuli. *American Journal of Psychology*, 1962, *75*, 401-410.

Spinelli, D.N., & Pribram, K.H. Changes in visual recovery functions and unit activity produced by frontal and temporal cortex stimulation. *EEG and Clinical Neurophysiology*, 1967, *22*, 143-149.

Sternberg, S., & Knoll, R.L. The perception of temporal order: Fundamental issues and a general model. In: S. Kornblum (Ed.), *Attention and performance IV.* New York: Academic Press, 1973, pp. 629-685.

Stevens, S.S. Issues in psychophysical measurement. *Psychological Review*, 1971, *78*, 426-450.

Stroud, J.M. The fine structure of psychological time. In: H. Quastler (Ed.), *Information theory in psychology.* Glencoe, IL.: Free Press, 1956, pp. 174-205.

Stubbs, A. The discrimination of stimulus duration by pigeons. *Journal of Experimental Analysis of Behavior*, 1968, *11*, 223-238.

Thomas, E.A.C., & Brown, I. B. Time perception and the filled-duration illusion. *Perception and Psychophysics*, 1974, *16*, 449-458.

Thomas, E.A.C., & Cantor, N.E. Interdependence between the processing of temporal and nontemporal information. In: J. Requin (Ed.), *Attention and performance VII.* Hillsdale NJ: Lawrence Erlbaum Associates, 1978, pp. 43-62.

Thomas, E.A.C., & Weaver, W.B. Cognitive processing and time perception. *Perception and Psychophysics*, 1975, *17*, 363-367.

Thompson, J.G., Schiffman, H.R., & Bobko, D.J. The discrimination of brief temporal intervals. Acta *Psychologica*, 1976, *40*, 489-493.

Treisman, M. Temporal discrimination and the indifference interval: Implications for a model of the internal clock. *Psychological Monographs*, whole nr. 576, 1963.

Warm, J.S., Stutz, R.M., & Vassolo, P.A. Intermodal transfer in temporal discrimination. *Perception and Psychophysics*, 1975, *18*, 281-286.

Warren, R.M. Auditory temporal discrimination by trained listeners. *Cognitive Psychology*, 1974, *6*, 237-256.

White, C.T. Temporal numerosity and the psychological unit of duration. *Psychological Monographs*, whole nr. 575, 1963.

Wiener, N. Time and the science of organization. *Scientia*, 1958, *93*, 199-205.

Wilsoncroft, W.E., Stone, J.D., & Bagrash, F.M. Temporal estimates as a function of difficulty of mental arithmetic. *Perceptual and Motor Skills*, 1978, *46*, 1311-1317.

Woodrow, H. The reproduction of temporal intervals. *Journal of Experimental Psychology*, 1930, *13*, 473-499,

Woodrow, H. Time perception. In: S.S. Stevens (Ed.), *Handbook of experimental psychology.* New York: Wiley & Sons, 1951, pp. 1224-1236.

Reference Notes

1. Blakely, W.A. *The discrimination of short empty temporal intervals*. Unpublished manuscript. University of Illinois, 1933.
2. Greenwood, P. *Contribution à l'étude des régulations temporelles acquises chez le chat.* Unpublished manuscript. Université de Liège, 1977.
3. Macar, F., & Besson, M. *The Contingent Negative Variation, a specific or aspecific index?* Manuscript submitted for publication, 1985.
4. Poynter, W.D. *Human time perception and memory processes: The role of retrieval in duration estimates*. Unpublished manuscript, Arizona State University, 1979.

Chapter 8. The Effects of Time Pressure on Duration Discrimination

Michelangelo Flückiger

Introduction

The study of reactions to periodic changes of duration in events that constitute event sequences is essential for the understanding of man's capacity to adapt to the flow of time. Most studies on the perception and discrimination of temporal durations are based, however, on single presentation paradigms of events (Allan, 1979; Macar, 1985, chapter 7 of the present volume). Many experimental variations on that theme have been used to address specific questions and all psychophysical studies adopting this approach share the basic tenet that the psychological mechanisms responsible for the processing of time can be satisfactorily investigated under static conditions. In comparison with these studies the principal results obtained with a repetitive presentation paradigm show a, by now well established, shift in differential sensitivity for duration (Michon, 1964; Fraisse, 1967). Somehow the repetition of stimulus intervals must therefore have a special effect on perceived duration.

The research reported here explores this aspect of duration perception with a dynamic approach which emphasizes the observer's orientation toward a continuously changing flow of discriminal information, that is, a flow in the context of which repeated discriminations are to be made. The basic conceptual framework which motivates this study can be conveniently described by using an analogy with the techniques for statistical inference. Conventional methods for deciding on the significance level of an observed difference are based on the calculation of mean and variance of two sample populations whose size is given and fixed. Sequential techniques differ radically in as much as they prescribe that discriminal information be accumulated only up to the point when a given criterion is met. Both techniques, of course, are designed to describe the same ›objective‹ reality, but the former adopts a static and the latter a dynamic point of view and consequently they offer a different interpretation of that reality.

My approach is conceptually analogous to the sequential one inasmuch as I shall consider the following situation: suppose that an observer is presented with a sequence of duration pairs which are included in a constant time period and that the difference between the members of these paired durations changes continuously in the course of the sequence. How long will it take before the observer has gathered enough discriminal evidence to conclude that the difference between the two durations is increasing: Is this detection time a function of the rate at which discriminal information is supplied? Is the differential sensitivity affected by this rate?

I have performed an experiment, to be reported below in brief outline, which addresses these questions within the described conceptual framework.[1] The principal independent variable in this experiment is the rate of change of the durations to be compared. The assumption is that the increase of this rate involves an effect I will call a time pressure on duration discrimination. A second point to be considered in this experiment concerns the role of an interstimulus interval between the perceived durations. Finally several lines of argument will be presented in the discussion to show that the dynamical approach affords an useful and hitherto insufficiently explored road to the understanding of time perception.

The Experiment

In the experiment described here, the objective temporal properties of the sequences used are determined by four parameters. The first of these defines the period of time within which a pair of successively presented lights is switched on. For all the situations used here, this time had a constant value of 300 ms. The second parameter determines the number of periods which make up an uninterrupted sequence, in this case a constant 600 periods. The third concerns the change, from one period to the next, in the size of the difference between two successive durations. This difference was automatically changed, for each successive period, by a step factor (ST) having three levels, 3, 6 and 9 ms respectively, constituting the principal independent variable. The fourth parameter determines the manner in which the lights are sequenced, and is the other independent variable, with two levels. In the condition, called WISI (Without Inter-Stimulus Intervals), pairs of lights follow each other immediately. For the other level of this independent variable, there is an Inter-Stimulus Interval (ISI) between the offset of one light and the onset of the next.

The *stimuli* in this experiment consisted of sequences of two alternating lights, with only the temporal parameters of these lights changing from one condition to another. Each pair of light flashes to be compared occurred within a fixed time period (P) lasting 300 ms. The duration of a given light was never the same from one period to the next. In the first period of each experiment the pair of lights have equal durations. Starting with the second period in a trial a difference in duration was introduced, whose value was equal to the step size (ST) chosen by the experimenter for that trial. Without subject intervention, the sequence continued in this linear fashion, the difference in duration between lights having a value of twice ST for the third trial, three times ST for the fourth, and so on. Thus the lights entertained an inverse linear relationship since an increase in duration for one light automatically implied an identical decrease for the other. Given this difference based algorithm, the effective change in the duration of a light from one period to the next is equivalent to twice the value of ST.

1 This research was supported by FNSRS grant nr. 1.353.0.81. and was carried out with the collaboration of A. Daccord and R. Maurer.

The *subjects* were 60 student volunteers aged between 20 and 27 years. They were randomly assigned in groups of 10 to six conditions of the experimental design. The subjects were tested individually in a quiet room with controlled lighting.

The *apparatus* consisted of a sequence generator implemented in a NORD-10 computer, which also recorded the subject's responses in real time, via a CAMAC interface. The overall temporal resolution of the system was 0.1 ms, the rise times of the lamps being negligible. These were light emitting diodes with a wavelength of 630 nm and an intensity of 6 mcd. They produced a red light, had a diameter of 5 mm and were placed horizontally in a plane perpendicular to the visual axis of the observer. At half the observation distance along the visual axis, the ambient lighting level was 15 lux. The lamps were 10 cm apart and at a distance of 1.5 m from the observer, the configuration thus subtending a visual angle of about 4 degrees. The subject was seated with his head resting on a rigid support.

The *design* comprised the six experimental conditions obtained by the combination of each of the two levels of the WISI-ISI variable with each of the three possible ST values, yielding a a 2x3 factorial design. At the beginning of the experiment, the increases in duration were applied to the first light to be presented for half the subjects of a group and to the second light for the others. In the three WISI conditions the lights were following each other without interruption. The independent variable was the step size (ST) with values of 3, 6, and 9 ms respectively for the three conditions. In the first period of every sequence, each lamp stayed on for 150 ms. In the three ISI-conditions, the lights did not follow each other immediately, each stimulus onset being separated by an ISI with a constant value of 80 ms from the preceding offset. Consequently in the first period the durations of the two lights to be compared were reduced to a value of 70 ms each, subsequently changing at the rate determined by the value of ST. As indicated, the period within which a pair of successive lights appeared was constant at 300 ms. The three ST values were identical to those used in the WISI conditions. We predict (1) that ST will influence the subjects' capacity to discriminate durations and (2) that this capacity is affected by the presence of an ISI.

The *instruction* to the subjects was to press a response key as soon as they noticed an increasing difference between the times each of the two lamps stayed on. Without subject intervention the difference in presented durations would increase linearly, up to the point where one of the lights would have a zero-duration.

The *procedure* followed included three phases: a demonstration, a pre-test and the 600-period experimental test.

The demonstration served to familiarize the subject with the experimental situation by using a stimulus pattern identical to the one used later. At this stage the function of the response key was not explained, the subject simply being told that he had to ›describe what happened‹ in the sequence of lights presented to him. In actual fact, after 100 periods with ST=3, one of the lights reached a duration of 30 ms and the computer stopped the sequence. For ST=6 or 9, the demonstration proceeded for 50 and 33 periods respectively. All subjects were able to describe correctly the physical, objective change in the durations which had taken place. Generally, their observations were expressed in a way suggesting something to the extent that the duration of one of the lights had ›completely gained‹ over the other. We used each subject's own words in the later phases of the procedure.

Prior to the pre-test the effect of the response key was explained. Pressing it sent a command to the computer to reverse the sign of the change and to bring the two light durations back to an objectively equal value and beyond. Detailed information about the precise moment the key was pressed was recorded and subjected to various analyses, as described in the results. The pre-test halted automatically after two responses and the subject was asked for an account of what had happened. Only one subject had to be rejected at this stage because, as he admitted spontaneously, he let one light ›gain completely‹ over the other. All the subjects' comments corresponded to the physical reality of what happened during the sequence, that is, subjects would report that the sequence continued after a response but that, if the difference had been increasing, it would now begin to decrease. When subjects detected a new difference, opposite to the first, they knew that they had to react immediately by pressing the response key.

The experimental phase could then begin. The subjective reports elicited after the experiment clearly confirmed that this procedure had directly oriented them towards the perception of a difference in durations.

The *results* of these experiments show the effect of the ST and ISI parameters on the capacity to discriminate durations in a dynamic context. In a complementary way, three kinds of measure characterize this capacity to react to an ongoing flow of events. The following analysis deals with (a) the number of interventions by subjects in the course of a sequence, that is, the number of changes (NC) in the direction of the difference between the on-durations of the lights; (b) the values for the points of subjective equality (PSE) and (c) the differential sensitivity to durations, as measured by the semi-interquartile range (SIQ) between the reversal points.

For each of these measures, the results are expressed in two forms, namely the global performance as expressed by calculations using responses over 600 periods in an entire trial (TOT), and performance on each of three equal 200 period parts (PART) of a trial, the latter allowing us to follow the evolution of performance in the course of a trial.

(a) The number of interventions or changes (NC) in a trial represents response rate. This is an important statistic since it allows us to appreciate subjects' response to the two independent variables using a simple numeric value. It will be seen from Table 1, in the two columns showing the overall (TOT) results, that NC increases with step size (ST) in the WISI as well as in the ISI condition. Furthermore, values of NC are always higher for ISI than for WISI, the difference being larger for the greater ST values. These observations were supported by a 2 x 3 Analysis of Variance (ANOVA). This confirms that the subjects increase their global response rate under the influence of ST, but clearly not in a directly proportional way. The results for NC will be further interpreted in relation with the performance with respect to the point of subjective equality (PSE).

When the PART results are considered, an interesting phenomenon emerges between the three subdivisions of the sequence (see Table 1). With small ST values, NC decreases over the trial, but for ST=9 response frequency increases. A

Table 1: Mean number of interventions for three different rates of change of events under ISI and WISI conditions of presentation (n = 60) for equal thirds of sessions (PART).

Steps (ms)	Without ISI				ISI			
	1–200	201–400	401–600	Total	1–200	201–400	401–600	Total
3	6.6	5.7	5.0	17.3	7.6	6.0	4.9	18.5
6	9.3	6.3	6.0	21.6	11.0	10.2	8.5	29.7
9	9.7	9.7	10.4	29.8	11.8	12.2	14.0	38.0

Multivariate Analysis of Variance (MANOVA) for repeated measures design, with ST and ISI as between subjects factors and PART as within subjects factor statistically confirms this effect, since the ST x PART interaction is significant ($p > 0.01$).[2]

(b) By considering the values of the points of subjective equality (PSE) it is possible to test for systematic bias in subjects' comparisons, in favor of one or the other of the two lights (left or right/first (odd) or second (even)). Analyses were carried out using both the mean PSE and the median PSE. These two statistics overlapped extremely well in the present results and only the latter was used in an ANOVA for TOT and a MANOVA for PART.

None of these tests was significant, showing that values of PSE are not affected by the experimental conditions. We conclude from this that the constant errors due to biases are negligible (ranging from 0.3 ms to 14 ms). This test shows good perceptual monitoring of the duration included in the temporal flow. It must be emphasized that this high level of performance must have been achieved by very different means (strategies) in the three experimental conditions, given the large range of NC values observed in those conditions.

(c) When differential sensitivity for duration is considered, there are several methods for analysis available. Choosing a measure is not problematic, however, since we are comparing results from different conditions obtained with the same apparatus. I have used the semi-interquartile range (SIQ) in view of its frequent use in the literature on the psychophysics of time. The results are shown in Table 2 and in summary form in Figure 1.

It is immediately apparent that the subjects' discriminative ability is far from being the same in the different experimental conditions. As in the NC, the results shows a noticeable and consistent effect of ST and ISI, especially on the overall values for the entire sequence (Figure 1).

Grouping the results into thirds shows how subjects evolve with respect to the task during the course of one trial (see Table 2). Interpretation of this evolution, however, is complex; only findings confirmed statistically will be discussed here. Introducing PART as a factor in a MANOVA with a significance criterion $p = 0.05$

2 In collaboration with L. Quennoz and G. Beck.

Table 2: Mean semi-interquartile ranges for three different rates of change of events under ISI and WISI conditons of presentations (n = 60) for equal thirds of sessions (PART).

Steps (ms)	Without ISI				ISI			
	1–200	201–400	401–600	Total	1–200	201–400	401–600	Total
3	35.6	34.3	39.0	36.2	30.2	36.8	33.8	33.6
6	41.0	53.2	56.3	50.7	40.0	40.1	47.8	42.6
9	57.5	56.8	60.0	58.1	47.5	44.0	44.2	45.2

reveals a main effect of all three variables. Of the interactions only STx PARTwas significant, which means that for the PART-analyses the apparent differences betweenWISI and ISI conditions are spurious.This allows us to pool the PARTdata across the WISI-ISI conditions but it also warns us to be careful and conservative when interpreting the results. Differential sensitivity decreases slightly for the ST= 3 and considerably more for ST = 6. In contrast, for the highest ST, differential sensitivity tends to increase slightly early in the trial.

One way of appreciating the psychological meaning of these effects is in their convergence with the results obtained for NC (Table 1). The most plausible interpretation of this analysis, in view of the slight improvement in performance during a trial when ST= 9, the slight decrement for ST= 3, and the relatively large performance deterioration when ST = 6, is that there is a kind of optimum in discrimination ability, as it is often found for duration sensitivity in other presentation paradigms.

The *discussion* of these results provides some new insights in the way time is perceived with ongoing events.

First of all I shall analyze the meaning of the number of changes (NC) that a subject triggers during a whole sequence. A constant number of changes

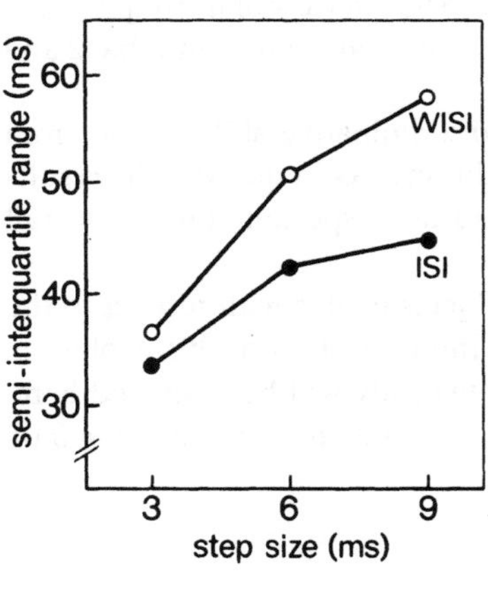

Figure 1. Semi-interquartile ranges as a function of step sizes 3, 6 and 9 ms for WISI and ISI conditions.

irrespective of step size (ST) would mean that subjects react only on the basis of a fixed interval timing not related to the periodicity of lights. Not surprisingly this was not the case. Instead the results show a variation of NC according to the rate of change in durations objectively dependent on ST. We assume that this rate corresponds subjectively to a specific time feature in a sequence with a given flow of events. It determines a perceptual effect on the responses that I shall call ›time pressure‹. Evidently, the higher ST is, the faster the subjects indicate that the difference between the two light durations is increasing. This behavior is, of course, adequate because an equal difference is attained with ST = 9 in a smaller number of periods than with ST=6 or 3. An increased response rate thus prevents an unwarranted overshoot in the duration difference at which action is taken, when ST is relatively large.

A second point concerns the effective consequences of this response strategy on the observed duration sensitivity as measured by the differential threshold. In order to maintain a constant level of discrimination performance, subjects would have had to increase NC in direct proportion to ST. In fact they were not able to do so, although NC did increase monotonically with step size. The results suggest that there is a relative loss in discriminating power when differences in duration build up more rapidly in a flow of events. It is as if subjects fail to keep up with what is going on. In fact ›time pressure‹ apparently leads to a deterioration in duration discrimination independently of the rate of occurrence of the events which, in this experiment, were presented (pairwise) at a constant rate with P = 300 ms. Other experiments are currently being carried out to shed further light on this issue. They attempt to isolate the effect of ISI by systematically varying both ISI and period length. For example, the same ›on‹ durations for the two lights are effectively obtained if P = 140 ms WISI and if P=300 ms with a 80 ms ISI. Preliminary results show a better discrimination under the former condition than in the experiments described here, but the effect of small ST as opposed to a larger ST appears to be the same. Thus, it may be that we are indeed dealing here with a limit in the perceptual processing of temporal information that is reached when the ST reaches a certain size. Incidentally, the time period being constant in both experiments, a differential effect of motor control load is unlikely.

Thirdly, the considerably better discrimination observed when an ISI is presented (compared to WISI conditions) should be considered in the light of the fact that the data have been expressed in terms of the absolute size of the SIQ. The result might imply that the ›off‹ periods between lights allow an enhanced processing of the information about light durations. This has, at least, been observed with physical dimensions other than time (Allan & Kristofferson, 1974). However, there is an alternative hypothesis which requires that we consider the observed discrimination performance in relative terms. In the ISI conditions, the ratio of the sensitivity index (SIQ) to the initial objective durations of lights (70 ms) has a value between 48 and 75 percent. In WISI conditions, this ratio value (relative to the initial 150 ms) ranges between 24 and 39 percent. Because this is not at all in agreement with Weber's law, we have to recognize the subjective importance of interstimulus intervals as an empty ›temporal background‹ (Michon, 1977).

Subjects must presumably process the duration of a pause that is perceptually integrated in a time sequence along with the task-relevant events (Flückiger, 1975).

Finally, the adaptation to the flow of time in the course of a trial is seen in the PART results. When time pressure is comparatively low (ST = 3 or 6) response timing slows down (see Table 1) and the discrimination tends to decrease (Table 2). In contrast for ST = 9, discrimination is relatively poor at the beginning of a presentation, but it can be improved somewhat (Table 2) by an acceleration of the response rate (Table 1).

The *conclusion* to be drawn from these experiments is that the capacity for temporal discrimination does not uniquely depend upon an instantaneous difference observed in a flow of modulated events. Subjects integrate the perceptual information related to the rate of change of durations (time intervals) to be compared. This leads me to some remarks about a possible definition of temporal information as related to a flow of events. A description in terms of frequency of events is not sufficient to psychologically characterize the flowing of time. The reactions to ongoing events are dependent on manifold aspects of a flow sequence, and perception concerns therefore ›time within time‹: periodicity as a cadence (or frame of reference) in which a given rate of change of event durations is embedded, similar to the way rhythmic structure is grafted upon the ›beat‹ or ›pulse‹ of a tonal pattern (see Povel, 1985, chapter 14 of the present volume).

In conclusion, some remarks with respect to future research and theory development can be made. First, the link with synchronization tasks has to be established more carefully. The theoretical avenue on ›temporal tracking‹ opened by Michon (1967) has not yet been fully explored. Such exploration would require a much more detailed consideration of the functional interdependencies in a chain of actions I have proposed in this chapter, as well as their relations to various duration judgment tasks.

A comparison between several perceptual modalities may eventually show model specificity in time processing. Some evidence in support of this expectation was obtained when the same paradigm was applied to the auditory domain (Flückiger & Osiek, Note 1). Alternating sequences of sound stimuli were also used by ten Hoopen & Akerboom (1983). Presently we are collaborating with these authors in order to compare auditory and visual results (see also ten Hoopen, 1985, chapter 9 of the present volume).

In summary, the repetitive presentation paradigm affords a way to characterize different strategies of combining timing performance and perceptual judgment of duration. Elsewhere I have proposed a typology for the classification of these strategies (Flückiger, 1983). A necessary step towards understanding the mechanisms involved in the processing of temporal information is, in my opinion, to link the results of rhythmic synchronization performance with those relative to direct judgments on event durations, in other words, an integration of perception and action.

References

Allan, L.G. The perception of time. *Perception and Psychophysics*, 1979, *26*, 340-354.

Allan, L.G., & Kristofferson, A.B. Psychophysical theories of duration discrimination. *Perception and Psychophysics*, 1974, *16*, 26-34.

Flückiger, M. Comparaisons de durées. Psychogenèse des perceptions de séquences sonores. *Archives de Psychologie*, 1975, *43*, 1-69.

Flückiger, M. Analyse de quelques stratégies d'autorèglage. *Archives de Psychologie*, 1983, *51*, 49-55.

Fraisse, P. Le seuil différentiel de durée dans une suite régulière d'intervalles. *l'Année Psychologique*, 1967, 67, 43-49.

Hoopen, G. ten. The detection of anisochrony in monaural and interaural sequences. In: J.A. Michon & J.L. Jackson (Eds.), *Time, mind, and behavior.* Heidelberg: Springer Verlag, 1985, pp. 140-150.

Hoopen, G. ten & Akerboom, S. The subjective tempo difference between interaural and monaural sequences as a function of sequence length. *Perception and Psychophysics*, 1983, *34*, 465-469.

Macar, F. Psychophysics in time. In: J.A. Michon & J.L. Jackson (Eds.), *Time, mind, and behavior.* Heidelberg: Springer Verlag, 1985, pp. 112-130.

Michon, J.A. Studies on subjective duration I: Differential sensitivity in the perception of repeated temporal intervals. *Acta Psychologica*, 1964, *22*, 441-450.

Michon, J.A. *Timing in temporal tracking*. Assen: Van Gorcum, 1967.

Michon, J.A. Holes in the fabric of subjective time: Figureground relations in event sequences. *Acta Psychologica*, 1977, *41*, 191-203.

Povel, D.J. Time, rhythms and tension. In: J.A. Michon & J.L. Jackson (Eds.), *Time, mind, and behavior.* Heidelberg: Springer Verlag, 1985, pp. 215-225.

Reference Note

1. Flückiger, M., & Osiek, C. *Une nouvelle méthode d'étude de l'autorèlage appliquée à la discrimination auditive de durées*. Paper presented at the 5. Tagung der Psychologischen Forscher, Bern, 1978.

Chapter 9. The Detection of Anisochrony in Monaural and Interaural Sequences

Gert ten Hoopen

It is a well documented phenomenon that listeners have a hard time processing auditory sequences that alternate relatively fast between the ears. This holds for several kinds of auditory material such as a series of unrelated words (Treisman, 1971), running speech (Cherry & Taylor, 1954; Wingfield, 1977), melodies (Deutsch, 1979) and click sequences (Guzy & Axelrod, 1972; Massaro, 1976; ten Hoopen & Vos, 1981). This kind of auditory stimulation, in which a sequence is alternated back and forth between the ears, is called interaural stimulation. Thus, interaural sequences are processed less efficiently than their nonalternating counterparts, that is, than the same sequences presented either monaurally (to one ear) or binaurally (to both ears at the same time). Recall of interaurally presented series of words is diminished, shadowing interaural speech is hampered, recognition of interaural melodies is worse, and the number of interaurally presented clicks is underestimated.

Nakao & Axelrod (1976) and van Noorden (Note 1) wondered whether the detection of anisochrony (that is, temporal inequality such as is found in galoping) in an ongoing sequence of sounds is also deteriorated by interaural presentation as compared to the nonalternating condition. Nakao & Axelrod presented their subjects with pairs of sequences, a pair consisting of an isochronous standard sequence and an anisochronous comparison sequence. Anisochrony was induced by offsetting all even-numbered sounds of an isochronous sequence in time towards their preceding (negative offset) or succeeding (positive offset) odd-numbered sounds (Figure 1). The amount of anisochrony of the comparison increased in each next pair by making the positive offset larger. The subjects had to indicate whether the rhythm or beat of the comparison was the same as or different from that of the isochronous standard sequence. The trial terminated after three successive ›different‹ judgments and then the positive offset in time of the even-numbered sounds decreased, went through ›zero offset‹ (= isochrony) into the negative offset range until the subject gave again three ›different‹ judgments. This cycle was repeated five times.

This procedure yielded a bandwith of times, centered around the zero time offset within which the listeners did not detect anisochrony. The values of this ›interval of tolerance‹ are depicted in Figure 2 as dependent on the Stimulus Onset Asynchrony (SOA) of the isochronous sequence and on the presentation mode (interaural vs. monaural)[1].

1 (Editors' note) The term ›Stimulus Onset Asynchrony‹ (SOA) has found widespread use in the type of research reported here. It is confusing because it suggests the possible occurrence of two (or more) synchronous or simultaneous stimulus onsets. A better term would be Onset-to-Onset Interval or Inter-Onset Interval.

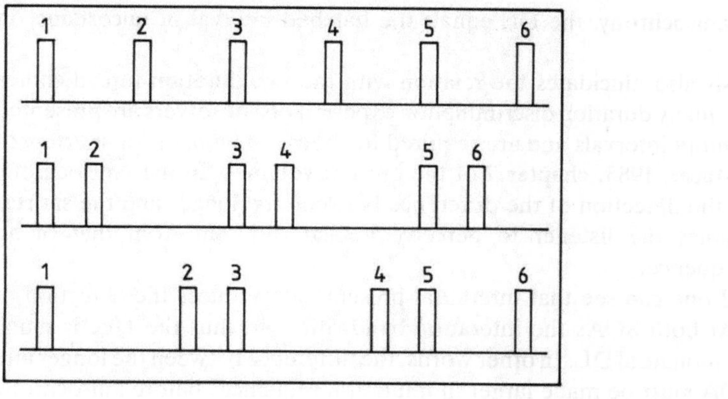

Figure 1. The *upper diagram* represents an isochronous sequence of sounds, numbered *1* through *6*. The repeating equal time interval between the onsets of the sounds is called the stimulus onset asynchrony (SOA). The *middle diagram* represents a sequence which is made anisochronous by a 'negative offset' in time of the even numbered sounds. The *lower diagram* represents a sequence which is made anisochronous by a positive effect in time of the even numbered sounds. Notice that the sum of two consecutive SOAs in the anisochronous sequences equals twice the base SOA of the isochronous sequence.

How should one define the Difference Limen (DL) for this particular paradigm? Normally the DL is defined as half the interval of uncertainty. However, it can be seen in Figure 2 that this interval, the hatched time interval in the Figure, equals the difference between the longer and shorter SOAs at which anisochrony becomes detectable. Since it is only by virtue of this difference that listeners are capable of

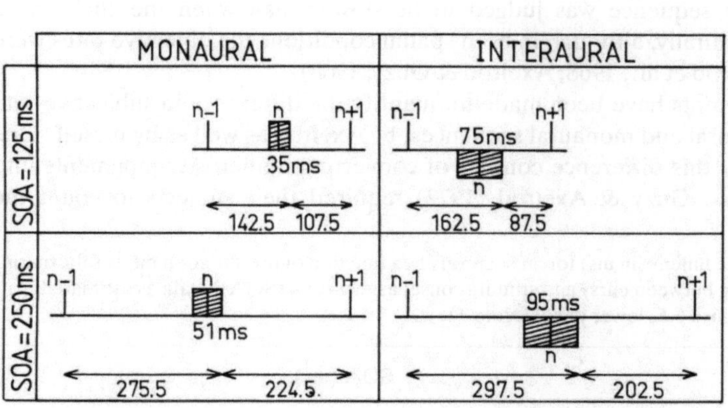

Figure 2. Nakao & Axelrod's (1976) results showing the interval of tolerance for anisochrony *(hatched area)* dependent on presentation mode (interaural vs. monaural) and stimulus onset asynchrony *(SOA)*. Notice that the difference limen for anisochrony, that is, the difference between the longer and the shorter SOA, equals the value of the bandwith.

detecting the anisochrony, the DL equals the hatched interval of uncertainty in Figure 2.

This analysis also elucidates the relation with the two-duration forced choice method used in many duration discrimination experiments: observers are presented two filled or empty intervals and are required to choose the longer (or shorter) of the two (see Macar, 1985, chapter 7 of the present volume). In the ›anisochrony task‹ it is also the detection of the difference between the longer and the shorter SOA that enables the listener to perceive a beat different from that of an isochronous sequence.

In Figure 2 one can see that interaural presentation reduces the detection of anisochrony. At both SOAs the interaural bandwith, and thus the DL, is much larger than the monaural DL. In other words, the difference between the longer and the shorter SOA must be made larger in interaural sequences before anisochrony can be detected.

The study by van Noorden (Note 1), in which a slightly different psychophysical procedure was used, confirmed this finding. However, as can be seen inTable 1, the difference between the alternating and nonalternating DLs tend to become smaller with increasing SOA, a pattern which is at variance with Nakao & Axelrod's results.

In the remainder of this paper I shall attempt to explain why such differences between the interaural and monaural DLs for anisochrony are to be expected. Secondly, I shall present the results of two experiments (1) to shed light on the difference in results between Nakao & Axelrod's and van Noorden's study, and (2) to test a quantitative prediction about interaural and monaural DLs for anisochrony.

In 1968 a curious auditory phenomenon was reported by Axelrod and coworkers. They established that the apparent rate of click sequences is dependent on the spatial structure of the sequence. When the clicks were presented interaurally, the rate of the sequence was judged to be slower than when the clicks were presented monaurally, although in both spatial conditions the objective rates were the same (Axelrod et al., 1968; Axelrod & Guzy, 1968).

Several attempts have been made to quantify the difference in subjective rate between interaural and monaural sequences, by Axelrod as well as by myself. One way to quantify this difference consists of converting numerosity judgments into ›counting times‹. Guzy & Axelrod (1972) required their subjects to count the

Table 1. Difference limens (in ms) for anisochrony as a function of presentation mode (alternating vs. non-alternating between ears) and stimulus onset asynchrony (SOA). Data as estimated from van Noorden's Figure 6.5, lower panel (Note 1)

	SOA (ms)							
	80	100	120	140	160	200	240	Mean
Alternating	44	56	62	70	48	44	40	*52*
Non-alternating	16	18	24	28	26	28	26	*24*
Difference	28	38	38	42	22	16	14	*28*

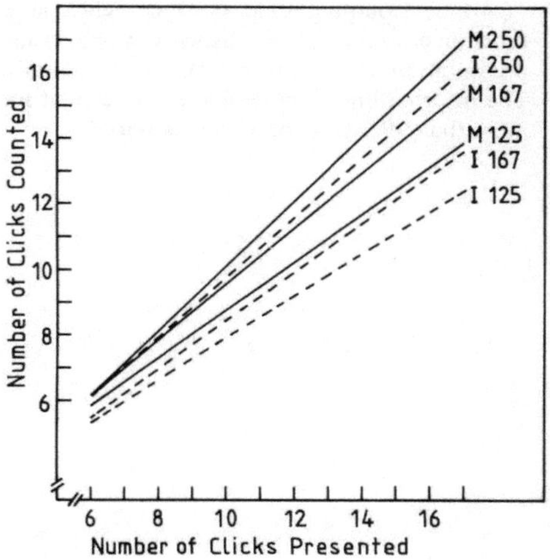

Figure 3. The number of clicks counted (N_c) as a function of presentation mode (interaural vs. monaural), stimulus onset asynchrony (SOA = 125, 167 or 250 ms) and number of clicks presented (N_p). Only the linear regression lines are drawn because they each explain at least 99.5 percent of the variance (M = Monaural, I = Interaural; 125, 167 and 250 denote the SOA-values). The equations of the linear regression lines are: $N_c = 1.48 + 0.723N_p$ (M125), $N_c = 1.18 + 0.838N_p$ (M167), $N_c = 0.24 + 0.986N_p$ (M250), $N_c = 1.43 + 0.645N_p$ (I125), $N_c = 1.07 + 0.734N_p$ (I167) and $N_c = 0.65 + 0.906N_p$ (I250). Based on the combined blocked group and neutral group data of ten Hoopen and Vos, 1981.

number of clicks in a sequence, presented either monaurally or interaurally. They found that interaurally presented clicks were ›undercounted‹ relative to those in monaural sequences. Unfortunately they applied a wrong analysis to their data as we demonstrated in ten Hoopen & Vos (1980).

Ten Hoopen & Vos (1981) replicated the result that interaural clicks are counted less well than monaural clicks. The pattern of results is shown in Figure 3. The number of sounds presented (N_p) in some time interval T equals T/SOA and the number of sounds that subjects can count (N_c) in T equals T/CT, where CT stands for counting time. Hence it follows that $N_c/N_p = SOA/CT$. Because N_c/N_p equals the slope b of a linear regression line in Figure 3, we computed the counting times at each SOA and for both presentation modes by $1/b \times SOA$. In Figure 4 these counting times are plotted and linear regression lines fitted through the interaural as well as through the monaural counting times. One sees that the counting times differ between both presentation modes by 24 ms, independent of the rate of the sequence.

Because the counting rate of the subjects is paced by the rate of the click sequence, the value of 24 ms reflects the subjective or perceptual difference between an interaural and a monaural SOA. Thus it seems as if the subjective time interval between two interaural clicks is *dilated* or *stretched* in comparison to the subjective time interval between two monaural clicks. Because these subjective time intervals exist between the *percepts* of the clicks and not between the clicks themselves, we refer in this case to Perceptual Onset Asynchrony (POA). Thus the difference between an interaural POA and a monaural POA, stemming from the *same* SOA, corresponds to 24 ms.

Of course it would be interesting when converging paradigms would support this supposition. Several years ago, Schaefer (Note 2) reported such a paradigm. His subjects were asked to react as quickly as possible when a sequence with an unknown number of clicks stopped. He found that it took his subjects 26 ms more

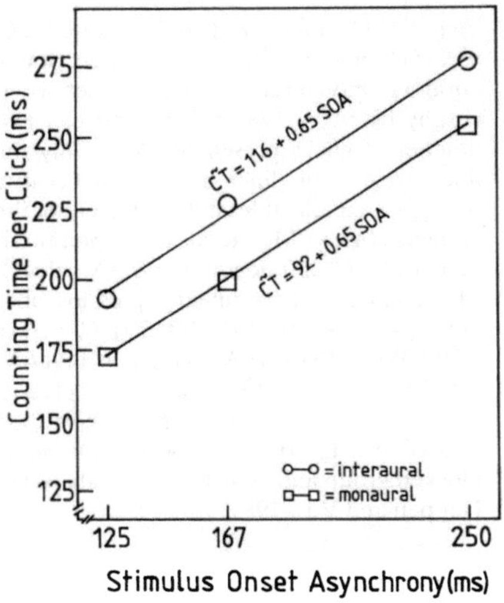

Figure 4. Counting time *(CT)* per click as a function of stimulus onset asynchrony (*SOA*) and presentation mode (interaural vs. monaural). The inserted linear regression lines account for more than 99.5 percent of the variance.

to react to the end of an interaural sequence than to the end of a monaural sequence, although the rate of these sequences was equal.

This so-called stop-reaction time (RT) paradigm has been utilized many times in my laboratory since then and it has been reported by ten Hoopen et al., (1982) that the stop-RT difference between interaural and monaural conditions amounts to a constant 24 ms throughout an SOA-range of 125-250 ms. Akerboom et al., (1983) found a stop-RT difference of 25 ms throughout a considerably larger SOA-range of 40-2130 ms.

What causes the stop-RT difference? Figure 5 diagrams the notion that, although in the real time domain interaural and monaural sounds have the same SOA, the ›projection‹ of the interaural SOA into the mental time domain is dilated with respect to the ›projection‹ of the monaural SOA.

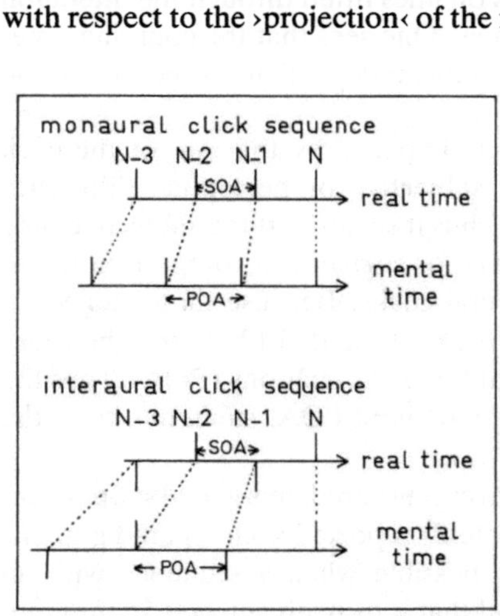

Figure 5. Although the monaural sequence *(upper diagram)* and the interaural sequence *(lower diagram)* have the same stimulus onset asynchrony (*SOA*) between the clicks (real time axis), the perceived onset asynchrony (*POA*) between the percepts of the interaural clicks is larger than that between the monaural percepts (mental time axis). Notice that bars (clicks or click percepts) drawn on the same side of the time axes represent the monaural condition, whereas bars drawn alternately on both sides of the axes represent the interaural condition.

What do listeners do in the stop-RTtask? Instructed to detect the unpredictable end of the sequence, they monitor mental time (*not* real time) to confirm whether a following sound arrives at the end of the POA. If this is the case they carry on monitoring. If no next sound occurs the stop response is initiated, which takes some additional amount of time (response initiation time, abbreviated RI). It is assumed that RI is not affected by the presentation mode of the sounds, because RI is a genuine motor component of the stop-RT. Under this assumption one may interpret the stop-RTdifference between the interaural and monaural conditions as the difference between their respective POAs. This is illustrated by Figure 6. Note that the estimates of the POA-difference in the counting experiment and the stop-RT experiments yield the same value (24 and 24-25 ms). However, it should be noted that our re-analysis of the data from the Axelrod studies did yield an estimate of about 40 ms. And it should also be mentioned that subsequent experiments also yielded somewhat smaller estimates of 19 ms (ten Hoopen & Akerboom, 1983) and 20 ms (Akerboom & ten Hoopen, 1983).

Until now it is not clear whether the time-dilation of interaural signals serves some ecological goal or what auditory mechanism causes it, but for the present purposes it is sufficient to know the amount by which interaural sounds are dilated in mental time. It allows predictions to be made about the detection of anisochrony in the interaural condition as compared to the monaural one: the difference between the DLs for anisochrony in interaural and monaural sequences should equal the POA-difference between those sequences and this difference should also be invariant with the rate of the sequence.

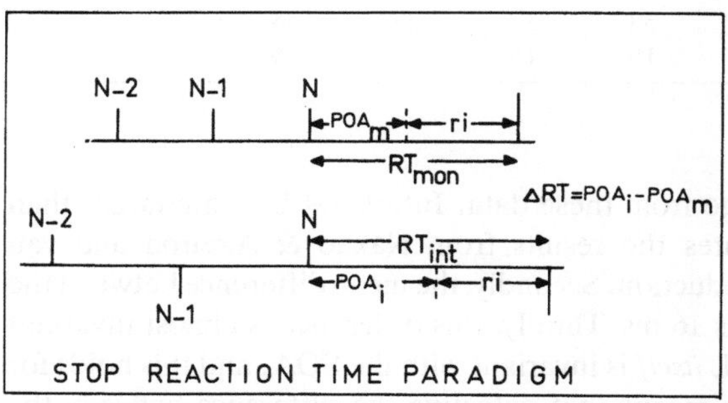

Figure 6. The *upper diagram* depicts the monaural mental time axis, the *lower diagram* the interaural mental time axis. When a sequence of unpredictable length (*N* clicks) stops, it takes the listener one extra POA before being able to detect that no further click arrives. The stop response is then initiated and response initiation (*RI*) time is assumed to be invariant with presentation mode. Hence the difference between the interaural and monaural stop reaction time (*stop-RT*) reflects the difference between the interaural and monaural POA (POA_i-POA_m).

Experiments

Method of Limits

Four subjects were presented with series of sequences which progressively became more anisochronous and with series which became progressively more isochronous. In the former case they had to respond when they detected that the series changed from regular to irregular and in the latter case vice versa. The basic SOAs were 100, 150, 200, 250 and 300 ms and presentation modes were interaural and monaural. The step size by which the even numbered sounds were offset in time was 3 ms, that is, in each consecutive sequence the difference between the longer and shorter SOAs increased (ascending series) or decreased (descending series) by 6 ms (only negative offsets were used, thus the even numbered sounds were shifted towards the preceding odd-numbered sounds). Table 2 gives the DLs for anisochrony, that is, the difference between the longer and shorter SOA at which the listeners judged the series of sequences to turn from isochronous into anisochronous or vice versa.

Table 2. Difference limens (in ms) for anisochrony as a function of presentation mode (interaural vs. monaural) and stimulus onset asynchrony (SOA)

	SOA (ms)					
	100	150	200	250	300	Mean
Interaural	53	54	54	55	56	*54*
Monaural	38	38	38	37	37	*38*
Difference	15	16	16	18	19	*16*

Four important facts emerge from these data. Interaural DLs are larger than monaural ones, which replicates the results from Nakao & Axelrod and van Noorden mentioned in the introduction. Secondly, the mean difference between the interaural and monaural DLs is 16 ms. Thirdly, this difference is almost invariant with the SOA. Fourthly, the DL *itself* is invariant with the SOA, and this holds for both presentation modes. This means that detecting the difference between the longer and the shorter SOAs, that is, detecting anisochrony, is not affected by the magnitude of the base SOA, at least not in range of 100-300 ms.

The four listeners regarded this method in which the sequences became more isochronous or more anisochronous gradually as rather difficult, probably because they had to compare each consecutive sequence with a mental template of (an)isochrony. I therefore decided to carry out a second experiment using an easier procedure.

Method of Constant Stimuli

Six subjects were confronted with a two alternative forced choice task. A trial consisted of successively presenting an isochronous standard stimulus and an anisochronous comparison stimulus (or vice versa) and the listeners had to judge whether the two sequences had the same or a different beat. Notice that this is the kind of procedure used by Nakao & Axelrod, but that they increased (or decreased) the offset in time in the comparison stimuli systematically whereas we randomized the amount of offset in time over the comparison stimuli.

Table 3. Difference limens (in ms) for anisochrony as a function of presentation mode (interaural vs. monaural) and stimulus onset asynchrony (SOA)

	SOA (ms)				
	150	200	250	300	Mean
Interaural	47	46	50	49	*48*
Monaural	20	23	27	27	*24*
Difference	27	23	23	22	*24*

In all other respects our second experiment was identical with the first one. The results are presented in Table 3. Because the results at the SOA of 100 ms were highly inconsistent, only the data for the other SOAs are given. As in the first experiment (see Table 2) the DL for anisochrony was larger in the interaural than in the monaural condition. In this case it was 24 ms larger and this amounts precisely to the POA-difference between interaural and monaural sequences. In addition, the difference between both DLs was almost invariant with the base SOA. Again, the DL *itself* was not much affected by the SOA-value.

Discussion

The first experiment yielded a difference of 16 ms between the DLs for interaural and monaural anisochrony. This value is smaller than the POA-differences between these spatial conditions than the other values we mentioned (19-25 ms). It should be stipulated, however, that the latter values were averages based on much larger samples of listeners than in the present study. The value found in the second experiment (24 ms) fits well within the range of POA-differences. Moreover, as the conception of the interaural time dilation predicts, the difference between the interaural and monaural DLs was invariant with the rate of the sequence in both experiments.

This pattern of results is in line with Nakao & Axelrod's (1976) study. They found (see Figure 2) a difference in DL of 40 ms between both spatial conditions at a SOA equal to 125 ms and one of 44 ms at a SOA of 250 ms. It is in contrast, however, with van Noorden's data (Note 1): there the difference in DL decreased with increasing SOA (see Table 1). It should be mentioned, however, that van Noorden's results were based on one subject, whereas Nakao & Axelrod's results are based on 15 subjects.

Although this was not the primary interest of the present study it is worthwhile to pay some attention to the pattern of DLs with respect to the rate of the sequence, or as we expressed it, the SOA. Inspecting Tables 2 and 3 one sees, as was already mentioned, that the DL does hardly, if at all, increase with increasing SOA. Intuitively we would have expected that with increasing SOA of the isochronous sequence the difference between two consecutive SOAs would also have to be made larger in order to detect anisochrony. But this notion, which would relate to Weber's law, is not supported by the data.

For example, in the monaural condition and with the method of limits (first experiment), the anisochrony in a 280-320 alternating SOA-pattern is detected equally well as in an 80-120 alternating SOA-pattern. In both temporal patterns the difference between the consecutive SOAs is 40 ms and in Table 2 one can infer that such a difference exceeds the DL at a SOA equal to 100 ms (DL = 38 ms) as well as at a SOA equal to 300 ms (DL = 37 ms).

Figure 7 illustrates this point further. For all data mentioned (Nakao & Axelrod, 1976; van Noorden, Note 1, and both experiments of the present study) I plotted the Weber fractions for anisochrony against the base SOA. The pattern, which holds for all these studies, is that the Weber fraction decreases with increasing base SOA (in the 80-300 ms range). Such a trend was also found in several duration discrimination studies (see Woodrow, 1951). Also Michon (1964), using an intermittency paradigm to investigate the differential sensitivity for repeated temporal intervals, found that the Weber fraction decreased with increasing SOA (in the 67-200 ms range). Beyond 200 ms, however, the Weber fraction sharply increased.

Of course it would also be interesting to study the detection of anisochrony at other SOA-values than those used in the present study; pertinent experiments are under way. Another question concerns the detection of anisochrony in other sensory modalities. Flückiger (1985, in Chapter 8 of the present volume) presented his subjects with two lights which were switched on in an alternating fashion. At the start of the experiment, the on-duration of the two lights was identical, namely

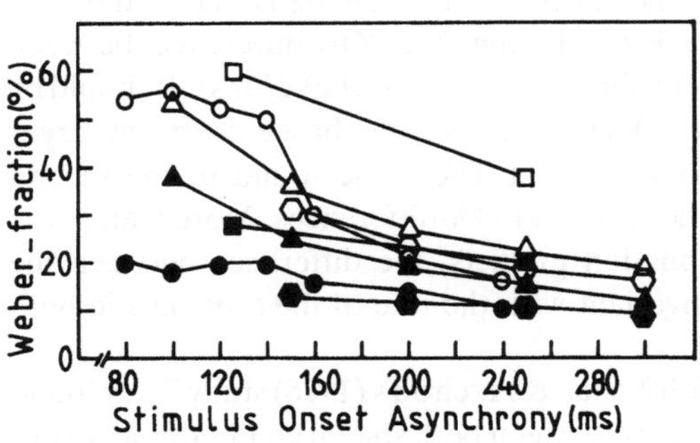

Figure 7. Weber fractions for anisochrony as a function of stimulus onset asynchrony for ear-alternating and non-alternating sound sequences *(open and closed symbols* respectively). *Squares:* Nakao & Axelrod (1976); *circles:* Van Noorden (Note 1), *triangles:* first experiment present study; *hexagons:* second experiment present study.

150 ms. But in each subsequent cycle the duration of one light was made longer and that of the other light shorter by the same amount of time (3, 6 or 9 ms). Subjects had to detect the difference between the two light durations. Therefore, Flückiger's data (at a step size of 3 ms) may be compared with our data obtained for the base SOA of 150 ms. It is interesting that Flückiger found a DL of about 35 ms which is not too different from our value of 29 ms (averaged over both experiments and at the monaural condition). This raises the question of whether there is a detection mechanism for anisochrony common to different sensory modalities. Further research is needed to answer this question.

References

Akerboom, S., & Hoopen, G. ten. The effect of a contralateral drone on the perceptual onset asynchrony of interaural tone sequences. *Perception and Psychophysics*, 1983, *33*, 571-574.

Akerboom, S., Hoopen, G. ten, Olierook, P., & Schaaf, T. van der. Auditory spatial alternation transforms auditory time. *Journal of Experimental Psychology: Human Perception and Performance*, 1983, *9*, 882-897.

Axelrod, S., & Guzy, L.T. Underestimation of dichotic click rates: Results using methods of absolute estimation and constant stimuli. *Psychonomic Science*, 1968, *12*, 133-134.

Axelrod, S., Guzy, L.T., & Diamond, I.T. Perceived rate of monotic and dichotically alternating clicks. *Journal of the Acoustical Society of America*, 1968, *43*, 51-55.

Cherry, E.C., & Taylor, W.K. Some further experiments upon the recognition of speech, with one and with two ears. *Journal of the Acoustical Society of America*, 1954, *26*, 554-559.

Deutsch, D. Binaural integration of melodic patterns. *Perception and Psychophysics*, 1979, *25*, 399-405.

Flückiger, M. Effects of time pressure on duration discrimination. In: J.A. Michon & J.L. Jackson (Eds.), *Time, mind, and behavior.* Heidelberg: Springer Verlag, 1985, pp. 131-139.

Guzy, L.T., & Axelrod, S. Interaural attention shifting as response. *Journal of Experimental Psychology*, 1972, *95*, 290-294.

Hoopen, G. ten, & Akerboom, S. The subjective tempo difference between interaural and monaural sequences as a function of sequence length. *Perception and Psychophysics*, 1983, *34*, 465-469.

Hoopen, G. ten, & Vos, J. Attention switching is not a fatigable process: Methodological comments on Axelrod and Guzy (1972). *Journal of Experimental Psychology: Human Perception and Performance*, 1980, *6*, 180-183.

Hoopen, G. ten, & Vos, J. Attention switching and patterns of sound locations in counting clicks. *Journal of Experimental Psychology: Human Perception and Performance*, 1981, *7*, 342-355.

Hoopen, G. ten, Vos, J., & Dispa, J. Interaural and monaural clicks and clocks: Tempo difference versus attention switching. *Journal of Experimental Psychology: Human Perception and Performance*, 1982, *8*, 422-434.

Macar, F. Time psychophysics and related models. In: J.A. Michon & J.L. Jackson (Eds.), *Time, mind, and behavior.* Heidelberg: Springer Verlag, 1985, pp. 112-130.

Massaro, D.W. Perceiving and counting sounds. *Journal of Experimental Psychology: Human Perception and Performance*, 1976, *2*, 337-346.

Michon, J.A. Studies on subjective duration I: Differential sensitivity in the perception of repeated temporal intervals. *Acta Psychologica*, 1964, *22*, 441-450.

Nakao, M.A., & Axelrod, S. Effects on bilateral alternation on perceived temporal uniformity of auditory and somesthetic pulse trains. *Perception and Psychophysics*, 1976, *20*, 274-280.

Treisman, A.M. Shifting attention between the ears. *Quarterly Journal of Experimental Psychology*, 1971, *23*, 157-167.

Wingfield, A. The perception of alternated speech. *Brain and Language*, 1977, *4*, 219-230,

Woodrow, H. Time perception. In: S.S. Stevens (Ed.), *Handbook of experimental psychology.* New York: Wiley, 1951, pp. 1224-1236.

Reference Notes

1. Noorden, L.P.A.S. van. *Temporal coherence in the perception of tone sequences*. Doctoral dissertation, Eindhoven (The Netherlands): University ofTechnology, 1975.
2. Schaefer, F. *Gerichtete und verteilte Aufmerksamkeit bei der Einschätzung richtungsalternierender Clicks* (Focused and divided attention when estimating interaurally presented clicks). Paper presented at the 21. Tagung experimentell arbeitender Psychologen, Heidelberg, West Germany, 1979.

Chapter 10. Memory for Temporal Information[1]

William K. Estes

The experimentally based study of the way human beings process temporal information has been approached from several distinct standpoints. Research stemming from the psychophysical tradition seems to be organized largely in terms of the conception that the human information processing system includes some kind of generator of temporal pulses or intervals and that an individual has access to counts, or the equivalent, of these pulses and can use them as a basis for temporal judgments. Studies on the internal clock and the way it gives rise to temporal judgments are covered elsewhere in this volume (Macar, 1985; Shaffer, 1985; and ten Hoopen, 1985; chapters 7, 15 and 9 respectively). Whatever the basis for perceptions or judgments of intervals, the way information about them is encoded and retained in memory is the subject of another major research tradition, and the one on which I focus in this paper.

In many instances in everyday life, as distinct from laboratory studies, memory for temporal information doubtless entails simply the remembering of clock times or dates or of propositional information that permits the inference of clock times or dates (Linton, 1975, 1978). This type of temporal memory must be of much practical importance, but it seems that it should be interpretable in terms of general principles of factual memory. In most laboratory research and also in much everyday life experience temporal memory does not, however, take the form of explicit verbal encodings of dates or times, but rather may be interpreted in terms of the encoding and retention of information about the temporal attribute of events or sequences of events, these processes presumably having much the same character as the encoding and retention of information about other attributes of memory (Estes, 1980; Michon, 1972, 1985, chapter 2 of the present volume; Underwood, 1969, 1977). In this paper I review the salient facts and principal theories about memory for temporal attributes.

Research Paradigms

The variety of experimental approaches that one finds in the literature can be usefully organized in terms of the tasks employed and the principal experimental parameters (see also Macar, 1985, chapter 7 of the present volume).

1 Preparation of this paper was supported in part by Grant BNS 80-26656 from the National Science Foundation.

Experimental Tasks

(1) Judgment of Elapsed Time. When subjects are fully instructed about the nature of temporal aspects of a task in advance, estimates of elapsed time are often found to decrease with the complexity or information load of the task (Hicks et al., 1976). Since the judgment is made immediately at the conclusion of the task, retention of temporal information is not importantly involved and the results may bear mainly on the degree to which concurrent information load interferes with encoding of the temporal attribute of the task. The sensitivity to interference suggests that the judgmental process involved is not automatic but, rather, depends on capacity-limited selective attention to appropriate aspects of the events being experienced.

(2) *Judgment of Relative Duration of Two Remembered Intervals.* Here, memory is more clearly implicated, since a representation of the first interval must be maintained in memory for comparison with the second. In connection with models for temporal memory (discussed in a later section), special interest attaches to the symmetry or asymmetry of judgments when two successive intervals are of equal duration (Block, 1978, 1985, chapter 11 of the present volume).

(3) *Absolute Judgments of Lag.* A sequence of items, usually letters, pictures, or words, is presented with some items repeated. The subject's task upon the occurrence of a repetition is to estimate the position of the earlier occurrence or the interval (often termed *lag* in the experimental literature) between the two occurrences (Hinrichs & Buschke, 1968; Lassen et al., 1974; Peterson et al., 1969). Alternatively, a sequence of items is followed by a test on which two items from the list are presented and the subject estimates the lag between them. An important parameter in this case is the similarity of the items (Hintzman et al., 1975).

(4) *Judgments of Relative Recency.* A sequence of items is presented to subjects as in category 3, but the subject is tested by being presented with a pair of items and asked to judge which of the two occurred more recently in the preceding sequence (Fozard, 1970; Yntema & Trask, 1963). Since the research objective is normally to get at properties of temporal memory, rather than merely performance, one might hope that the tasks in categories 3 and 4 could be shown to tap the same memorial processes. In a study by Lockhart (1969) two groups of subjects were presented with the same sequence of items, but one group made absolute and the other relative recency judgments; the main result, a comfort for those hoping for theoretical simplicity, was that it proved possible to make accurate predictions of relative recency judgments between pairs of items on the basis of the absolute judgments obtained for items in the same list positions separately. Further support for this conclusion has been provided by Underwood & Malmi (1978).

(5) *Positional Recall.* Following sequential presentation of a list of items, subjects recall the items, usually after a filled retention interval, and indicate the position of each item in the list or on a time scale representing the duration of list presentation

(Conrad, 1964; Michon & Jackson, 1984; Shiffrin & Cook, 1978). It is important to distinguish this task, which explicitly requires a recall of positional or temporal information about the occurrence of an item, from the more common task of simple ordered recall. The implications of this distinction have been spelled out by Lee & Estes (1977). The positional recall task has been the principal source of information about the way in which memory for temporal attributes of events changes during retention intervals.

(6) *Dating of Long-Term Memories*. Subjects attempt to recall the dates or estimate relative recencies of events that have occurred over a time span of years in everyday life, for example television programs or news events (Linton, 1975, 1978; Squire et al., 1975; Underwood, 1977). In the nature of things, studies of this type are not well controlled in comparison to laboratory studies, but nonetheless it is of interest to see whether some of the functional relationships obtained in the laboratory do extend to naturalistic situations.

Major Experimental Dimensions

Experimental dimensions that have proven to be of major theoretical significance have to do with the duration of the interval over which temporal information has to be retained and the complexity of the items or events about which memory is tested. These factors may exert important constraints on the processes that contribute to temporal judgments.

In studies using retention intervals within the short-term memory range, a critical parameter is the presence or absence of measures to control rehearsal. It seems likely that short-term memory for temporal positions of events or intervals between them can be directly assessed only when rehearsal is precluded by a shadowing task or the equivalent. In studies following the general procedure initiated by Conrad (1964), for example, the events to be remembered are presentations of letters on a display screen, and rehearsal is prevented by using short exposure durations (typically 2 to 2½ letters per second) and requiring the subject to pronounce the name of each letter as it appears (Healy, 1974; Lee & Estes, 1977). An alternative technique, used by Shiffrin & Cook (1978), is to instruct the subjects to attend to a concurrent task (e.g. signal detection) as their primary responsibility and not to rehearse the stimulus letters. When such techniques are not used subjects are likely to rehearse the items, and not necessarily in the input order, thus generating memory representations in which temporal relations may be different from those of the original event sequence.

The greater part of research on temporal memory has been conducted with moderate retention intervals, ranging from minutes to hours. Most often words are used as the stimuli, though a few studies have compared words and pictures (Fozard & Yntema, 1966; Lassen et al., 1974). The heavy reliance on words may be unfortunate, since word presentations are not very representative of the kinds of events for which temporal memory is significant in ordinary life. Not surprisingly, failures to obtain robust or systematic results for such measures as estimates of lags between stimulus occurrences (Underwood, 1977; Underwood & Malmi, 1978) have occurred with words as stimuli. One problem may be the common tendency to cluster semantically related words in rehearsal, another that in the absence of instructions to do so, subjects may not attend to temporal relations between words. In some as yet unpublished research in my laboratory, we have used brief descriptions of episodes as stimuli, and results suggest that events are more likely than word stimuli to engage habits of temporal encoding.

Much longer retention intervals, ranging from days to years, have so far been explored only in observational studies of memory for complex events (Linton, 1975, 1978; Squire et al., 1975; Underwood, 1977). The extension and testing of models for temporal memory beyond the very restricted range of current laboratory paradigms may not be able to go far until we begin to see research conducted at longer retention intervals with at least some of the controls that have proven important in the short-term studies.

Major Experimental Results

My principal guideline for sifting the considerable accumulation of empirical studies on temporal memory is theoretical relevance, and with this criterion in mind it is not difficult to select a small number of phenomena that merit close consideration. A few robust findings of some generality have been the focal points on which the development of models in this area have turned. Perhaps foremost among these is the relationship between absolute judgments of intervals between item occurrences and the actual intervals. In experimental studies with simple items and involving short- to moderate-term memory, a common finding is that schematized in Figure 1; lag estimates are a smooth negatively accelerated function of actual lags, with short lags being overestimated and long lags underestimated (Hinrichs & Buschke, 1968; Hinrichs, 1970; Lassen et al., 1974; Lockhart, 1969; Wells, 1974). This function is characteristic of experiments in which the items are rehearsable and the subject's task is to estimate the interval between a probe given at the end of a sequentially presented list and the position at which the indicated item occurred during list presentation. In the case of very long term memory for everyday life events, the same deviations from direct proportionality are present, lags of recent events being overestimated and remote events underestimated, but the function is much closer to linear than that in Figure 1 (Underwood, 1977). It may well be that in the naturalistic studies estimates of lags are based only in part on the same processes responsible for the laboratory findings and in part also on memories of dates.

A more complex pattern of results emerges from studies in which subjects are called on to judge from memory, not the absolute lag of a list item, but rather the lag between the list positions of two items presented on a test. If the test pair comprises two semantically related words, or two instances of the same word, then lag estimates increase with lag, but if the members of the test pair are unrelated words,

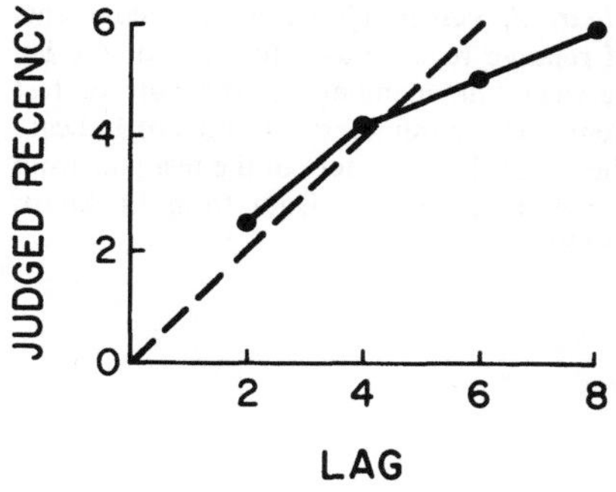

Figure 1. Judged recency of items as a function of the lag from item occurrence to the end of the list (Data from Peterson et al., 1969).

the lag function is virtually flat, as depicted in Figure 2 (Hintzman et al., 1975). These two groups of findings on estimates of absolute lag pose two prime questions for theoretical interpretation: Why should short lags tend to be overestimated and long lags underestimated and why should the lag function differ so sharply for related versus unrelated items? We shall return to these questions in a later section.

A related group of findings arises from experiments on relative recency judgments, the almost uniform result being a negatively accelerated increasing function relating percentage of correct relative recency judgments to lag between the items of the test pair, as schematized in Figure 3. Functions of this form have been reported, for example, by Fozard (1970) with either words or pictures as stimuli, and by Lockhart (1969) with words as stimuli. In this paradigm, correct recency judgments decrease with the lag between the test and the list occurrence of the more recent member of the test pair (Yntema & Trask, 1963).

The third major group of phenomena involves what has been termed the *repetition effect*. In theories based on the idea that individuals judge the recency of occurrences of events on the basis of the strength of their memory traces, there is necessarily a trade-off between recency and repetition of items. That is, the strength of the memory trace must be expected to decrease with time following the occurrence of an item but to increase with each repetition of the item. Consequently, other things equal, a repeated item tends to be judged as more recent than an unrepeated item. Suppose, for example, that two items of a list occur in the pattern ABB, that is, the occurrence of item A being followed after an interval by

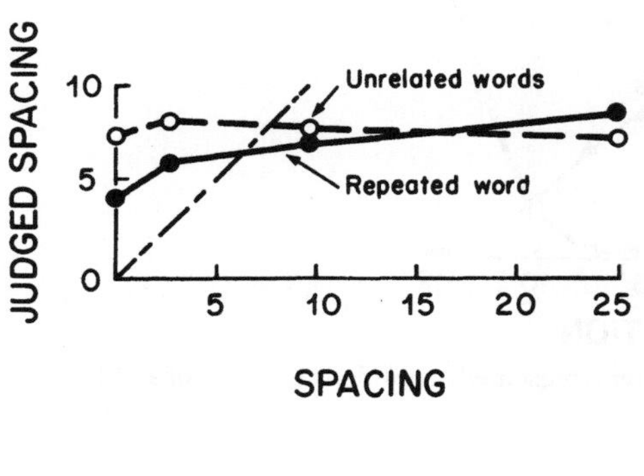

Figure 2. Judged spacing between items as a function of their actual spacing in a list (Data from Hintzman et al., 1975).

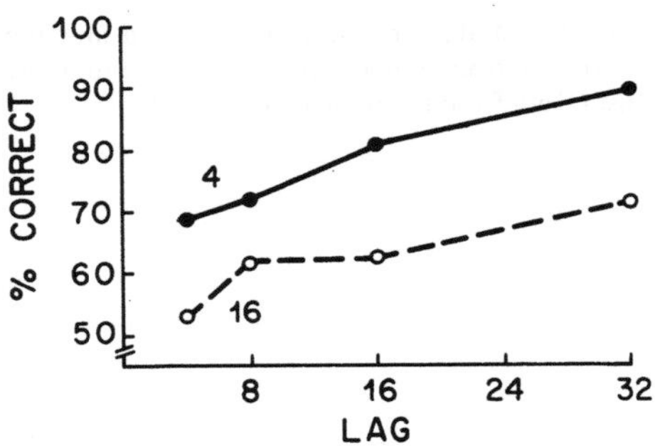

Figure 3. Percentage of correct judgments of relative recency as a function of the lag between the members of the pair of test items. The parameters (*4* and *16*) indicate the lag of the later item of the test pair from the end of the list (Data from Lockhart, 1969).

item B, and, following another interval, by a repetition of B. On a test at the end of the list subjects should be expected to judge B as more recent than A with higher probability than they would if the repetition had not occurred. Similarly, following the sequence AAB, the likelihood of correct judgments of B being more recent should be reduced. These expectations have been confirmed by Fozard & Yntema (1966), Galbraith (1976), Morton (1968) and Peterson et al. (1969) among others. This effect has been one of the principal sources of support for strength theories of memory.

The final cluster of results I shall focus on centers around uncertainty gradients. In experiments in which subjects are asked, following presentation of a list, to recall items in their temporal or sequential positions, accuracy of memory for the temporal or positional attribute can be represented by what has been termed an uncertainty gradient (Estes, 1972; Healy, 1974; Lee & Estes, 1977; Shiffrin & Cook, 1978). Characteristically the gradient is very steep, that is closely centered around the correct position, for items late in a list, spreads in an orderly manner to the maximum variance for items in the interior of the list, then steepens for items at the beginning of the list, though not as sharply as for those at the end, the characteristic pattern being as schematized in Figure 4. Accounting for the form of these gradients and the orderly way in which they change during a retention interval provides yet another challenge to models of memory for temporal attributes.

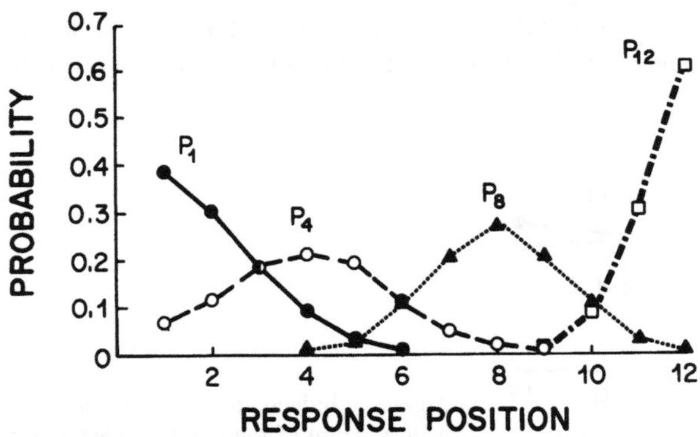

Figure 4. Observed uncertainty gradients for items presented in positions 1, 4, 8, 12 of a 12-item list (Data from Lee & Estes, 1977).

Models and Interpretations

Trace Strength Models

The first kind of model put forward for the interpretation of temporal memory, and the motivation for a substantial part of the research that has been done on the topic, is termed *strength theory.* The central idea, well recommended on considerations of parsimony, is that a person's judgment of the recency of an event is based on the strength of a memory trace established when the event occurred. All theories of memory include the assumptions, in one form or the other, that a memory trace is established when an event occurs and that strength or availability of the trace tends to decay or weaken with the passage of time, especially when the retention interval includes events or activities that are similar in some way to those of the original episode. All that needs to be added to have a basis for interpreting judgments of recency or temporal duration is that an individual has access to the strength of the memory trace, either directly or by virtue of its relation to such indicators of availability as the time required to recall or recognize an event (compare the notion of *availability* in the work of Tversky & Kahneman (1973) on judgments of probability).

Some properties predicted from strength theory with regard to absolute and relative judgments of recency are illustrated in Figure 5. In the illustration it is assumed that three items or events, A, B, C, have occurred successively and that the observer is called on for the estimates of absolute or relative recency at the time *t*. If the task is to estimate the interval between the occurrence of each event and the time, *t*, of testing, the estimate for A will be longest because its memory trace at the time *t* is the weakest. Similarly, if a judgment of relative recency were called for, say between B and C, C would be likely to be judged the more recent because its memory trace is the stronger. The model accounts for the well documented observations that absolute judgments of duration increase directly with the actual interval and that judgments of relative recency between two events tend to increase with their temporal separation. The picture is not entirely simple, however, since the probability of a correct recency judgment is predicted (and observed) also to decrease as the interval between the more recent of two events and the time of testing increases. The repetition effect is also accounted for, on the supposition,

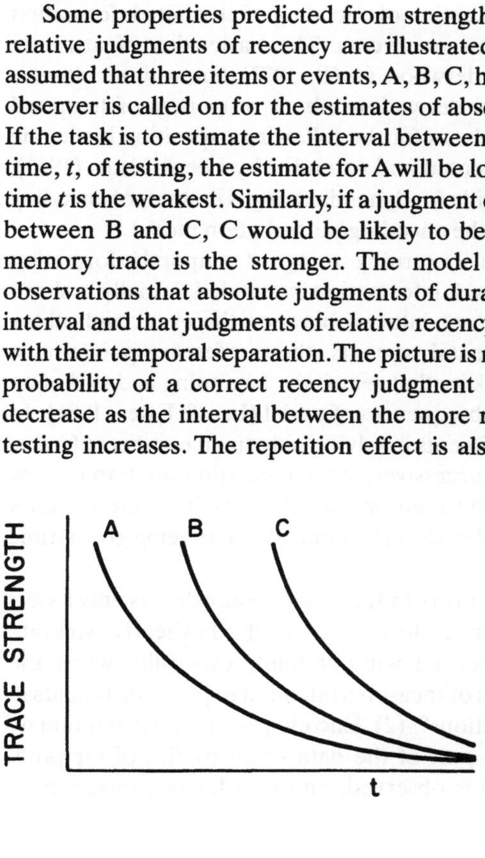

Figure 5. Schematic representations of the curves of declining memory trace strength as a function of time for three successively presented items, *A*, *B* and *C*, with a memory test being given at time *t*.

common to most strength theories, that repetition on an item or event increases the strength of its memory trace. Thus, if event B in Figure 5 were repeated, its trace would be strengthened and thus the likelihood available to the individual as a correlate of temporal relations. If the trace of an event A is still active when an event B occurs, and this relation is attended to by the individual, then the information that A preceded B is available to be recorded in memory. Also, the magnitude of the trace provides an index of the duration of the interval between the events. The idea that this process may provide the basis for the initial encoding of information about temporal intervals in memory is compatible with the »study-phase retrieval« interpretation of judgments of temporal spacing proposed by Hintzman et al. (1975) and seems even closer to the extension of that proposal by Tzeng et al. (1979). Thus there may be a basic continuity between the once dominant but now unfashionable trace theory and the currently more influential class of encoding models, to which we now turn.

Encoding Models

The type of model that might prove most widely applicable to the phenomena of temporal memory is based on the conception of encoding of attribute information concerning relations or intervals between events. The basic idea of attribute encoding has been developed by Underwood (1969, 1977) but without process assumptions that would enable actual derivations about properties of temporal memory. The version I shall take as a basis for the following exposition is the one introduced by Estes (1972) and augmented and extended by Lee & Estes (1981), which does include processes that provide for dynamic changes in states of encoding over time. For brevity I shall speak of the encoding/perturbation model.

In this model, the basic assumption concerning storage of temporal information is that when an event or item is perceived, information is encoded, initially in short-term memory, concerning the relation or interval between the event and others which serve as referents. In experimental contexts, these reference points most often take the form of distinctive markers that set off the beginning and end of an experimental trial, or in some cases subsequences of a trial (Lee & Estes, 1981). In short-term recall or recognition experiments involving only small numbers of items presented in regular sequence, the successively presented stimulus frames may provide adequate reference points, and then one speaks loosely of the memory representations or items being entered in slots (Conrad, 1964) or being given ›time tags‹ (Flexser & Bower, 1974).

The assumption of encoding in the form of a time tag or equivalent is only a start on a model. Several major problems need to be addressed. (1) Clearly, accurate tagging of item positions cannot proceed without limit, especially when the individual is subject to time constraints or measures intended to preclude rehearsal. What is the basis of capacity limitations? (2) Encoding cannot be unfailingly precise, for errors are a conspicuous part of the data of all studies of temporal memory. (3) Generally retention loss is observed, and an adequate model must include some mechanism to account for increasing imprecision of temporal memories over retention intervals.

Regarding the first problem, the model for memory span proposed by Drewnowski (1980) simply includes the assumption that only a small number of item positions can be accurately encoded; an estimate of between 2 and 4 positions proved satisfactory for his data. A theoretical basis for the capacity limitation is provided by the idea, discussed above, about positional encoding of relations between the trace of the start signal for a trial and the first one or two items. The negatively accelerated form of the function relating trace strength to time (Figure 5) would not allow distinctively different encodings much beyond this limit. However, when item presentations have an auditory component, allowing recycling of the representations of several items in sequence for a few seconds in an ›echo box‹ or rehearsal loop (Baddeley, 1976; Crowder & Morton, 1969; Sperling, 1967), similar encoding should be possible for the relations between the traces of the last one or two items and the signal for the end of the list. This assumption would account for the typically symmetrical form of the serial position curve for 4-item lists in short-term recall (Healy, 1974).

In principle, the same type of encoding could be accomplished for the temporal relations between successive items within a list, but evidently it does not occur when rehearsal is precluded (Lee & Estes, 1977, 1981). This observation is not surprising in view of the fact that the idea of automatic encoding of temporal information (Hasher & Zacks, 1979; Tzeng et al., 1979) has been severely discounted by results indicating that little encoding of this kind occurs in the absence of conditions conducive to selective attention to temporal relations (Jackson, 1985, chapter 12 of the present volume; Michon, 1972; Michon & Jackson, 1984). I suspect that the picture may actually be somewhat mixed. It seems likely that temporal relations between reference points such as the start and end of a list and adjacent items are normally encoded automatically but that relations between items or events within a sequence are only encoded if they are explicitly attended to and rehearsed (Estes, 1972).

The second and third problems, having to do with the occurrence of errors and the course of forgetting, are handled in part by the related proposals of Flexser & Bower (1974) and Hinrichs (1970). In these models, it is assumed that values of ›time tags‹, in the former case, or trace strength, in the latter, are encoded, but as normal distributions of values rather than fixed points on time or strength continua. Readout from these distributions at the time when recall or temporal judgments are called for generates distributions of errors that accord quite well with data (Flexser & Bower, 1974; Hinrichs & Buschke, 1968). The chief limitation of these models is that they do not provide any principled way of accounting for retention loss over time. This deficiency is overcome in the encoding/perturbation model.

The dynamic aspect of the model is that once a representation of an item or event and its temporal position are stored in memory, it is periodically reactivated in a rehearsal mode (which may be either an automatic or a controlled process) along with other events from the same episode or sequence of episodes, and its position is re-encoded on each reactivation. However, the coding process is subject to error. Specifically, it is assumed that during any brief interval of time following the initial encoding of an item, there is some probability that a perturbation of encoded position will occur, and when one does occur, it is equally likely to displace the representation of the item toward the beginning or the end of the list or sequence (Estes, 1972). This assumption is basically the same as one incorporated by Michon (1967) in a model for temporal tracking in skilled motor performance. Michon assumed that immediately following the occurrence of an event in a movement sequence, the interval between it and the preceding event is recorded in memory exactly, but during each subsequent brief interval the remembered duration is incremented or decremented by a unit amount with some probability, p_+ or p_-. The relation between these probabilities is not specified, but the finding that the first of two equal intervals is remembered as longer than the second (Block, 1985, chapter 11 of the present volume), suggests that increments have higher probability than decrements.

The perturbation process generates errors in remembered relative temporal positions of events in much the same way that small irregularities in terrain and behavior cause a group of horses, even though originally well matched for speed, that start a race in a particular order gradually to drift out of the original positions. Over a period of time following presentation of a list of items to a human observer, the temporal information tends to become increasingly fuzzy, the result being the generation of what is termed an *uncertainty gradient*. An illustration of the uncertainty gradients that would be expected after an interval following the presentation of a 12-item list in a short-term memory situation is illustrated in Figure 6. Formal derivations of uncertainty gradients are given in a number of sources (Estes, 1972; Lee & Estes, 1977, 1981), and here I shall simply mention a few of the salient properties. The heart of the process is that as time passes following the encoding of an event in memory, recallable information about its position becomes increasingly uncertain, the uncertainty being describable by a curve, as represented

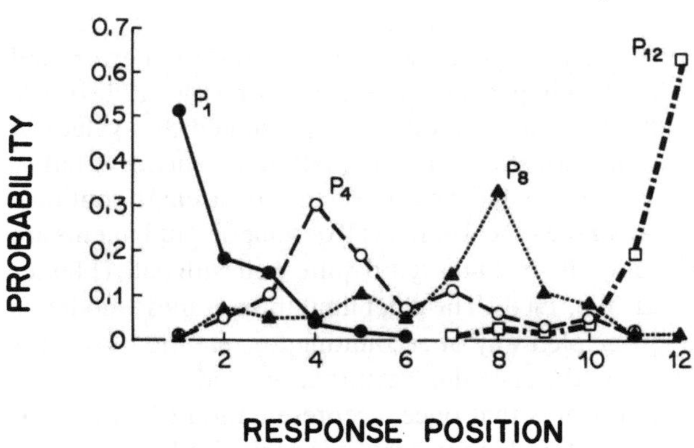

Figure 6. Predicted uncertainty gradients from the encoding/perturbation model for items presented at positions *1, 4, 8* and *12* of a 12-item list. The forms of these gradients may be compared to the observed gradients in Fig. 4.

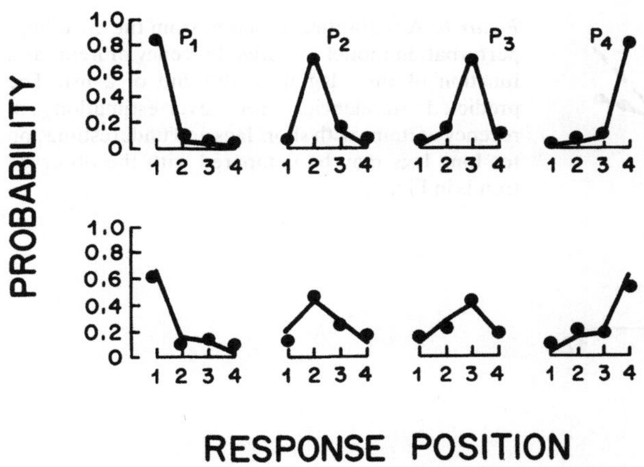

Figure 7. Observed uncertainty gradients for the items of a 4-item list at short and long retention intervals (an interval of 3 item presentations for the upper panels and 18 for the lower panels) together with predictions from the encoding/perturbation model, represented by the connected lines (Data from Healy, 1974).

in Figure 6, plotting the probability at any given point that the event would be recalled in its original position or in other positions earlier or later in time. Gradients near the beginning and end points of the trial are steepest, since uncertainty typically does not spread beyond the trial boundaries, and becomes greatest in the interior of the list. Also, on the average, uncertainty increases more for earlier than for later positions, since for the former there is more time for the perturbation of temporal information to occur before the point of a test following the trial.

Initial applications of the encoding/perturbation model were to short-term recall experiments of the type initiated by Conrad (1964) and Healy (1974), described in a preceding section. The main outcome is that the model generates predicted uncertainty gradients and serial position functions that yield good quantitative accounts of the corresponding empirical functions (Estes, 1972; Lee & Estes, 1977, 1981; Cunningham et al., 1984), a typical result being exhibited in Figure 7.

Implications of the encoding/perturbation model for some of the kinds of studies of temporal judgments discussed in the preceding review are readily obtained by combining the assumptions of the model with the idea of a memory search as developed by Murdock (1972). The idea is that when an individual is called on for an absolute or relative recency judgment, he or she searches mentally through the sequence of memory representations of items of the list until the representations of the indicated item or items are located and responds on the basis of the temporal relations between these and the reference points; thus uncertainty in temporal judgments is the consequence of uncertainty in remembered positions of items following a retention interval. As may be seen in Figure 8, the function

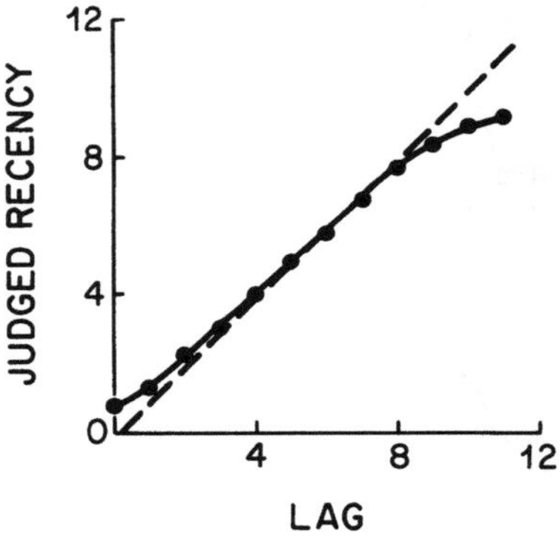

Figure 8. A theoretical function from the encoding/perturbation model for judged recency of items as a function of their lag from the end of a list. The predicted tendencies for over-estimation of recency of items with short lags and underestimation for long lags may be compared with the observed trends in Fig. 1.

predicted by the model for estimates of absolute lag has the characteristic form of observed functions, short lags being overestimated and longer lags underestimated. Referring back to the relevant empirical function in Figure 2, however, we recall that the observed relationship between judged and actual lag between items of a list depends strongly on the similarity of the items involved. The lag between two items is judged with greatest accuracy if the second item is a repetition of the first and with least accuracy if the items are unrelated (Hintzman et al., 1975). My interpretation is similar to that of Tzeng et al. (1979) and goes back to the suggested basis of coding in relations between memory traces and event occurrences. Once an item has been perceived during presentation of a list, a later presentation of the same item or one similar to it is likely to direct attention to the trace of the earlier occurrence, thus providing the basis for encoding of the interval between the two presentations, the encoded duration being a function of trace strength or availability. But if the two items are unrelated, this process would generally be unavailable, and judgment of the interval between them would have to be based on their separate encodings relative to end points of the sequence.

It may be seen in Figure 9 that predictions concerning relative recency judgments also accord with the major results reviewed earlier, the probability that

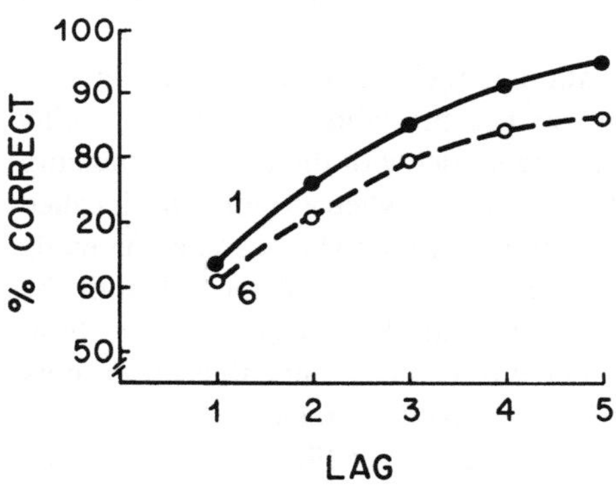

Figure 9. Theoretical functions from the encoding/perturbation model for percentage of relative judgments as a function of the lag between the members of the test pair, the numerals 1 and 6 denoting lag of the more recent item of a pair from the end of the list. The predicted trends may be compared with those of the observed data in Fig. 3.

the more recent member of a pair is more correctly judged to be more recent increasing with the lag between them and decreasing with the lag between the later of the two items and the point of reference at the end of the trial.

An important issue not yet touched on in this review is the role of the frame of reference within which encoding of temporal relations occurs. As indicated above, I think there is adequate reason to believe that memory for temporal aspects of events depends as strongly on selective attention and other control processes as do other forms of memory. Thus, we should expect the encoding of information about temporal intervals to be significantly influenced by the individual's orientation toward the situation or task in which events are experienced. It has been assumed in recent developments of the encoding/perturbation model, that relatively accurate encoding of positional information is possible only if the observer or experimental subject has a prior conception, or schema, of the length of a trial and the way it is demarked from other trials (Lee & Estes, 1981). In short-term recall experiments, it typically is made extremely easy for subjects to form and use trial schemata, and under these circumstances temporal properties of recall data are characteristically well accounted for by the model. In contrast, the conditions under which such phenomena as increasing proportions of correct recency judgments as a function of lag fail to appear are those in which the formation and use of such schemata are precluded by experimental arrangements as, for example, using relatively long lists of items and excluding the early and late items of a list from those entering into recency tests (Hintzman et al., 1975; Underwood, 1977).

Organization in Temporal Memory

To my mind the most significant recent shift of emphasis in theories of temporal memory is the emerging concern with organizational factors. In all of the earlier theories, memory for events or episodes was conceived to consist, in effect, of a linear sequence of representations extending from the present into the past. The salient properties of that conception were captured nicely in the ›conveyor belt‹ model of Murdock (1974), in which the depositing of representations of events in memory was likened to the depositing of items of baggage on a conveyor belt. The ordering of items on the belt corresponds to the ordering of items in memory and the visually perceived size of the items on the belt, diminishing as they move away from the observer, corresponds to the decrease in strength of a memory trace with the passage of time.

This starkly simple conception of memory for temporal aspects of events was in sharp contrast to the rather elaborate hierarchical network models for semantic memory that developed during the 1970s. However, there is reason to believe that memories for events or episodes also takes the form of an organized hierarchical system, with information concerning temporal attributes of events and intervals being incorporated into the representation just as are other kinds of categorical information in semantic networks (Estes, 1982).

The beginnings of hierarchical organization of memory for temporal positions of events can be detected even in short-term recall situations. Thus, a study by Lee & Estes (1981) set subjects the task of recalling a series of items that had been grouped by means of different types of markers into subgroups, recall taking the form either of a full report of the sequence presented on a trial or partial report of a cued subgroup. Substantial evidence was adduced for an interpretation of recall performance in terms of the memory structure illustrated in Figure 10. It is assumed that memory for the sequence of events that occurred on a trial is represented, not by a linear sequence, but rather by a hierarchical network in which the items of a subgroup are grouped together, or chunked, by means of their associations to a common control element and these chunks are in turn associated with a higher order node or control element of the network. Information concerning relative temporal position of the items within a chunk is encoded at the lowest level, information concerning the relative temporal positions of the chunks at the next higher level, and so on; and it is assumed that uncertainty gradients develop at each level just as in the simple version of the encoding/perturbation model discussed earlier.

A number of implications of this model are borne out in the data of Lee & Estes (1981). In one condition of that study, subjects were given maximal advance information about the structure of an experimental trial. The stimuli were two sets of four letters and one set of four digits, always presented sequentially in the order letter set 1, digits, letter set 2, with only the order of items within the sets varying randomly from trial to trial. Highly symmetrical serial position curves were obtained within segments, only the level of the curve varying with temporal remoteness from the recall test. On some trials, the subjects attempted to recall the whole 12-item list, but on other trials they were cued after list presentation to report only one of the three segments. The pattern of relationships between serial position curves for the full and partial report conditions was predictable on the assumption that, given a partial report cue, that subjects could search only the segment-level

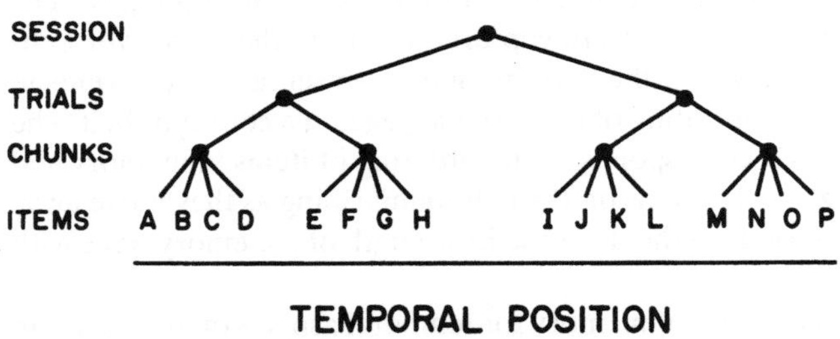

Figure 10. Schematic representation of a hierarchical network representation of memory for items in experimental lists. Clusters of individual items, *A*, *B*, *C*, etc., are associated with nodes in the network that represent chunks of episodic information; these nodes are associated with higher order nodes representing trials, and so on. Uncertainty gradients are assumed to develop at each level during a retention interval so that memory becomes imperfect with regard to the relative positions of items within a chunk, of chunks within a trial, and of trials within a session.

control elements (Figure 10) until the cued group was located, then search within the segment for the individual items.

In a second condition, the stimuli were three sets of four mixed letters and digits. On each trial, the groups were assigned randomly to the first, second, and third segments of the 12-item list, the groups being set off by markers and order of items varying randomly within segments. The data showed the predicted patterns of errors of recalled order of items within groups and also errors resulting from perturbations of the remembered order of the groups. In a third condition, groups of four items were again set off by markers, but individual items were assigned randomly to the 12 list positions on each trial. Nonetheless, subjects were expected to organize their memory representations hierarchically as in the first two conditions. Evidence for the predicted structure was yielded by the observed tendency for an item transposed in recall from the segment in which it was presented to another segment to carry along information about its intra-segment position. That is, the uncertainty function was of the same form, with its mode at the correct relative position for items recalled in their correct segments and items transposed between segments.

Other evidence supporting a hierarchical organization of the kind illustrated in Figure 10 has been reported by Tzeng & Cotton (1980) in a study in which presentations of categorized word lists were followed by relative recency judgments on intra- en inter-category pairs (the former yielding higher accuracy, as predicted by the hierarchical model but not by alternative models considered).

There would seem to be no reason to believe that the hierarchical organization of information about temporal position of events is peculiar to short-term situations. Rather, I conjecture that long-term memory for episodes in an individual's experience and temporal relations among them is similarly represented in a hierarchical network (Estes, 1982). Individual events or episodes would be clustered or chunked on the basis of both temporal proximity and similarity with regard to other attributes. Access to items in the network would be achievable by either sequential search, which however would become slow and uncertain when extended time intervals were involved, or by direct access to higher order control elements by means of retrieval cues related to the semantic or categorical properties of the events involved, followed by search of the segment of memory constituting the particular event representations associated with the accessed control element. This notion of retrieval by means of a combination of direct assess and local search is similar to one that has been put forward with some support by Oliver & Ericsson (Note 1) with regard to performance of actors recalling particular cued segments of script from a play. Little has yet been accomplished toward working out the detailed properties of long-term episodic networks, but it seems reasonable to expect that progress may be achievable by approaches similar to those that have been fruitful in the case of semantic networks.

Finally, it might be noted that the theoretical ideas I have developed in this paper are entirely compatible with the ›equivalence postulate‹ of Michon (1972). The gist of this concept is that temporal attributes of events are processed in the memory system in basically the same way as sensory or other attributes. Given that the assumption of equivalence is sound, we should expect theoretical results from other areas of cognitive research to prove applicable to phenomena of temporal memory, - a motif that has been frequently illustrated in previous sections of this paper. The other side of the coin is that theoretical results arising from the study of temporal memory should prove extendable to other areas. For one example, I have been seeking in current research to test the idea that the perturbation process common to my model of temporal memory (Estes, 1972) and that of Michon (1967) applies similarly to encoding on other attributes. It is possible that the extended model, allowing for concurrent perturbation on multiple attributes of items or events, may provide a replacement for the ›decay theory‹ of memory, which retains intuitive appeal even though it has lapsed from favor among memory investigators during decades of almost exclusive focus on ›interference theory‹. The currently widely held view that memories once stored are never truly lost could be reconciled with the observation that memories almost universally lose availability during intervals of disuse. For, in the perturbation model, memories are not lost, they simply become increasingly imprecise, and hence difficult to retrieve, with the passage of time.

References

Baddeley, A.D. *The psycholoy of memory.* New York: Basic Books, 1976.

Block, R.A. Remembered duration: Effects of event and sequence complexity. *Memory and Cognition*, 1978, *6*, 320-326.

Block, R.A. Contextual coding in memory: Studies of remembered duration. In: J.A. Michon & J.L. Jackson (Eds.), *Time, mind, and behavior.* Heidelberg: SpringerVerlag, 1985, pp. 169-178.

Conrad, R. Acoustic confusions in immediate memory. *British Journal of Psychology*, 1964, *55*, 75-84.

Crowder, R.G., & Morton, J. Precategorical acoustic storage (PAS). *Perception and Psychophysics*, 1969, *5*, 365-373.

Cunningham, T.F., Healy, A.F., & Williams, D.M. Effects of repetition on short-term retention of order information. *Journal of Experimental Psychology: Learning, Memory and Cognition*, 1984, *10*, 575-597.

Drewnowski, A. Attributes and priorities in short-term recall: A new model of memory span. *Journal of Experimental Psychology: General*, 1980, *109*, 208-250.

Estes, W.K. An associative basis for coding and organization in memory. In: A.W. Melton & E. Martin (Eds.), *Coding processes in human memory.* Washington DC: V.H. Winston & Sons, 1972, pp. 161-190.

Estes, W.K. Is human memory obsolete? *American Scientist*, 1980, *68*, 62-69.

Estes, W.K. Multiple coding and processing stages: A review. In: F. Klix, J. Hofmann, & E. van der Meer (Eds.), *Cognitive research in psychology.* Berlin: VEB Deutscher Verlag der Wissenschaften, 1982, pp. 14-21.

Flexser, A.J., & Bower, G.H. How frequency affects recency judgments: A model for recency discrimination. *Journal of Experimental Psychology*, 1974, *103*, 706-716.

Fozard, J.L. Apparent recency of unrelated pictures and nouns presented in the same sequence. *Journal of Experimental Psychology*, 1970, *86*, 137-143.

Fozard, J.L., & Yntema, D.B. The effect of repetition on the apparent recensy of pictures. Paper read at 37th meeting of the Eastern Psychological Association in New York, NY, April, 1966. *American Psychologist*, 1966, *21*, 879 (title only).

Galbraith, R.C. The effects of frequency and recency on judgments of frequency and recency. *American Journal of Psychology,* 1976, *89*, 515-526.

Hasher, L., & Zacks, R.T. Automatic and effortful processes in memory. *Journal of Experimental Psychology: General*, 1979, *108*, 356-388.

Healy, A.F. Separating item from order information in short-term memory. *Journal of Verbal Learning and Verbal Behavior*, 1974, *13*, 644-655.

Hicks, R.E., Miller, G.W., & Kinsbourne, M. Prospective and retrospective judgments of time as a function of amount of information processed. *American Journal of Psychology,* 1976, *89*, 719-730.

Hinrichs, J.V. A two-process memory-strength theory for judgment of recency. *Psychological Review,* 1970, *77*, 223-233.

Hinrichs, J.V., & Buschke, H. Judgment of recency under steady-state conditions. *Journal of Experimental Psychology,* 1968, *78*, 574-579.

Hintzman, D.L., Summers, J.J., & Block, R.A. Spacing judgments as an index of study-phase retrieval. *Journal of Experimental Psychology: Human Learning and Memory,* 1975, *1*, 31-40.

Hoopen, G. ten. The detection of anisochrony in monaural and interaural sequences. In: J.A. Michon & J.L. Jackson (Eds.), *Time, mind, and behavior.* Heidelberg: Springer Verlag, 1985, pp. 140-150.

Jackson, J.L. Is the processing of temporal information automatic or controlled? In: J.A. Michon & J.L. Jackson (Eds.), *Time, mind, and behavior.* Heidelberg: Springer Verlag, 1985, pp. 179-190.

Lassen, G.L., Daniel, T.C., & Bartlett, N.R. Judgment of recency for pictures and words. *Journal of Experimental Psychology,* 1974, *102*, 795-798.

Lee, C.L., & Estes, W.K. Order and position in primary memory for letter strings. *Journal of Verbal Learning* and *Verbal Behavior,* 1977, *16*, 395-418.

Lee, C.L., & Estes, W.K. Item and order information in short-term memory: Evidence for multi-level perturbation processes. *Journal of Experimental Psychology: Human Learning and Memory,* 1981, *7*, 149-169.

Linton, M. Memory for real-world events. In: D.A. Norman & D.E. Rumelhart (Eds.), *Explorations in cognition*. San Francisco: Freeman, 1975, pp. 376-404.

Linton, M. Real world memory after six years: An in vivo study of very long term memory. In: M.M. Gruneberg, P.E. Morris & R.N. Sykes (Eds.), *Practical aspects of memory.* London: Academic Press, 1978, pp. 69-76.

Lockhart, R.S. Recency discrimination predicted from absolute lag judgments. *Perception and Psychophysics*, 1969, *6*, 42-44.

Macar, F. Psychophysics in time. In: J.A. Michon & J.L. Jackson (Eds.), *Time, mind, and behavior.* Heidelberg: Springer Verlag, 1985, pp. 112-130.

Michon, J.A. *Timing in temporal tracking*. Soesterberg: Institute for Perception RVO-TNO, 1967.

Michon, J.A. Processing of temporal information and the cognitive theory of time experience. In: J.T. Fraser, F.C. Haber, & G.H. Müller (Eds.), *The study of time*. Heidelberg: Springer Verlag, 1972, pp. 242-258.

Michon, J.A. The compleat time experiencer. In: J.A. Michon & J.L. Jackson (Eds.), *Time, mind, and behavior.* Heidelberg: Springer Verlag, 1985, pp. 20-53.

Michon, J.A., & Jackson, J.L. Attentional effort and cognitive strategies in the processing of temporal information. In: J. Gibbon & L. Allan (Eds.), *Timing and time perception*. Annals of the New York Academy of Sciences, Vol. 423, 1984, pp. 298-321.

Morton, J. Repeated items and decay in memory. *Psychonomic Science*, 1968, *10*, 219-220.

Murdock, B.B. Jr, Short-term memory. In: G.H. Bower (Ed.), *The psychology of learning and motivation: Advances in research and theory,* Vol. 5. New York: Academic Press, 1972, pp. 67-127.

Murdock, B.B. Jr. *Human memory: theory and data*. Potomac, MD: Lawrence Erlbaum Associates, 1974.

Peterson, L.R., Johnson, S.T., & Coatney, R. The effect of repeated occurrences on judgments of recency. *Journal of Verbal Learning and Verbal Behavior,* 1969, *8*, 591-596.

Shaffer, L.H. Timing in action. In: J.A. Michon & J.L. Jackson (Eds.), *Time, mind, and behavior.* Heidelberg: Springer Verlag, 1985, pp. 226-241.

Shiffrin, R.M., & Cook, J.R. Short-term forgetting of item and order information. *Journal of Verbal Learning and Verbal Behavior*, 1978, *17*, 189-218.
Sperling, G. Successive approximations to a model for short-term memory. *Acta Psychologica*, 1967, *27*, 285-292.
Squire, L.R., Chace, P.M., & Slater, P.C. Assessment of memory for remote events. *Psychological Reports*, 1975, 37, 223-234.
Tversky, A., & Kahneman, D. Availability: A heuristic for judging frequency and probability. *Cognitive Psychology*, 1973, 5, 207-232.
Tzeng, O.J.L., & Cotton, B. A study-phase retrieval model of temporal coding. *Journal of Experimental Psychology: Human Learning and Memory*, 1980, *6*, 705-716.
Tzeng, O.J.L., Lee, A.T., & Wetzel, C.D. Temporal coding in verbal information processing. *Journal of Experimental Psychology: Human Learning and Memory*, 1979, *5*, 52-64.
Underwood, B.J. Attributes of memory. *Psychological Review*, 1969, *76*, 559-773.
Underwood, B.J. *Temporal codes for memories: Issues and problems*. Hillsdale, NJ: Lawrence Erlbaum Associates, 1977.
Underwood, B.J., & Malmi, R.A. An evaluation of measures used in studying temporal codes for words within a list. *Journal of Verbal Learning and Verbal Behavior*, 1978, *17*, 279-293.
Wells, J.E. Strength theory and judgments of recency and frequency. *Journal of Verbal Learning and Verbal Behavior*, 1974, *13*, 378-392.
Yntema, D.B., & Trask, F.P. Recall as a search process. *Journal of Verbal Learning and Verbal Behavior*, 1963, *2*, 65-74.

Reference Note

1. Oliver, W.L. & Ericsson, K.A. *Actors' memory for their parts*. Program of Psychonomic Society 24th Annual Meeting, San Diego, CA, 1983, p. 339.

Chapter 11. Contextual Coding in Memory: Studies of Remembered Duration[1]

Richard A. Block

Two metatheories have dominated psychological studies of memory and of time during this century. Although the historical pattern of adoption of these metatheoretical approaches overlaps somewhat, one of them is clearly older. The older metatheory has roots in the behavioristic and neobehavioristic psychology of the earlier part of this century, but it extends it; branches into some recent information processing psychology. I refer to this as a *stimulus-based* approach, because it emphasizes memory for stimulus events per se. The newer metatheory originates in some cognitive psychology of the later part of this century. I refer to this as a *context-based* approach, because it emphasizes contextual coding in memory.

Consider first the stimulus-based paradigm. It is implicitly based primarily on a mechanistic root metaphor, or ›world hypothesis‹, discussed so well by Pepper (1942), although it also partially relies on what he called a formistic root metaphor. The prototypical example of the adoption of this approach by memory researchers was the verbal learning tradition of the 1940s, 1950s, and early 1960s, which emphasized stimulus-response associations. It was initially assumed that memory could be adequately characterized in terms of associations between relatively meaningless stimuli and relatively meaningless responses. Later, many researchers realized that this framework was insufficient to describe the complexities of human memory.

In the last decade or two, the mechanistic flavor of this approach has receded somewhat, but the emphasis on memory for stimulus information remains, even if it is sometimes tempered by phraseology such as ›stimulus-as-coded‹. Information processing models of human memory, which implicitly or explicitly adopt a computer metaphor, can be seen to be primarily mechanistic, or perhaps a mixture of mechanistic and formistic root metaphors (Hoffman & Nead, 1983; Pepper, 1942). In theorizing on the experience of duration, Ornstein (1969) explicitly relied on a computer metaphor. He referred to the ›storage size‹ in memory taken up by encoded and retrievable stimulus information as determining the remembered duration of a time period. He assumed that if more stimuli occurred during a time period, or if the stimuli were coded in a more complex way, the experience of duration would lengthen. Indeed, Ornstein conducted several cleverly designed experiments that seemed to support this particular embodiment of a stimulus-based metatheory, his ›storage size metaphor‹ of remembered duration. Fraisse (1984)

1 Some of the research described here was work supported by National Science Foundation Grant ISP-8011449.

stated that this is the leading account of effects on the estimation of duration in retrospect, although his view would be questioned by some researchers.

Now consider an alternative conceptualization of memory, one which is context-based. It is derived from contextualism, another root metaphor discussed by Pepper (1942). One of the earliest proponents of this approach among memory theorists was Jenkins (1974), who concluded: »What is remembered in a given situation depends on the physical and psychological context in which the event was experienced, the knowledge and skills that the subject brings to the context, the situation in which we ask for evidence for remembering, and the relation of what the subject remembers to what the experimenter demands« (p. 793). In a somewhat similar way, Estes (1980) asserted that human short-term memory is »oriented toward events and their attributes rather than toward the retention of items as units« (p. 65). A contextualist approach, such as that advocated by Jenkins and a few other cognitive scientists, divides the world into two basic ontological categories – events and changes. Contextualism emphasizes that change and novelty are inherent in observer-environment interactions and that every event has a unique quality and texture. It rejects the notion of permanent structures; instead, it emphasizes the »salient or important aspects of events« that depend upon »the observer's particular purposes« (Hoffman & Nead, 1983, p. 519).

In the past I have not adopted a ›pure‹ contextualist approach to remembered duration, but rather a combination of mechanism and contextualism. I wish to explore a more purely contextualist approach in the present selective review and report of my research. To do so entails a rejection of the mechanistic assumption of relatively static memory structures, such as those that comprise Ornstein's (1969) ›storage size‹. Instead, dynamic terms must be used to describe the encoding of contextual information and its use in judging the duration of a time period in retrospect.

Review of Previous Studies

A selective review of some of my previous studies on remembered duration is needed as background for the present study. These studies are all characterized by a common methodology. Events occurred during each of two moderately long periods of equal duration (between 54 and 280 s). Then subjects unexpectedly were asked to judge the relative duration of the two time periods. Half of the subjects judged the first time period relative to the second, and the other half judged the second time period relative to the first (see Ornstein, 1969). Table 1 presents a summary of these studies, including a characterization of independent variables as well as main effects[2]. All data reported here are expressed in terms of the ratio of the apparent duration of the first time period (D_1) to that of the second time period (D_2), even though a different ratio was used in some of the original articles.

2 The analysis of main and residual (i.e., interaction) effects is based on the methods which have recently been described clearly by Rosenthal & Rosnow (1984, Chapter 21).

Table 1. Summary of main effects in previous studies

	Description of treatment	Treatment effect[a]	Time-Order effect[b]
Block (1978)			
Experiment 1	Complex (C) vs. simple (S) stimuli	$C \approx S \pm 0.00$	$D_1 > D_2 + 0.21$*
Experiment 2	Complex (C) vs. simple (S) sequence	$C > S \pm 0.11$*	$D_1 > D_2 + 0.14$*
Block & Reed (1978)			
Experiment 1	Deep (D) vs. shallow (S) processing	$D \approx S \pm 0.01$	$D_1 > D_2 + 0.12$*
Experiment 2	Mixed (M) vs. unmixed (U) processing	$M > U \pm 0.09$*	$D_1 > D_2 + 0.15$*
Block (1982)			
Experiment 1	Disruption only		$D_1 > D_2 + 0.12$*
	Disruption & change		$D_1 \approx D_2 + 0.03$
Experiment 2	No Disruption		$D_1 > D_2 + 0.19$*
	Disruption only		$D_1 > D_2 + 0.07$*
	Disruption & change		$D_1 \approx D_2 - 0.04$
Experiment 3	Mixed (M) vs. unmixed (U) processing		
	Disruption only	$M > U \pm 0.09$*	$D_1 \approx D_2 + 0.05$
	Disruption & change	$M \approx U \pm 0.01$	$D_1 \approx D_2 + 0.04$

Note. Significant effects are indicated by an asterisk (*). [a]Treatment effects are shown as plus or minus the amount by which the mean is each condition differed from the overall mean. An effect of ± 0.11, for example, indicates that *remembered duration* was 22 percent longer in the first condition listed than in the second condition listed. [b]*Time-order effects* are shown as the amount by which the mean D_1/D_2 judgment ratio differed from 1.00 (the theoretical mean).

Event and Sequence Complexity

Two experiments on effects of complexity (Block, 1978) gave me the first indication that Ornstein's (1969) stimulus-based storage size metaphor was fundamentally incorrect. In the first experiment, the complexity of individual stimuli did not affect remembered duration, as would be predicted by Ornstein's hypothesis. In the second experiment, a more complex *sequence* of stimuli was remembered as being about 22 percent longer in duration than a less complex sequence. The nature of the entire sequence of events – the ›texture‹ of the whole experience – seemed to be much more important than the nature of the individual stimuli.

In addition, a positive time-order effect was observed in both experiments: with all other factors equal or controlled by counterbalancing, the first duration was remembered as being longer than the second duration. Discussion of this important effect is postponed until a later section.

Levels of Processing and Changes in Processing

The results of another study (Block & Reed, 1978) also supported a context-based, rather than a stimulus-based, account. In the first experiment, subjects spent a 64 s duration processing 32 words at either a *shallow*, structural level or a *deep*, semantic level. Then they performed the other kind of processing during a second 64 s duration. The remembered duration of the deep processing task was not greater than that of the shallow processing task. However, recognition of the stimulus words was considerably greater for those that were deeply processed. Thus, a measure of the ›storage size‹ of stimulus information »remaining in storage« showed that more stimuli were retained as a result of the deeper, semantic processing. Ornstein's (1969) hypothesis –as well as other similar hypotheses like those of Block (1974) and Fraisse (1984) – simply cannot explain these data.

In order to clarify these findings, we conducted a further experiment. Subjects either performed structural or semantic processing (as in Experiment 1) or alternated between the two kinds of processing during the first of two 80 s durations; they performed the other kind of task during the second time period. The duration containing alternating, or *mixed*, kinds of processing was remembered as being about 18 percent longer than the *unmixed* duration, regardless of whether the unmixed processing was structural or semantic. Again, a stimulus-based storage-size hypothesis would not predict and cannot easily explain this finding, because later recognition of words from the mixed task was intermediate between recognition of words from the shallow processing task and recognition of words from the deep processing task.

These findings seemed to support a *contextual-change hypothesis*, a more dynamic explanation which asserts that remembered duration is based on »memory for the overall change in cognitive context« during a time period (Block & Reed, 1978, p. 664; see also Block, 1979). In these experiments, a major kind of contextual change was assumed to be produced by variations in process context; that is, because the mixed-processing task required different kinds of cognitive processes, dynamic changes in the internal context resulted (see B.J. Underwood, 1977). The quality of the mental events that occurred during the whole duration apparently caused changes in memory in a fairly direct way, perhaps as a byproduct of the particular processing of stimulus information. It is on this quality – what we referred to as the amount of contextual change rather than on memory for stimulus information *per se* that subjects apparently rely in order to judge the duration of a time period in retrospect. If this is the case, judgments of duration may be expected to play a central role in any contextualist account of memory: »Retrospective judgment of duration may serve... as an index of the overall amount of change in cognitive context« during a time period (Block & Reed, 1978, p. 665).

Changes in Environmental Context

Three more recent experiments investigated environmental context as another potential source of contextual changes (Block, 1982). In the first two experiments, 15 stimulus words were processed in the same way during each of two 150 s durations. Thus, both the number of words and the complexity of coding of the words were held constant. The experimental manipulation simply concerned the room, or environmental context, in which the two tasks were performed. If subjects spent the second time period in a room which was different from that in which they spent the first time period – that is, if the environmental context was both disrupted and changed – the remembered duration of the second time period was relatively longer (i.e., about 23 percent longer than it would have been judged had the environmental context not been changed). The remembered duration of the second time period was also slightly lengthened if subjects simply were asked to leave the room (and then to return to the same room) in between the two time periods – that is, if there was simply a disruption of context between the two durations. Thus, the remembered duration of the second time period was relatively longer to the extent that the encoding of the environmental context during the second time period was assumed to be changed in some way. As before, these experiments reveal effects on remembered duration that simply cannot be explained by a stimulus-based hypothesis, such as Ornstein's (1969) storage size hypothesis or Fraisse's (1984) adaptation of it.

A third experiment explored the possibility of interacting effects of environmental- and process-context changes. The environmental context manipulation was the same as that in the first two experiments; the process context manipulation was similar to that used by Block and Reed (1978, Experiment 2), which was discussed earlier. This experiment revealed a main effect of process context which replicated that found in the earlier experiment.

More importantly, there was an interaction of the two kinds of contextual factors: If there was no change in environmental context between the two durations, the effect of process context was substantial; but if the environmental context was changed, the effect of process context was eliminated. One possible explanation is that the changes in environmental context were more salient aspects of the experimental situation than the changes in process context. In other words, different kinds of contextual factors do not necessarily produce additive effects on remembered duration, so that the person's subjective reaction to the quality – the »total meaning« (Jenkins, 1974, p. 786) – of the situation is of critical importance.

Results of Present Study

The results of four recent experiments (Block, Note 1) enable us to distinguish between two versions of a contextual-change hypothesis. The findings also rule out any sort of explanation in terms of the storage size of stimulus information. In these experiments, subjects performed a different kind of imagery task involving the 32 words presented during each of two equal durations between 224 and 280 s in length. One task, the *environmental imagery* task, was assumed to require the deliberate encoding of environmental stimuli. It involved imagining the referent of each presented word interacting in some way with a unique object or location in the room. The other task, the *internal imagery* task, was assumed to restrict the deliberate encoding of environmental stimuli. It involved imagining the referent of each word either in a single location (Experiment 1) or interacting in an internal image with the referent of the preceding word (Experiment 2). Table 2 shows that the results in either case were the same: The internal imagery task was remembered as about 18 percent longer in duration than the environmental imagery task.

Two additional experiments (Experiments 3 and 4) clarify the possible cause of this difference. Before performing the imagery tasks, subjects either wrote a description of the experimental room or a description of a familiar room, relying on internal imagery. The combined results shown in Table 3 reveal that the main effect of imagery task on remembered duration was replicated. More importantly, there was an interaction between the preceding description condition and the subsequent imagery task. Simply stated, the kind of cognitive processing assumed to be required in order to perform the preceding description task caused a relative lengthening of the remembered duration of the imagery task which did require a *different* kind of cognitive processing.

Thus, the results of Experiments 1 and 2 rule out a contextual explanation in terms of the encoding of varied environmental associations, because the environmental imagery task was remembered as being shorter in duration than the internal imagery task. The results of Experiments 3 and 4 suggest another kind of contextual explanation: the factor that is critical in determining remembered duration in this situation involves the more holistic changes in process context that resulted from the performance of an imagery task.

Table 2. Summary tables of means and effects in present study, experiments 1 and 2

Experiment	Imagery-task order		
	Environmental-internal	Internal-environmental	Average
	Table of means		
Experiment 1	1.04	1.22	1.13
Experiment 2	1.02	1.20	1.11
	Table of effects		
Experiment 1	− 0.09*	+ 0.09*	+ 0.13*
Experiment 2	− 0.09*	+ 0.09*	+ 0.11*

Note. Significant effects are indicated by an asterisk (*). The overall means (column at right) reveal the time-order effect. Data are from Block (Note 1).

Table 3. Summary tables of means and effects in present study, experiments 3 and 4 (combined data)

Description task	Imagery task order		
	Environmental-internal	Internal-environmental	Average
	Table of means		
Environmental-imagery	0.94	1.18	1.06
Internal-imagery	1.05	1.05	1.05
Average	1.00	1.11	1.05
	Table of effects		
Environmental-imagery	− 0.06*	+ 0.06*	+ 0.00
Internal-imagery	+ 0.06*	− 0.06*	+ 0.00
Average	− 0.06*	+ 0.06*	+ 0.05

Note. Significant effects are indicated by an asterisk (*). The grand average (1.05, or an effect of + 0.05) reveals the time-order effect. Data are from Block (Note 1).

A subsequent free recall test revealed that the task requiring environmental imagery produced a higher level of recall of words than did the task requiring internal imagery. This is additional evidence rejecting a storage-size hypothesis.

Time-Order Effects

The results of some of these recent experiments (Block, 1982, Note 1) also reveal the likely origin of time-order effects in retrospective duration judgments of relatively long time periods. Contrary to what is implied by Ornstein (1969), a positive time-order effect was found in all of the experiments discussed here. With all other factors equal or controlled by counterbalancing, the first of two equal time periods is remembered as being 12 percent to 21 percent longer in duration than the second time period. This reliable finding of a positive time-order effect has played an important role in the present contextualist account, as well as in exploring effects of various contextual factors. The positive time-order effect is eliminated if the environmental context prevailing during the second of two durations is changed (Block, 1982; see Table 1) or if the changes in emotional context that might ordinarily occur during the first duration do, instead, occur during an experimental task that precedes it (Block, Note 1; see Table 3). Thus, the finding of a positive time-order effect in previous experiments is explained in terms of the greater contextual changes that ordinarily occur during the first duration.

A Contextualist Model of Temporal Experience

The findings of the studies reviewed here, especially the interaction effects in the later studies (Block, 1982, Experiment 3; Block, Note 1), suggest that the remembered duration of a time period is a product of the combined influence of many factors. Jenkins (1979) proposed a general contextualistic model of memory, which he applied specifically to research on levels of processing. Bransford (1979, pp. 6-9) presented a similar model of »learning, understanding, and remembering«. Figure 1 shows an adaptation of this general model to the understanding and explanation of interacting factors that affect temporal experience. Each of the four vertices of the tetrahedron represents a cluster of factors that are commonly found to affect temporal experiences. (The studies reviewed here involved manipulations of variables on all four vertices). Each of the six edges of the tetrahedron represents a two-way interaction of different kinds of factors, and each of the four planes represents a three-way interaction. (In the research described here, two- and three-way interactions were routinely found.) Finally, the whole, solid tetrahedron represents the complex four-way interaction of the different kinds of variables. Just

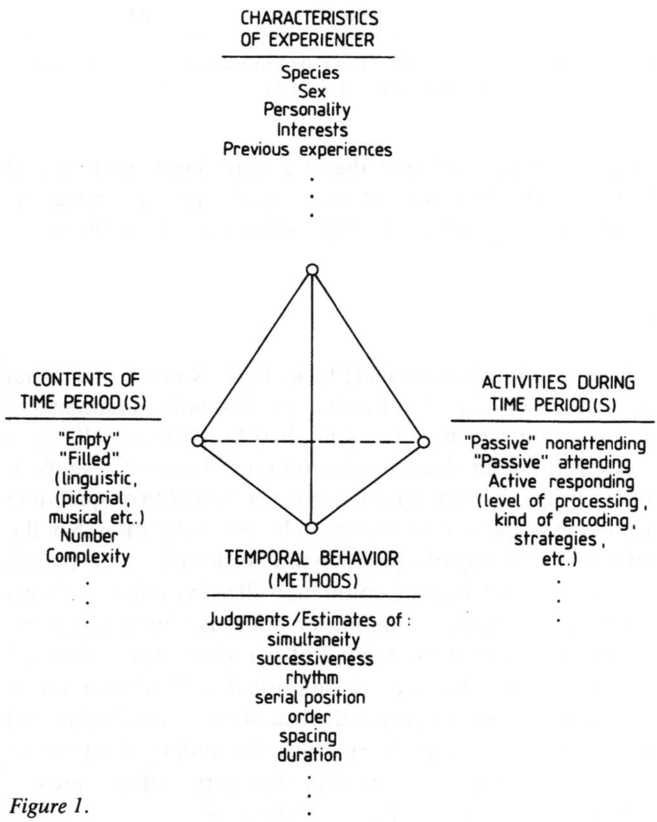

Figure 1.

as such a contextualistic model seems to be necessary to capture the complex interactions of different factors in the research reported here, it also seems to be helpful in understanding the research reported elsewhere in this volume.

Summary and Conclusions

All of the findings discussed here are consistent with a context-based approach, but strongly reject a stimulus-based approach like that advocated by Ornstein (1969) and Fraisse (1984). Other investigators (e.g. Poynter, 1983) using somewhat different methods have arrived at a similar conclusion. In addition to being able to explain diverse findings, a contextual hypothesis can also parsimoniously subsume other kinds of explanations, such as Vroon's (1970) informational hypothesis and G. Underwood's (1975) attentional hypothesis (see Block, 1979). In experiments directly comparing predictions of these rival hypotheses and those of a contextual hypothesis, the results have supported the latter. In contrast to other hypotheses, a contextual hypothesis may also be able to explain effects on the experience of duration in passing, or prospective duration experience (Block, 1979; Block et al., 1980). Finally, a contextualistic approach to remembered duration is consistent with current theories of memory dynamics.

References

Block, R.A. Memory and the experience of duration in retrospect. *Memory & Cognition*, 1974, *2*, 153-160.

Block, R.A. Remembered duration: Effects of event and sequence complexity. *Memory & Cognition*, 1978, *6*, 320-326.

Block, R.A. Time and consciousness. In: G. Underwood & R. Stevens (Eds.), *Aspects of consciousness: Vol. 1. Psychological issues*. London: Academic Press, 1979, pp. 179-217.

Block, R.A. Temporal judgments and contextual change. *Journal of Experimental Psychology: Learning, Memory, and Cognition*, 1982, *8*, 530-544.

Block, R.A., George, E.J., & Reed, M.A. A watched pot sometimes boils: A study of duration experience. *Acta Psychologica*, 1980, *46*, 81-94.

Block, R.A., & Reed, M.A. Remembered duration: Evidence for a contextual-change hypothesis. *Journal of Experimental Psychology: Human Learning and Memory*, 1978, *4*, 656-665.

Bransford, J.D. *Human cognition: Learning, understanding, and remembering*. Belmont, CA: Wadsworth, 1979.

Estes, W.K. Is human memory obsolete? *American Scientist*, 1980, *68*, 62-69.

Fraisse, P. Perception and estimation of time. *Annual Review of Psychology*, 1984, *35*, 1-36.

Hoffman, R.R., & Nead, J.M. General contextualism, ecological science and cognitive research. *Journal of Mind and Behavior*, 1983, *4*, 507-560.

Jenkins, J.J. Remember that old theory of memory? Well, forget it! *American Psychologist*, 1974, *29*, 785-795.

Jenkins, J.J. Four points to remember: A tetrahedral model of memory experiments. In: L.S. Cermak & F.I.M. Craik (Eds.), *Levels of processing and human memory*. Hillsdale, NJ: Lawrence Erlbaum Associates, 1979, pp. 429-446.

Ornstein, R.E. *On the experience of time*. Harmondsworth, England: Penguin Books, 1969.

Pepper, S.C. *World hypotheses*. Berkeley, CA: University of California Press, 1942.
Poynter, W.D. Duration judgment and the segmentation of experience. *Memory & Cognition*, 1983, *11*, 77-82.
Rosenthal, R., & Rosnow, R. L. *Essentials of behavioral research: Methods and data analysis*. New York: McGraw-Hill, 1984.
Underwood, B.J. *Temporal codes for memories: Issues and problems*. Hillsdale, NJ: Lawrence Erlbaum Associates, 1977.
Underwood, G. Attention and the perception of duration during encoding and retrieval. *Perception*, 1975, *4*, 291-296.
Vroon, P.A. Effects of presented and processed information on duration experience. *Acta Psychologica*, 1970, *34*, 115-121.

Reference Note

1. Block, R.A. *Remembered duration: Imagery processes and contextual encoding*. Manuscript in preparation, 1984.

Chapter 12. Is the Processing of Temporal Information Automatic or Controlled?[1]

Janet L. Jackson

Introduction

It has been suggested that temporal properties of events are functionally identical to other stimulus properties (e.g. Michon, 1972; Underwood, 1977; Estes, 1980). Cognitive processing of temporal properties, such as encoding and remembering relative or absolute positions and durations, can therefore be viewed as being essentially equivalent to the processing of other stimulus elements.

An increasing amount of recent experimental literature has demonstrated that temporal coding does occur (e.g. Yntema & Trask, 1963; Zimmerman & Underwood, 1968; Hinrichs, 1970; Block & Reed, 1978; Tzeng et al., 1979). Such results have been cited by Hasher & Zacks (1979) as strong evidence for their view that temporal attributes of events are encoded automatically into memory. They describe automatic encoding as being unintentional and as absorbing little or no capacity. This type of processing is in contrast to effortful operations, such as rehearsal and elaborative mnemonic activities, which do demand attentional capacities.

Hasher & Zacks specified several criteria for determining through which information processing mode a certain attribute is actually encoded. They argued that varying instructions, level of practice, state variables, divided attention requirements and developmental trends ought not to affect task performance if the information is processed in the automatic mode.

Some Experimental Results

Using a line of reasoning similar to that of Hasher & Zacks we have, in a recent series of experiments, come to a somewhat different, less straightforward, conclusion. I shall first consider a selection of these experiments which explore performance on order, lag and position judgment tasks.

(1) If temporal information is processed automatically, similar temporal judgments should be expected for words that have different semantic characteristics. The concrete-abstract distinction was explored in two experiments (Jackson & Michon, 1984) which employed a directed-forgetting paradigm (Bjork, 1972) in which half of the words presented were followed by a cue to remember (R-cue) and

1 This research was supported by a grant from the Netherlands Organization for the Advancement of Pure Research under project number 15-23-15.

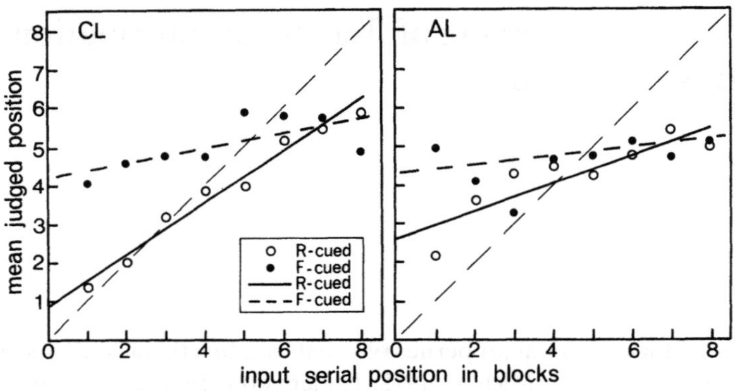

Figure 1. Mean judged position for the R-cued and the F-cued words as a function of their actual input serial position (in blocks); within the concrete *(CL)* and abstract *(AL)* word lists. (R = to-be-remembered word; F = to-be-forgotten words).

the other half by a cue to forget (F-cue). Results, summarized in Figure 1 show that temporal order retention is higher with concrete R-cued than with abstract R-cued lists.

(2) In two experiments exploring the effect of incidental versus intentional acquisition of temporal order information for both concrete (CL) and abstract (AL) word lists (Jackson, Note 2), the results shown in Figure 2 suggest that the type of instruction, that is, recognition (RI) or temporal position (TPI) instructions does indeed appear to affect temporal position recall. This effect is particularly clear with abstract word lists.

(3) A further way of examining the performance of temporal judgment tasks following an incidental learning task was carried out in an experiment using a levels-of-processing approach (Craik & Lockhart, 1972). We used either *shallow* (physical structure) level or *deep* (semantic) level tasks (see Block, 1985, chapter 11 of the present volume) and three types of judgment task (order, lag and position). Comparing temporal judgment tasks is not easy since the questions such tasks normally ask yield very different types of data. To allow us to compare performances we therefore chose to frame the questions in a novel manner – namely by turning all three types of judgment task into binary decision tasks. Care was taken to balance separation and position of test items. A more detailed explanation of the procedure will be found in Jackson (Note 2). Examples of the tasks and types of question asked are shown in Table 1. The use of binary questions allowed us to test performance in all conditions against a random guessing level of performance. The results, presented in Table 2, show that subjects who performed the physical structure orienting tasks were very poor indeed in making temporal judgments. In no instance did their performance exceed a random guessing level. Such guessing behavior was also evident in the lag judgments made by subjects after the semantic orienting task. Performance in the other two judgment tasks following this orienting

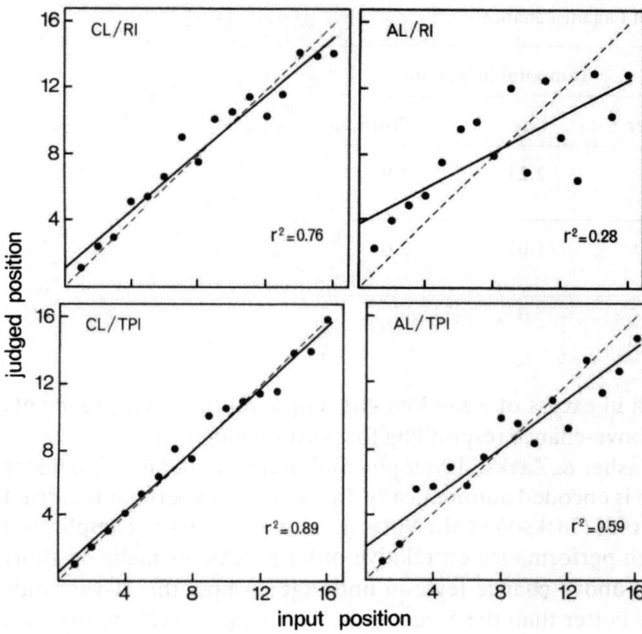

Figure 2. Mean judged position as a function of actual input serial position; within the concrete *(CL)* and abstract *(AL)* word lists. (*RI* = recognition instructions; *TPI* = temporal position instructions).

Table 1. Examples of questions and tasks used in the levels-of-processing experiment

Level of processing	Orienting questions	Correct response Yes	No
(1) Physical structure	Is there an a in the word?	hand	tree
(2) Semantic features	Would the word fit the sentence: »A ... has a tail«?	horse	flag
Type of Judgment task	**Test questions**	**Binary**	**choice**
(1) Order judgments	Which of these two words appeared earlier in the original series?	b or	a
(2) Lag judgments	How many words appeared between these two in the original series?	8 or	14
(3) Position judgments	What was the exact position of this word in the original series?	11 or	7

Table 2. Observed values of t against chance

	Temporal judgments		
Condition	Order	Lag	Position
physical structure	1.52	2.21	1.93
semantic feature	3.05*	1.07	2.67

* Significant at 5% level.

task was, however, well in excess of a random guessing level for order judgments and was approaching above-chance responding for position judgments.

(4) According to Hasher & Zacks, developmental trends should not be evident if temporal information is encoded automatically. In a series of experiments carried out with 5- and 11-year olds (Jackson et al., Note 3), a somewhat more complicated picture emerges: though performance on relative order judgments made on short lists (7 items) was well above chance level in both age groups, the 11-year olds performed significantly better than the 5-year olds. With long lists (28 items) both age groups were again above chance but the developmental effect disappeared. In a second experiment the temporal task selected involved judging an item's position in a serial list. With this task, significant differences were found between the performances of 5- and 11-year olds, the latter group always producing more correct judgments in short as well as long lists. Such significant developmental effects argue against the claim that all temporal information is automatically encoded in memory. That the improvement in performance arises from deliberate processing, is suggested by the effects of induced rehearsal: when young children are encouraged to adopt such rehearsal strategies their performance, though still not matching that of older children, does improve. When long lists are presented, however, this improvement disappears (see Figure 3). Though no effects are found with older

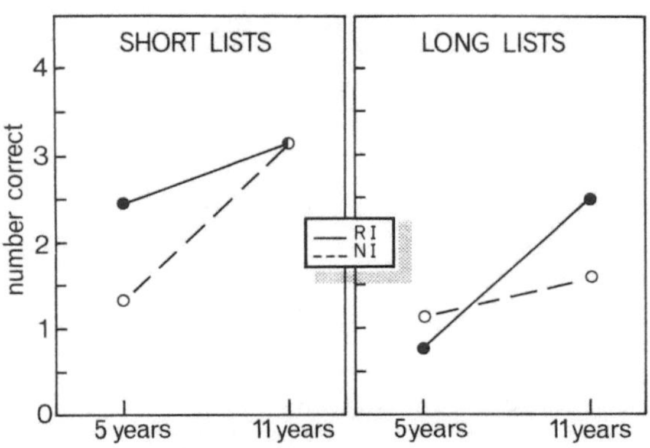

Figure 3. Average correct responses as a function of age and instruction (*RI* = Rehearsal Instructions; *NI* = No Rehearsal Instructions).

children given short lists, an improvement in performance occurs as a result of rehearsal instructions when long lists are presented (see Figure 3). This result suggests that although 11-year olds may spontaneously rehearse, such rehearsal strategies may be far from optimal. By stressing the use of particular rehearsal procedures, these 11-year olds may have been encouraged to develop a more efficient, task-related strategy.

This sample of experiments suggests that conclusions such as those drawn by Hasher & Zacks (1979) are too simple. Subjects who were asked to perform shallow tasks at the level of physical structure, which probably only involve scanning procedures, were very poor at making all types of temporal judgments. This suggests that it is necessary for complete items to appear in an input buffer before satisfactory temporal judgments can be made. Such a result is basically in agreement with the hypothesis put forward by Tzeng et al. (1979). These authors were among the first to use the principle of intrinsic order in a process model of temporal information coding. Their results suggested that being present in a rehearsal buffer may indeed be a necessary condition for temporal coding to take place, allowing old-new relations to become established. We have shown, however, that recalled concrete items produce better temporal order judgments than do recalled abstract items. We must conclude therefore that although being present in an output buffer may be a necessary condition to establish efficient temporal order coding, it is not a sufficient condition. Something extra, possibly related to selective attentional demands or to the processing strategies used by subjects (see Michon & Jackson, 1984) does also play a role in making temporal judgments. Doubts also arise when we consider the results of subjects who performed temporal judgment tasks following ›deep‹, semantic level tasks. Though words in such conditions can be assumed to enter some sort of buffer presumably resulting in the establishment of old-new relations, this relationship was not sufficient to ensure above-chance performance on all temporal judgment tasks. In fact, only relative order judgments reached this level. Michon & Jackson (1984) and Underwood & Malmi (1978) have also suggested that temporal judgment tasks do not all imply the same level of difficulty: the ability of subjects to make correct judgments depends on the questions asked. Though relative order judgments may be performed well, irrespective of age or instruction and may indeed reflect some automatic encoding of intrinsic order, such coding is not sufficient to allow subjects to perform more complex temporal judgment tasks adequately. We have found that, while position judgments present difficulties for some of our subjects, lag judgments for such unrelated items are performed very poorly by all. In chapter 10 of the present volume Estes (1985) has suggested that such lag judgments would have to be based on the separate encoding of each item relative to end points of the sequence. Our results suggest that such judgments require deliberate processing.

Individual Differences

So far, in the experiments we have been discussing, we have concentrated our attention on group data. Although this approach has successfully identified many interesting variables, it has also tended to obscure individual differences in processing. An examination of individual results showed subjects to be behaving in quite varied, unexpected ways. To tap the source of these differences we had to choose between two methods. One was to try to give more explicit and precise instructions, taking deliberate measures to encourage subjects to stay with one strategy and the second was to make use of verbal protocols. Since we were not at that time able to predict the efficiency of particular strategies when temporal judgments are required, the second solution offered the better chance of gaining more insight into both the number and nature of available strategies and into the likelihood of success which comes from using such strategies.

The first experiment was an in-depth study of 4 subjects. They were trained to verbalize concurrently with the task, both in the encoding/storage phase and the retrieval/test phase. With such instructions to verbalize a direct trace of the information in working memory is obtained, and hence indirect evidence for the internal cognitive processing steps. The results of the protocol analysis have been summarized in Table 3, which shows the various strategies adopted by our subjects. (See Michon & Jackson, 1984 for a more detailed description). One general finding from our results suggests that the exclusive use of simple rehearsal strategies does not yield correct judgments of temporal position. If we look at the performance of Subject 1 shown in Figure 4, a subject who made extensive use of simple rehearsal,

Table 3. Summary of coding strategies obtained from protocol analysis

Objective Organization	
1. Physical characteristics	
2. Functional and/or category associations	
Subjective Organization	
A. Simple rehearsal	B. Elaborative rehearsal
3. Repetition	7. Number pegs
a – single words	8. Combining
b – blocks of words	a – 2 words into 1 word
c – corresponding number	b – 2 words into 1 sentence
4. First letter of word	9. Story
5. Before – After	a – elaboration of a word
*6. Begin – Middle – End	b – one connected
	c – two unconnected
	d – several unconnected
	*10. Visual interacting

* Strategies frequently found in retrieval protocols but never in coding protocols.

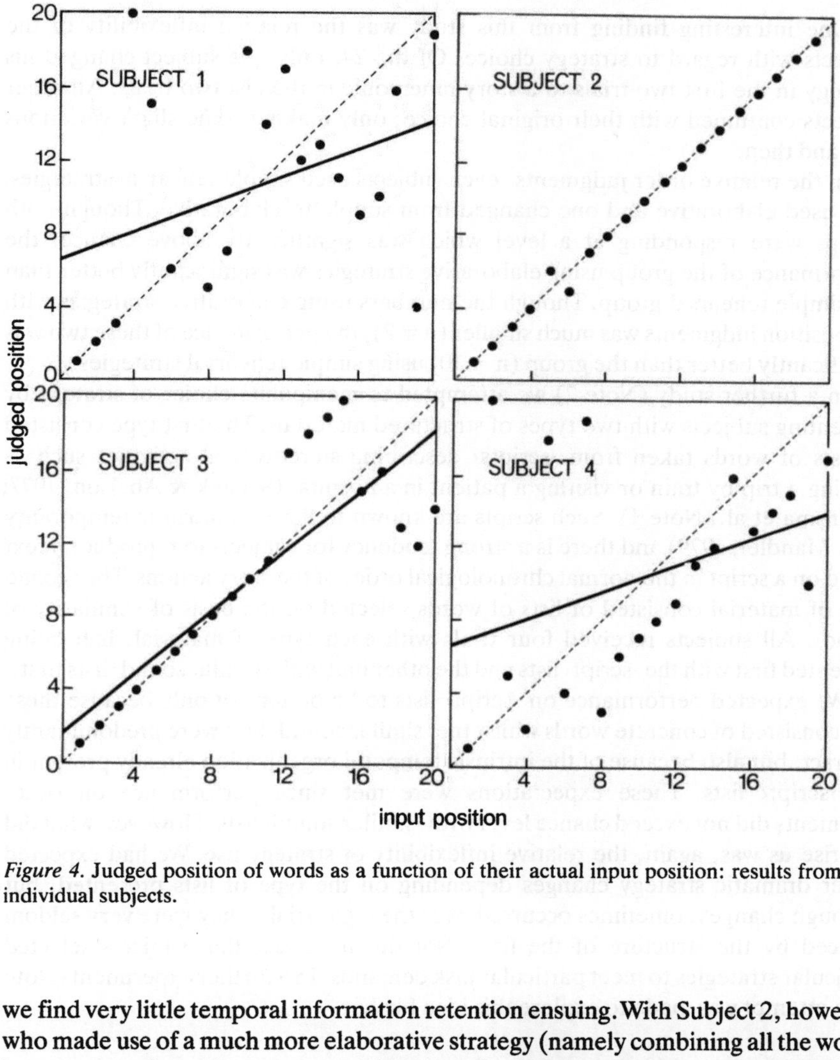

Figure 4. Judged position of words as a function of their actual input position: results from two individual subjects.

we find very little temporal information retention ensuing. With Subject 2, however, who made use of a much more elaborative strategy (namely combining all the words into one story), performance was almost perfect. Though Subject 1 persisted in his use of simple rehearsal strategies, the other three subjects all developed fairly elaborate strategies.

A further study using verbal protocols was then run to examine both the efficacy of the coding system we had devised from our in-depth study, and also whether different strategies did indeed produce varying levels of performance. Twenty-four language students were asked to participate, 12 performing four series of relative order judgments and the other 12 four series of exact position judgments. The reasoning behind the selection of such a student population sample was the hope that, since learning lists of words is for them a fairly common occurrence, a fair number of them would be found to engage in simple rehearsal strategies.

One interesting finding from this study was the relative inflexibility of the subjects with regard to strategy choice. Of the 24, only one subject changed his strategy in the first two trials to a story mnemonic in the last two trials. All other subjects continued with their original choice, only making some slight variations now and then.

In the relative order judgments seven subjects used simple rehearsal strategies, four used elaborative and one changed from simple to elaborative. Though both groups were responding at a level which was significantly above chance, the performance of the group using elaborative strategies was significantly better than the simple rehearsal group. Though the numbers using elaborative strategies with the position judgments was much smaller (n = 2), the performance of these two was significantly better than the group (n = 10) using simple rehearsal strategies.

In a further study (Note 2) we attempted to manipulate choice of strategy by presenting subjects with two types of structured materials. The first type consisted of lists of words taken from ›scripts‹ describing stereotypical activities such as making a trip by train or visiting a patient in a hospital (Schank & Abelson, 1977; Bruinsma et al., Note 1). Such scripts are known to have an intrinsic temporality (e.g. Mandler, 1979) and there is a strong tendency for subjects to reproduce a text based on a script in the normal chronological order of the story actions. The second type of material consisted of lists of words selected on the basis of ›similarity of sound‹. All subjects received four trials with each type of material, half being presented first with the ›script‹ lists and the other half with ›similar sound‹ lists first.

We expected performance on ›script‹ lists to be better not only because these lists consisted of concrete words whilst the ›similar sound‹ lists were predominantly abstract, but also because of the intrinsic temporal organization already present in the ›script‹ lists. These expectations were met since performance on order judgments did not exceed chance level with ›similar sound‹ lists. However, what did surprise us was, again, the relative inflexibility of strategy use. We had expected rather dramatic strategy changes depending on the type of lists presented, but although changes sometimes occurred over the eight trials, they were very seldom induced by the structure of the lists. Nor did it appear that subjects selected particular strategies to meet particular task demands. In a further experiment (Note 2) an attempt was made to explore this idea further.

Table 4. Design of Experiment exploring the effect of an unexpected change of temporal task on the fifth trial

Number of subjects	Judgment task on trials 1 – 4	Judgment task on trials 5 and 6
10	Position	Order
10	Lag	Order
10	Order	Lag
10	Position	Lag
10	Lag	Position
10	Order	Position

All 60 subjects in this experiment were asked to make a series of six sets of temporal judgments, the main manipulation being an unexpected change of task on the fifth trial (see Table 4 for experimental design).

On the basis of their verbal protocols, the results of subjects in each condition were split into two groups those using a simple rehearsal strategy and those using a more elaborative strategy. In the order as well as the position judgment condition, 6 subjects employed elaborative strategies and 14 simple rehearsal; while in the lag judgment condition 5 subjects employed elaborative strategies and 15 simple rehearsal.

For all three temporal tasks choice of strategy had a very significant effect on performance with subjects who used an elaborative strategy always performing better. These results are summarized in Figure 5. For the order judgments performance was well above a chance level for all subjects, independently of strategy choice and there was no effect of practice. These results strengthen our view that order judgments may indeed be less dependent on deliberate processing. Although an analysis of variance revealed no significant practice effect with lag judgments, Figure 5 reveals with elaborative strategies there is certainly a tendency for improvement with practice. Only in trial 4 was there a lack of improvement. This can in part be traced back to the protocol of one subject who became confused and muddled up his ›story‹ during the acquisition stage, could therefore not retrieve it at time of making his judgments and consequently scored zero correct. With position judgments[2], there was a significant effect of practice and this was found for both types of strategy use. These results must certainly raise problems for those who view temporal coding as an automatic process. It is clear that there is some sort of hierarchy of difficulty among temporal judgment tasks and they do not all require

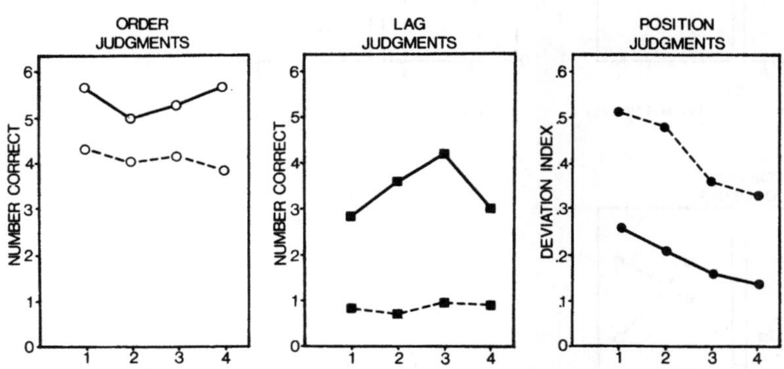

Figure 5. Mean judgment scores as a function of trial and strategy. (Solid lines represent simple rehearsal strategies; dotted lines represent elaborative strategies).

2 It should be noted that the performance measure used here for temporal position judgments is the Deviation Index (D). The deviation index score is based on absolute deviations from true positions (d_i) as well as an a value (k_i) which adds a bonus for partially correct segments. Its value varies between zero and one, with a score of zero signifying perfect performance. Further discussion of the measure can be found in Jackson (Note 2).

the same degree of processing. Performance on lag and position tasks is not only closely related to the strategies used, but also reveals a considerable effect of ›level of practice‹, a criterion which according to Hasher & Zacks (1979) is indicative of the deliberate information processing mode.

Another set of interesting results is revealed when the unexpected‹ trial 5 of the last experiment is examined. These results are shown in Figure 6.

It appears that for lag and position judgments there are no significant differences relating to *which* task subjects performed in the previous four trials, but instead, the important factor is *how* that task was carried out. Those subjects who had used elaborative strategies on trials 1-4 performed significantly better on either of these unexpected tasks than did subjects who had chosen a simple task independent rehearsal strategy for tasks 1-4. In the unexpected‹ order condition, however, a different picture emerges: in this condition it makes no difference either *which* task was performed in the previous four trials or *how* it was carried out, performance in all instances being equally good. This again suggests that order judgments are not too dependent on any deliberate form of processing.

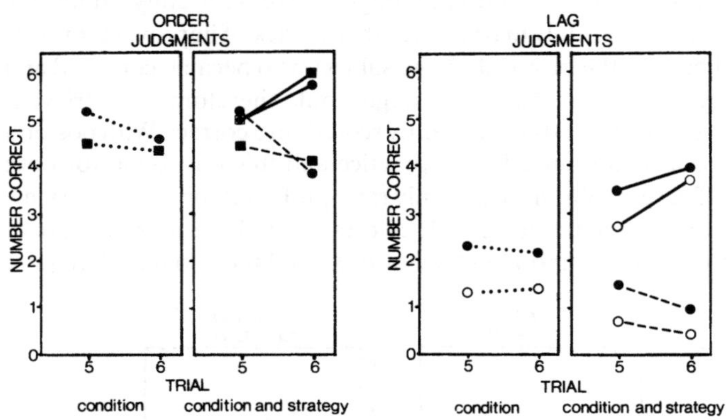

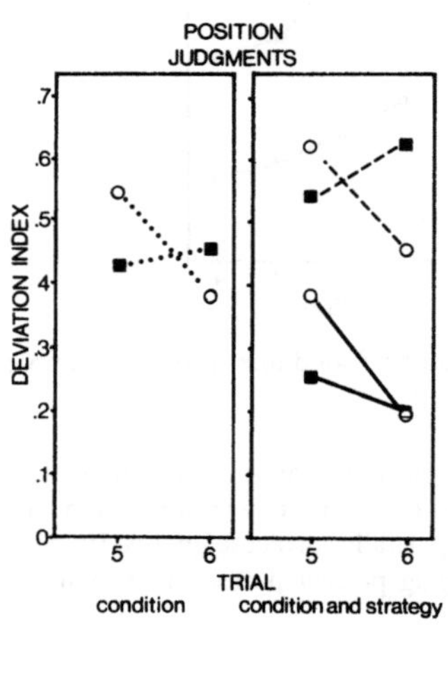

Figure 6. Mean judgment scores on trials 5 and 6 as a function of previous task and strategy. (○ represents order judgments on trials 1–4; ■ represents lag judgments on trials 1–4; ● represents position judgments on trials 1–4; *Solid lines* represent simple rehearsal strategies; *dotted lines* represent elaborative strategies).

Conclusions

The conclusion to be drawn from this series of experiments seems to be clear. We must be careful about how we define temporal information. It is not enough to examine one restricted interpretation of such information and then proceed to generalize from it into other domains. Also it is it less than elegant to generalize too much from group data. Employing a verbal protocol analysis framework has revealed individual strategies to be a prominent feature of temporal information processing.

To simply qualify temporal coding as being an automatic byproduct is a gross understatement of the possibilities inherent in the processing of temporal information. Results instead suggest that different stimulus materials as well as different temporal judgment tasks demand different levels of processing.

Since the initial writing of this chapter, Zacks et al. (1984) have also reported data which challenge two of their original automaticity criteria. They show that retention of temporal position does increase with practice and also that reliable individual differences can be found on judgments of temporal information, findings which support the conclusions we have drawn here.

References

Bjork, R.A. Theoretical implications of directed forgetting. In: A.W. Melton & E. Martin (Eds.), *Coding processes in human memory.* Washington, DC: Winston & Sons, 1972, pp. 217-235.

Block, R.A. Contextual coding in memory: Studies of remembered duration. In: J.A. Michon & J.L. Jackson (Eds.), *Time, mind, and behavior.* Heidelberg: Springer Verlag, 1985, pp. 169-178.

Block, R.A., & Reed, M.A. Remembered duration: Evidence for a contextual-change hypothesis. *Journal of Experimental Psychology: Human Learning and Memory,* 1978, *4,* 656-665.

Craik, F.I.M., & Lockhart, R.S. Levels of processing: A framework for memory research. *Journal of Verbal Learning and Verbal Behavior,* 1972, *11,* 671-684.

Estes, W.K. Is human memory obsolete? *American Scientist,* 1980, *68,* 62-69.

Estes, W.K. Memory for temporal information. In: J.A. Michon, & J.L. Jackson (Eds.), *Time, mind, and behavior.* Heidelberg: Springer Verlag, 1985, pp. 151-168.

Hasher, L., & Zacks, R.T. Automatic and effortful processes in memory. *Journal of Experimental Psychology: General,* 1979, *108,* 356-388.

Hinrichs, J.V. A two-proces; memory-strength theory for judgment of recency. *Psychological Review,* 1970, *77,* 223-233.

Jackson, J.L., & Michon, J.A. Effect of item concreteness on temporal coding. *Acta Psychologica,* 1984, *57,* 83-95.

Mandler, J. Categorical and schematic organisation in memory. In: C.R. Puff (Ed.), *Memory organisation* and *structure.* New York: Academic Press, 1979, pp. 259-299.

Michon, J.A. Processing of temporal information and the cognitive theory of time experience. In: J.T. Fraser, F.C. Haber & G.H. Mueller (Eds.), *The study of time.* Heidelberg: Springer-Verlag, 1972, pp. 242-258.

Michon, J.A., & Jackson, J.L. Attentional effort and cognitive strategies in the processing of temporal information. In: J. Gibbon & L.G. Allan (Eds.), *Timing and time perception.* Annals of the New York Academy of Sciences, Vol. 423, 1984, pp. 298-321.

Schank, R.C., & Abelson, R. *Scripts, plans, goals and understanding.* Hillsdale, NJ: Lawrence Erlbaum Associates, 1977.

Tzeng, O.J.L., Lee, A.T., & Wetzel, C.D. Temporal coding in verbal information processing. *Journal of Experimental Psychology: Human Learning and Memory*, 1979, *5*, 52-64.

Underwood, B.J. *Temporal codes for memories: Issues and problems*. Hillsdale, NJ: Lawrence Erlbaum Associates, 1977.

Underwood, B.J., & Malmi, R.A. An evaluation of measures used in studying temporal codes for words within a list. *Journal of Verbal Learning and Verbal Behavior*, 1978, *17*, 279-293.

Yntema, D.B., & Trask, F.P. Recall as a search process. *Journal of Verbal Learning and Verbal Behavior*, 1963, *2*, 65-74.

Zacks, R.T., Hasher, L., Alba, J.W., Sanft, H., & Rose, K.C. Is temporal order encoded automatically? *Memory and Cognition*, 1984, *12*, 387-394.

Zimmerman, J., & Underwood, B.J. Ordinal position knowledge within and across lists as a function of instructions in free-recall learning. *Journal of General Psychology*, 1968, *79*, 301-307.

Reference Notes

1. Bruinsma, A., Michon, J.A., & Jackson, J.L. *Contextual cues in temporal information processing*. Internal report, Institute for Experimental Psychology, University of Groningen, undated.
2. Jackson, J.L. *The Processing of Temporal Information*. To appear in 1985.
3. Jackson, J.L., van Schagen, I., & Michon, J.A. *Developmental effects in temporal coding*. In preparation.

Part III.
Patterns: The Structure and Organization of Time

Chapter 13. Structural Organization of Events in Time

Mari Riess Jones

Our experienced environment is not a static one. It consists of energy that is patterned over time. And so a fair description of the organizational structure of things in time must include time and temporal relationships. Here I consider three approaches to this issue in terms of broad categories that reflect my own assessment of the way different theorists incorporate time and temporal relations. These categories are termed *Rate-Relational, Coding*, and *Dynamic Serial Transformations*.

The focus of this review is upon the way people respond to temporal structure as indexed by judgments or reproductions of serial (temporal) order. Theoretical differences arise with regard to the portrayal of time and rhythm as determinants of perception and retention of this organization. Thus, what I term *Rate-Relational* theories commonly propose that perception of serial patterns follows general processing principles, but that pattern rate constrains this processing. In some cases these approaches reflect a search for a time-limited or threshold process. Therefore pattern rate is often an important variable. In contrast, *Coding* models do not tend to focus upon pattern rate. Instead, order retention is explained in terms of rule-governed serial codes. Usually, these rules reflect relationships between events arising from non-temporal dimensions (alphabets). The third approach emphasizes the function of *Dynamic Serial Transformations* in the environment. Here the serial structure is portrayed as a nested hierarchy of changes over and in time. Consequently a pattern's rate and rhythm become important parts of the whole structure that governs perception and retention of serial order.

These approaches share a common interest in understanding how people perceive and remember the ordered arrangement of things in time. In the following sections, I emphasize this commonality by focusing upon ways in which these approaches study perception and retention of temporal order of events within sequences of varying degrees of structure, that is, within serial patterns. A serial pattern is taken to be a coherent (non-random) arrangement of things in time[1]. Thus, many environmental patterns qualify as serial patterns. The visual pattern created by a walker (Johansson, 1975), for example, is a serial pattern. So too are speech utterances, music and other natural sound patterns (e.g. Howard & Ballas, 1982). In light of this breadth, the focus of this review is relatively narrow, although I believe the ideas and findings do have significant generality. I restrict my discussion to experiments directly related to testing hypothesis about order

1 More generally, a serial pattern is any ordered arrangement of things in space or time. Visual sequences can be serial patterns.

retention from one or another of the three general categories already suggested. To a degree this means that the patterns of interest are auditory sequences, although notable exceptions exist (e.g. Restle, 1970).

Rate-Relational Theories

Rate-relational approaches posit processes that people use to perceive organization in time. These processes may be more or less directly tied to the structure of an event series, but in general they are conceived as consuming time. Consequently, pattern rate becomes theoretically important in precipitating a breakdown or restructuring of processing under time constraints. The basic idea is that time, in some absolute sense, limits processing activity. Consequently, perception of serial order changes and/or becomes more difficult as rate increases.

Historically, these theories combine two distinct traditions. One is the Gestalt tradition, in which structural principles which govern perceptual organization are emphasized. The other arises from Helmholtzian psychophysics and it reflects a search for an absolute temporal order threshold. Together these traditions offer a backdrop against which several rate-relational approaches make sense.

Garner

Garner and his colleagues have investigated the perception and learning of repeating temporal sequences presented at various rates (Garner, 1974; Garner & Gottwald, 1967, 1968; Preusser, 1972). Garner distinguishes perceptual activities that are passive and immediate responses to structure in fast patterns from conceptual activities that are active and intellectualized responses to slow patterns. With slow patterns verbal encoding of pattern relationships may play an important role and this encoding may be time limited (Garner, 1974, p. 69). Yet, regardless of time constraints, Garner maintains that Gestalt organizing principles direct the psychological response to pattern structure at all rates; people simply ›use‹ these principles differently at different rates.

Garner proposes two Gestalt organizing principles appropriate primarily for binary patterns, that is, patterns composed of two different events (e.g. A, B): the Run and Gap principles. He assumes that people respond to one of these as a figure (e.g. A) and the other as a ground (e.g. B). The Run principle has an individual reorganizing a recurrent sequence so that the longest string of figure events (that is, a run) begins the pattern as in ...AAAAB B AB.... According to the Gap principle a different organization may obtain where the sequence ends with the longest run of ground events as in ...AB AAAAB B.... Thus, Run and Gap principles separately applied do not *necessarily* produce the same organization. However, if they do agree, this should result in fewer ways of organizing the sequence and consequently such patterns should be perceptually simple. Garner proposed that complex sequences are those which afford larger sets of alternative organizations.

Complex binary sequences, by this criterion, turn out to be more difficult in tapping and other tasks (Garner & Gottwald, 1967; Royer & Garner, 1966). Yet serial organization effects appear to have their most prominent effects on pattern difficulty at slower rates (at least as measured by observation time and verbal descriptions) (Garner & Gottwald, 1968). Also Preusser (1972) found that people relied more upon the Run principle at slow rates and the Gap principle at faster rates if they had to verbally describe serial structure. Such findings contributed to Garner's proposal for a two-process model incorporating rate-limiting effects. At fast rates, an immediate passive perceptual process obtains whereas at slow rates an active conceptual process governs serial expectancies.

Warren

Warren also assumes that two processes are involved in discrimination and retention of temporal order within complex event sequences. In contrast to Garner, Warren addresses sequences based on more than two different events. However this approach is less explicit about the function of Gestalt organizing principles than it is about the function of temporal limits on processing. His ideas have been strongly influenced by research on temporal order thresholds in which critical time periods between events are assumed to limit order perception (Warren, 1974; Macar, 1985, chapter 7 of the present volume).

Ideas of both Stroud (1955) and Hirsh (e.g. Hirsh, 1959; Hirsh & Sherrick, 1961) paved the way for Warren's theorizing. Stroud proposed that discrete psychological time quanta package the continuous physical time dimension, whereas Hirsh and his colleagues sought to experimentally pin down the actual value of a time interval between two events within which order information was lost to a perceiver (that is, a temporal order threshold). Curiously, values originally proposed for these limiting time quanta do not agree (e.g. 100 ms for Stroud and 20 ms for Hirsh). In fact, subsequent research by Hirsh and others suggests that the more widely cited 20 ms temporal order threshold value must be qualified due to influences of event type, context and training among other things (Allport, 1968; Fay, 1966; Hirsh, 1974; Hirsh & Fraisse, 1964; Divenyi & Hirsh, 1974).

Warren, specifically rejects the 20 ms order threshold value in his description of order discrimination between events in complex sequences (Warren, 1974). He notes that order judgments about arrangements of ›unrelated‹ items (e.g. hisses, tones, etc.) that were embedded within the context of recycled sequences were not accurate unless event durations were at least 200 ms (Warren, Note 5; Warren et al., 1969). According to Warren, event order is directly perceived only at durations in excess of 200 ms because this is the requisite time to verbally encode (label) each item; memory for event order depends on verbal codes.

Warren's model of temporal order discrimination is finally a two-factor one. Serial order depends on rate limited verbal encoding and upon training. While listeners can learn to recognize very rapid sequences this skill is assumed to depend on a different, holistic pattern recognition process which does not involve direct perception of individual event order. Sequences presented at rates that break the 200 ms limit of verbal encoding are holistically perceived. Order information in these cases occurs when people learn to associate informative verbal labels with particular patterns.

Support comes from studies in which people learned to recognize isochronous sequences'in which four events were permuted (that is, tone, square wave, noise, tone *versus* tone, noise, square wave, tone). Individual event durations varied from 5 ms to 500 ms, and with enough trials people could accurately distinguish order even when events were only 10 ms long.

There is little doubt that changes in rate alter one's experience of a sequence. Both Garner's and Warren's research testify to this. Furthermore, both theoretical accounts propose a transition in processing mode close to a rate corresponding to inter-event times of 200-250 ms. For Garner this shift represents limits due to verbal encoding of Gestalt principles, whereas for Warren it reflects a more abrupt shift from verbal or serial encoding to holistic pattern recognition. Particularly in Warren's model, the failure to specify the way in which serial structure itself dictates perception at various rates means that any lawful interactions of serial structure with rate cannot be parsimoniously explained. It is not clear, for example, how Warren's model can explain the data of Royer & Garner (1966) or Preusser (1972) cited earlier. However, an even more compelling phenomenon that reveals certain interactions of serial structure with rate threatens the explanatory generality, not only of Warren's threshold oriented approach, but also of Garner's serial organization approach. This is the phenomenon known as auditory pattern streaming.

Auditory Streaming and Perception of Serial Order

The term ›streaming‹ refers to a phenomenal effect wherein a temporal event sequence (e.g. of sounds) perceptually ›break apart‹ into related sub-patterns. The most characteristic feature of this effect is that the perceived subpatterns seem to co-occur or temporally overlap (Dannenbring & Bregman, 1976). The result is that if, say, listeners must report the actual temporal order of events within the whole sequence they cannot do it. This does not necessarily mean that they cannot discriminate between two different patterns that decompose phenomenally into streams. It does mean that people will have difficulty reporting precisely the order of events within a given ›decomposed‹ pattern (Bregman, 1978): order confusions abound. Instead, people accurately report event order only *within* one or another subpattern or stream (Bregman & Campbell, 1971).

The phenomenon of streaming is important because it highlights the fact that perceived serial organization depends upon *interactions* of rate with event structure. In other words, streaming cannot be explained by simple rate limited processes such as, for instance, verbal encoding. To illustrate, consider the classic study of streaming by Miller & Heise (1950). They used auditory trills (that is, alternations: A B A B ...) presented at 10 tones/s, and showed that when one frequency is gradually shifted, listeners report no longer hearing a serial trill pattern of A B A B ..., at approximately a 3 semitone pitch interval[2]. Instead they report hearing two co-occurring streams: A - A - A ... against B - B - B .... In short, the serial integration of the trill breaks up if pitch distance is too great, provided the rate is fast enough.

From the perspective of theories such as Garner's and Warren's the lawfulness of streaming presents difficulties. Breakup depends not simply upon rate but also upon *degree* of relatedness (namely pitch distance) between putatively nominally distinct events. A further problem is that when the breakup occurs it results in a simultaneous, figure-ground percept, not a serially integrated one (see also Dannenbring & Bregman, 1976; McNally & Handel, 1977).

Streaming is by now much researched. A classic study of Bregman & Campbell (1971) required that people report the serial order of six tones each lasting 100 ms within a recurrent sequence. Three different high frequency tones (H_1, H_2, H_3) were combined with three low frequency tones to form patterns such as H_1 L_1 H_2 L_2 H_3 L_3. Listeners accurately reported the order of the three high tones and/or of the low tones. But they could not order the events overall to reconstruct the actual sequence (H_1 L_1 H_2 L_2 ...). Others have replicated this effect (Handel et al., 1983; McNally & Handel, 1977; Jones et al., 1978; van Noorden, Note 4).

The general tenor of research on streaming suggests that numerous structural variables interact with pattern rate to determine whether a pattern will phenomenally fragment and at what rate it will be judged to restructure from serial integrity to serial simultaneity. Some of the variables which influence streamability are:

(1) *Pitch Interval Size*. If pure tones are involved, then patterns with relatively large pitch interval between adjacent tones are more likely to stream at a given rate than other (Dannenbring & Bregman, 1976; Jones et al., 1978; van Noorden, Note 4). A tradeoff of sorts between pitch change and time period occurs so that Jones (1976a) proposed that the critical temporal order threshold is not independent of rate but depends on a pitch-time ratio.

(2) *Context*. The larger serial pattern context determines streamability so that patterns with jagged contours, for instance, are more likely to stream than those with smooth contours (Divenyi & Hirsh, 1974; Handel et al., 1983; Jones et al., 1978).

(3) *Rhythm*. Elements within the larger pattern context are more likely to segregate into separate streams if rhythm renders them temporally unpredictable within the whole serial pattern (Handel et al., 1983; Jones et al., 1981). Work on polyrhythmic perception is undoubtedly also relevant here in suggesting effects of certain time ratios (Handel & Oshinsky, 1981; Oshinsky & Handel, 1978).

2 A semitone is defined to be a frequency ratio $\Delta f/f = 0.059$. It is the smallest pitch interval in traditional Western music.

(4) *Instructions and perceptual learning* also influence the degree to which a perceiver will hear an integrated or parallel pattern percept (Bregman, 1978; van Noorden, Note 4).

Bregman attempted to explain stream formation in terms of Gestalt heuristics which parse incoming sequences, allowing perceivers to infer different originating sources. He and others (e.g. Handel et al., 1983) suggest that the phenomenal stream experience *mediates* judgments of temporal overlap and serial pattern recognition in addition to guiding actual serial productions. These ideas lead to at least two questions regarding the relationship between streaming and performance. The first is »What observable pattern conditions determine serial ordering and hence phenomenal streams?« and the second is »Can we assume that perception determines productions?«

With regard to the first question, reliance on Gestalt principles of similarity, proximity and continuity is often problematic because these rules relate observable pattern structure to phenomenal experience in multiple and, sometimes, conflicting ways. For instance, within any sequence serial order retention may derive from invariant pitch changes (continuity), invariant contour changes (continuity), small pitch changes (proximity), similar quality (similarity) and so forth. Violation of any one of these invariances is likely to disrupt experiences of serial ordering. Yet it is not clear which violations are more serious nor is it clear how these structural features interact to determine phenomenal reports. Finally, explanations of streaming require more rigorous hypothesis about higher-order interactions of structural variables that go beyond Gestalt principles.

The second question asks whether the perception of a stream influences pattern reproduction. Here a more straightforward resolution is possible. The work of Bregman & Campbell (1971) and others (Handel et al., 1983; Jones et al., 1978) indicates that when people judge patterns to yield overlapping streams they also tend to order events only within those streams in productions. These findings also square with Garner's research which shows that the way a pattern is perceptually organized influences reproduction.

Massaro

Massaro (1975) has proposed a stimulus-based approach to pattern structure. It differs from other rate relational approaches in its attention to the role of serial relationships such as relative pitch (music-like) relations in auditory sequences. Massaro proposed people encode information about pitch intervals and contour in sequences and this encoding takes time. In this context, the phenomenon of streaming results from a breakdown of pitch processing.

Idson & Massaro (1976) trained listeners to identify arrangements of patterns of three (test-) tones. Subsequently, these people had to identify the learned three-tone patterns in six-tone contexts wherein each of the test tones was followed by a new masking tone. Mask tones were either from the same octave as test tones or from a different one. According to Idson & Massaro, when the masks fall in a different octave listeners should fail to process the pitch information and auditory streaming will occur. Presumably, if this happens trained listeners will more readily identify a familiar test tone subpattern as a stream. In fact, this occurred. People were quite good at test pattern identification when masks were very different from test tones, and they performed poorly when masks were similar. Related findings are found in Divenyi & Hirsh (1975) and Dowling (1973). (While this is taken as support for the hypothesis that large pitch intervals prevent pitch processing, it should be pointed out that, among other things, pattern contour also covaried with mask octave).

To study processing time, Idson & Massaro also manipulated certain time intervals between test and mask tones within the six-tone sequences. Within a condition this time period could take one value (that is, between 70 ms and 260 ms), while the time interval between test tones themselves was fixed at a different value. Results of these manipulations led the authors to conclude that the amount of time required for extraction of relative pitch information is much less than the 250 ms limit proposed for absolute pitch recognition of single tones (Massaro, 1975): »... all of the information needed to make such a judgment can be extracted from the tone within a period of time as short as 70 ms from onset...« (p. 166, Idson & Massaro, 1976).

In sum, this approach begins to address structural effects deriving from contour and interval relationships about tonal elements within subpatterns (that is, three-tone streams) and within larger patterns. Thus, it has, in a sense, broader applicability than the models of Garner and Warren. It does not, however, directly formulate hypotheses regarding possible interactions of structural variables such as contour, pitch intervals and pattern rate. In fact, there remains an emphasis upon identifying processing time thresholds associated with amounts of information. While it is possible that such absolute time limits exist, the experiments of Idson & Massaro cannot finally be taken as unequivocal evidence for their existence. This is because their manipulations of available processing time within a sequence necessarily produced correlated changes in pattern rhythm. Since rhythms arising from variations of time intervals within a sequence can directly facilitate stream formation (Handel et al., 1983) it is clear that relative, and not absolute, time may have been responsible for these results.

Summary

Rate-relational approaches rightly stress the importance of time as pattern rate. A person's percept of serial relations does change with rate. Here these changes are explained by time constraints on psychological processes, such as run or pitch encoding at faster rates.

Finally, the fact that serial order retention declines with increments in rate is undeniable. However, these approaches tend to overlook another fact, namely that as serial ordering declines with rate different configurational relationships may come to control perception (e.g. simultaneous figure-ground relations). A strict rate limiting notion therefore can place undue emphasis on order errors alone. The research on streaming is instructive in illustrating the complexity of interactions involving rate and other pattern relationships, interactions which contribute to different configurational percepts (serial versus simultaneous). Nevertheless some rate-relational theories remain committed either tacitly or explicitly to models involving absolute time. In other words, the possibility that the psychological effects of time as rate are *relative* to aspects of pattern structure (pitch, interval, contour) and even to other time intervals (rhythm) has not been seriously considered.

Serial Organizations in Time and Coding Theory

A line of theorizing originating with Simon's detailed description of pattern relationships (Simon, 1972; Simon & Kotovsky, 1963; Simon & Sumner, 1968) has culminated in powerful ideas relating to the processing of serial structure. One important assumption is that processing consists of three putatively independent stages: encoding, memory representation, and retrieval. Another is that when confronted with manifest structure, people will attempt to encode relationships inherent in a pattern in some relatively efficient way. Patterns that lend themselves to more economical encoding will be perceived and recalled more readily. Consequently, this approach maintains that order retention resides, in part, in memory codes that express serial event relations.

Issues which direct research in this area are formal ones that deal with nomenclature and assessment of code complexity. One is the issue of structural ambiguity. In the context of theoretical assumptions about the independence of encoding, storage and retrieval, the structural ambiguity issue has led to the proposal that a given memory code may give rise to many different reproductive strategies, strategies that are not necessarily identical to encoding activities. Closely associated with this topic are questions surrounding the representation of certain well-formed serial patterns by rule recursive memory codes.

Serial coding models have focused upon descriptions of temporal organization in which order is derived from rules defined on non-temporal dimensions. For example, the temporal digit sequence: 1-2-3-6-5-4 may be described by rules such as ›+1‹ of ›–1‹ that dictate moves along a number line. The point here is that neither absolute time, as rate, nor relative time, as rhythm, enter in to modify the codes. There are notable attempts to specify rhythm directly in terms of coding principles (e.g. Povel, 1981), but these have not yet resulted in the incorporation of rhythm as a part of the structure of patterns that have non-temporal rule codes.

Simon's Approach

Initially Simon & Kotovsky (1963) proposed that people solve Letter Series Completion problems by inducing a serial concept, fixating it and then using it to extrapolate unfinished letter patterns. The concept, in this case, is reflected in combinations of Next (N), Backward Next (BN), or Identity (I) rules on the English alphabet. They assumed that people brought to these tasks knowledge of alphabets and elementary rules and through interaction with a visual sequence acquired a specific abstract description of its structure in terms of rule-based moves originating from one (or more) loci on the appropriate alphabet. Thus, for example, the sequence a a a b b b c c c ... is described by fixing the originating position ml at a [m1= a] in the English alphabet, and the serial concept is coded as [3m1, N(m1)]* (where * means repeat)[3]. A key feature involves the assumption that people discover and represent periodicities found within a sequence in terms of alphabetic moves (rules).

Because their descriptions of salient serial relations are so explicit, Kotovsky & Simon (1973) could generate a program that predicted letter series solutions under the assumption that certain things in the coding process led to error (e.g. unfamiliarity of alphabet or rules, memory load, etc.). Empirical tests with human problem solvers were consistent with the hypothesis that people initially discover periodicities and then create rule based pattern concepts that influenced their overt extrapolations. Solution predictions (solved or not) by the computer program yielded point-biserial correlations ranging from 0.16 to 0.78 with various performance summaries involving solution time and errors.

These ideas, developed on visual letter sequences, have a straightforward extension in Simon & Sumner's (1968) approach to musical pattern structure where alphabets are musical scales and chords. Recently they have been elaborated further by Deutsch & Feroe (1981). Another noteworthy extension of the alphabet notation, developed by Simon & Sumner themselves, involves the incorporation of note durations and stress alphabets (three levels of stress) to incorporate rhythm. Descriptions of a variety of wellknown tunes therefore is possible.

Simon (1972) has integrated research from other serial coding models (e.g. Vitz & Todd, 1969) showing that most can be subsumed by codes that address different alphabets but yet exhibit common relationships (e.g. ›same‹, ›next‹) and repetitive redundancies. He proposed a general assessment of pattern complexity resting upon the number of symbols required to encode a pattern (that is, code length), that should, in many tasks, reliably predict pattern difficulty.

Restle's Approach

Restle introduced structural trees and the attendant, highly economical, rule recursive encodings that underlie certain kinds of serial patterns (Restle, 1970). He proposed that many temporal sequences (including musical patterns) could be learned by acquiring rules that related sequence subparts to one another in a hierarchical fashion. His claims were supported with data gathered from transfer tasks and from serial anticipations of visual patterns created by temporal arrangements of six equally spaced lights (this alphabet is represented numerically as 1 2 3 4 5 6) (Restle & Brown, 1970a, 1970b).

3 Simon has used various notations for this idea.

Restle's most influential proposal involves hierarchical (rule recursive) encoding. He proposed that people encode rules such as Transposition (T_k) Mirror Image (M) and Repeat (R) which operate through transformation and concatenation (o) of some subpattern within a sequence. For example, T_1 (12) = 12 o t_1 (12) = 1223 where lower case letters indicate nonconcatenated transformations and upper case letters imply an argument-plus-transformation concatenation (symbolized by o). In this case the transposition, t_k (where k = + 1) advances 12 by 1. Similarly, for mirror-imaging the m complements (e.g. on base 6) a subpattern: M (12) = 12 o m(12) = 1265, Repeat is an identity operator: R (12) = 12 o r (12) = 1212. A strictly nested binary structural tree is composed of recursive combinations of such rules to yield hierarchical (well-formed) serial patterns. Thus M(T(R(12)) = 12 12 23 23 65 65 54 54 arises from systematic application of rules to ever-expanding arguments[4]. Figure 1 a shows the binary structural rule tree corresponding to this pattern. Serial anticipation profiles of such sequences support Restle's predictions that events initiating higher level rule transitions (e.g. between 12 and 23) will be more difficult to anticipate.

Restle's theory has been highly influential. It often yields useful descriptions of performance errors and serial latencies in a variety of tasks including motor tracking, predictions and reproductions (Restle, 1976; Restle & Brown, 1970b; Rosenbaum et al., 1983). And in some cases hierarchical patterns are more quickly learned than nonhierarchical versions, supporting Restle's initial claim that the former result in more concise memory codes (Restle & Brown, 1970b; Simon, 1972). However, other research questions the latter prediction particularly when nonhierarchical patterns themselves have significant (but not recursive) structure (Jones, 1976b; Jones & Zamostny, 1975; Boltz et al., in press).

Although Restle's ideas tap important elements of the way people respond to events organized in time, they also raise some unresolved issues. For example, much support for multilevel structural trees is inferred from serial position anticipation errors or latencies. However, these profiles can be misleading. Higher level tree transitions are usually confounded with serial position; therefore more errors will normally accrue to higher order transitions in more central serial locations. Correlated effects of pattern contour and recurrent identities within rule generated patterns may also account for observed recall effects of hierarchical patterns and/or for certain serial error profiles (Jones et al., 1978). In short, there remain questions about Restle's conception of structural trees in terms of static interval rule codes, codes which do not incorporate the timed placement, for example, contour changes or recurrent identities.

Leeuwenberg's Approach

Leeuwenberg (1969) and his associates also contributed to coding theory by proposing rules that capture certain Gestalt properties in terms of simplifying

4 Restle's notation can also be expressed in terms of Complement (C_i) rules for M, and Next (N^j) rules for T_k (see Jones, 1981b).

structural redundancies and efficient memory codes. Typically, a serial pattern is initially encoded as a concatenated string of symbols that represents, on an element-by-element basis, the event sequence. This yields a *primitive code*. In patterns with many redundancies (that is, repeated elements, symmetries, recursive properties), Leeuwenberg proposes that reduction principles apply to a primitive code to yield increasingly shorter and more economical, *reduced codes*. It is the reduction principles which summarize iterative pattern features in terms of good continuation rule recursion and so forth.

In applying his theory to the Letter Series Completion patterns studied by Simon & Kotovsky (1963), Leeuwenberg found that his description of pattern complexity, defined as length of a reduced code, correlated with people's letter series solutions somewhat better than did Simon & Kotovsky's measure.

Leeuwenberg also addressed the problem of structural ambiguity. Recall that this refers to the fact that a particular can be encoded in different ways. Thus, the sequence 12 23 45 56 can be recursively encoded as $T_3(T_1(T_1(1))$ or as a linear string of rules, each applied to the preceding argument: (12) o t_1 (12) o t_2 (23) o t_1 (45) (again o is concatenation). The solution to the structural ambiguity problem as it applies to perception, according to Leeuwenberg, involves a minimum principle: perceived structure will correspond to the shortest reduced code (that is, that with fewest symbols). Truly, ambiguous patterns are ones with exactly equivalent final reduced codes lengths (see also Collard & Buffart, 1982).

Other Advances in Coding Theory

Other advances come primarily from work of Greeno & Simon (1974) and Collard & Povel (1982). This work focuses upon encoding strategies that guide overt serial productions during retrieval. According to coding theory, encoding, representation and retrieval stages can be independent of one another; therefore retrieval output strategies may only loosely reflect encoding constraints. Essentially this means the serial pattern *perception* can be decoupled from overt serial *production*. Perception does not necessarily determine action (Collard & Povel, 1982; Greeno & Simon, 1974).

The structural basis for this decoupling returns us to the issue of structural ambiguity: sequences afford multiple ways of perceiving and/or producing them. Since rule recursive patterns, especially, lend themselves to multiple descriptions, this issue has special meaning for these patterns (Jones, 1981b).

Insofar as perceiving has been studied, most coding approaches predict that people should perceive *hierarchical serial relations* (e.g. Jones, 1974; 1981b; Restle, 1976) since the most economical encoding of ambiguous patterns with rule recursive relations is one that capitalizes on the recursiveness. Tests of this prediction that are uncontaminated by serial production strategies are rare, but recently Boltz et al. (in press) found only conditional support for this prediction.

From a different perspective, Deutsch & Feroe (1981) proposed that musical pattern perception depends upon encoding relations between adjacent and nonadjacent notes within a given melody according to rules defined upon different musical alphabets (e.g. chromatic, diatonic, chordal, etc.). Different levels of musical structure reflect selected encoding from one or another alphabet. The issue of structural ambiguity in perceiving is expanded here since at any musical level there is not only the potential for serial different rules from a selected alphabet, but there is also some potential for still different rule expressions based on alternative alphabets. Furthermore, serial anchoring points for these alphabets can vary as well, and this moves the ambiguity issue into questions about the perceiver's choice of reference frame. In music, such questions concern tonality and key identification (see Butler & Brown, 1984; Deutsch, 1980; Jones, 1981b).

Much current research arises from the implications structural ambiguity holds for serial *production* strategies. A basis for production accounts is that rule recursive expressions, in fact, characterize the memory code of hierarchical patterns (Collard & Povel, 1982; Greeno & Simon, 1974). Yet, because this code an be ›interpreted‹ in multiple ways, various production strategies having characteristically different serial error and latency profiles are possible. Order retention, summarized in a memory code, may not necessarily determine serial order recall as evidenced in one's overt productions. Greeno & Simon (1974) proposed three different output strategies, namely Push-Down, Recompute and Doubling that reflect different applications of various rules to stored sequence elements. These strategies are graphically suggested in Figure 1, panel b, where it is evident that the Doubling strategy is closest to Restle's initial analysis.

The analysis of Greeno & Simon is dependent, in part, upon their use of rules that form a mathematical group. Group properties enable an elegant justification for the authors' demonstration of different interpreters. For example, according to group properties any two rules within one interpretive description of a particular sequence can be expressed by a different group rule thus forming the basis for another interpretive strategy (see also Jones, 1974, 1976a). Collard & Povel (1982), also relying upon group properties, add a fourth interpretive strategy, called Tree Traversal. According to this strategy, if rules form a commutative group (a so-called, Abelian group, see Jones, 1974) the structural tree of recursive codes can be adapted to reflect interpretive processes acting on the memory code. This strategy exploits additional group properties, namely that each rule, r_i, of a group has an inverse rule, $\hat{r}_i$, in the set such that $r_i \circ \hat{r}_i = e$ where e is the identity rule (also in the set). The relationship between r_i, $\hat{r}_i$ and e forms the core of the tree traversal strategy which is shown in Figure 1, panel c. Here the *i*th serial element is generated by transforms wrought on the immediately preceding element. Thus by applying r_i, $\hat{r}_i$ and/or e depending on the particular tree location, different mental loads occur at different serial transitions; this leads to predictions of hierarchically determined latency profiles in serial productions.

Collard & Povel (1982; Povel & Collard, 1982) tested these ideas in a serial finger tapping task wherein pattern elements (1, 2, 3, 4) were equated respectively with fingers of one hand (see also Rosenbaum et al., 1983). As predicted by the Tree Traversal model, they found that finger latencies were not consistent with motor transition difficulty but rather with ›mental load‹ in hierarchically codable finger patterns.

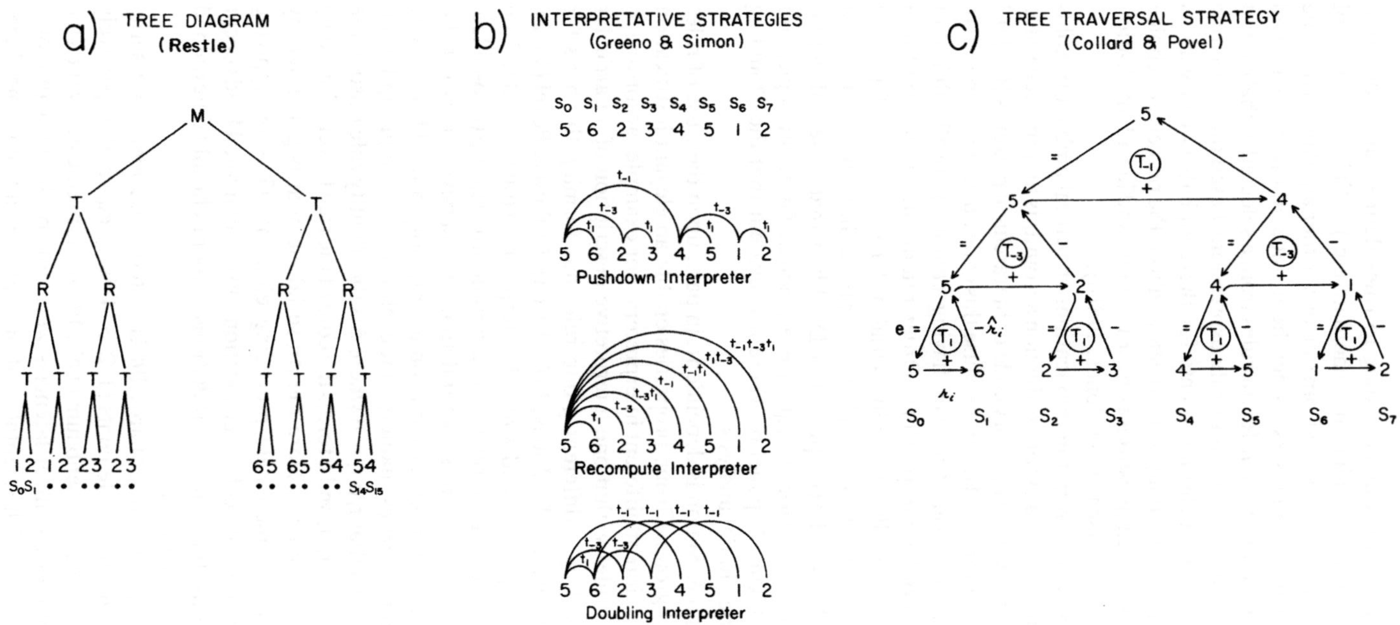

Figure 1. (*a.*) Restle's binary tree diagram of a long (S_0-S_{15}) regular digit pattern (from Restle, 1970). (*b.*) Three different interpretative strategies posited by Greeno & Simon for a rule recursive memory code for a different digit pattern (S_0-S_7): $(T_{-1}(T_{-3}(T_{+1}(5))))$. Arrows indicate inputs, outputs and operators for successive digits (from Greeno & Simon, 1974). (*c.*) A tree traversal interpreter posited by Collard & Povel for the same digit pattern examined by Greeno and Simon (1974) (from Collard & Povel, 1982).

Povel's Approach

Povel's (1981) rhythmic coding model described rhythmic productions in terms of time ratios. Inter-event intervals were varied to create simple serial time ratios (e.g. 1:2, 1:3, 1:4) and complex ones (2:3, 2:5, 3:4) denoting thereby rhythms of unequal difficulty. According to Fraisse's theory (1956), those rhythms with two durations with a ratio short:long of 1:2 should be simplest. Povel, in fact, found this to be true. However, he also found evidence that people could accurately reproduce more complex rhythms depending upon the patterning of durations.

In contrast to Simon & Sumner's approach, Povel proposed a hierarchical model of rhythm involving temporal nesting. According to this model, listeners first encode a time pattern by segmenting it equally in time following prominent accent points. Subdivisions of this so-called beat interval, must also be equal in time (or abide by a 1:2 ratio). Furthermore, simple (natural) codings are reflected in an equal number of subdivisions throughout the pattern. Thus, if a rhythm with the following inter-event durations (ms): 250 - 250 - 250 - 250 - 500 - 500 occurs, then two codings are possible. The preferred (good) coding identifies two beat intervals of 1000 ms. Then, natural subdivisions of this interval must be of 500 ms so that the first two subgroups match the final two subgroups in number. Ultimately, a three-tiered hierarchy results. A bad hierarchical model would initially subdivide the 1000 beat into a two-tiered hierarchy of 4 and 2 groups. These ideas have been extended in the form of temporal grids (Povel, 1984a) and internal clocks (Povel & Essens, 1984) to explain rhythmic complexity judgments and reproduction profiles of temporal sequences. However, most recently Povel (1985, chapter 14 of the present volume) found evidence for rhythmic expectations and surprise and these phenomena are most difficult for static coding hypotheses (but see Jones, 1981a).

Much of Povel's research reveals an influence of Fraisse's studies on rhythmic tapping (Fraisse, 1956). Fraisse's theory and much of the voluminous research on perception of rhythmic structure and production (for reviews see Ehrlich, 1960; Fraisse, 1982, 1984; Gabrielsson, Note 2; Jones, 1976a, 1978; Martin, 1972) has not been formulated in terms of coding theory. There is emphasis upon Gestalt principles, many of which could enter into a coding formulation, but the fact that Gestalt principles express reorganizational or distorting tendencies may pose problems for some models, including Povel's (1981) model. Fraisse, for example, has emphasized the importance of distinguishing between long and short time intervals having, say, 2:1 ratios and of attendant distortions revealed in regularizing similar durations (assimilation) and exaggerating differences between dissimilar ones (differentiation). In rhythmic productions maximum contrasts between long and short intervals are evident in tapping at the individual's preferred (spontaneous) tapping rate (see Fraisse, 1982). Others have shown that people group elements according to certain relative time properties (e.g. Collard et al., 1981). Furthermore, people prefer sequences that end with longer durations, and may even distort serial reproduction to achieve end-lengthening (Handel, 1974; Martin, 1972).

An old problem in the study of rhythm involves whether time patterning arises from a sequence of variable durations or whether it is based on accented elements, for in many rhythms both accents and durations can be found (e.g. Povel & Okkerman, 1981). Simon & Sumner emphasized stress and duration alphabets. On the other hand, Martin (1972) has suggested that accent level and duration are inextricably related such that stronger accents are placed farther apart in time than are weaker ones. He combines durational differences reciprocally with accent strength in proposing an accent-rhythm hierarchy based on the principle of isochrony. Features of this hierarchy then are responsible for good and distortable (poor) rhythms.

Summary

Many powerful descriptive techniques have evolved to explain order retention in terms of encoding and decoding strategies. A basic assumption has been that encoding and decoding are to a significant degree dependent and therefore can be studied separately. The issue of structural ambiguity has been a motivating factor, the idea being that there are multiple ways to encode and decode the same sequence.

Generally models addressed to serial order retention/production agree that errors evident in serial ordering arise from multiple task-specific circumstances and these affect a person's keeping track within an alphabet and his/her memory load for different rule combinations. Nevertheless, certain kinds of patterns should be easier to encode and remember, although people may differ in the way they ultimately reproduce them in time (e.g. hierarchical sequences).

Encoding of rhythm which emphasizes, respectively, stress alphabets (e.g. Simon & Sumner, 1968) or nested durations (Povel, 1981) has also been proposed. In the latter, while a given arrangement of durations can, theoretically, be encoded in several different ways, simple codes are those which incorporate good ratios.

In contrast to rate-relational approaches, these models have not been extended to predict changes in serial ordering as a function of variation in rate. Theoretically, time constraints could limit encoding, but the additional possibility that such constraints might *selectively* limit encoding of certain serial relations has not been considered. A consequence is that coding models fail to accomodate perceptual changes that occur with rate variations. The most significant of these is auditory streaming. Streaming phenomena pose potential problems for coding theory as currently formulated (see Jones, 1976a for a discussion). For example, some hierarchical patterns, which according to coding models have very economical codes, will stream (that is, perceptual ordering fails) at faster rates selectively as function of contour and pitch distance (Jones et al., 1978)[5].

5 Rule recursive formulas for auditory patterns in the Jones et al. study are based on semitone distances and an enlarged group of rules including a Mirror Imaging rule (M or C) and Next rules (T_k or N^j) for a chromatic scale. Recursive formulas for all of the patterns studied were not explicitly presented as such in the published manuscript but it can be easily verified that all sequences had rule recursive generative formulas.

Streaming phenomena pose yet another challenge to coding models. This involves the coding theory presumption that perception and production are independent. As indicated earlier, people's percepts of figure-ground aspects of ›streamed‹ sequences guide their productions. This means that action may not be independent of perceiving *within* a given individual. Perhaps structural ambiguity applies only to the various interpretations of a pattern offered by different perceivers. If so, this qualifies a basic assumption of coding theory.

Dynamic Serial Transformation

A striking feature of many environmental time patterns is the way their structure seems to invite us to follow and participate in changing relationships over time. The approach which I label dynamic and tranformational focuses upon descriptions of sequences that incorporate the idea that people acquire a sense of patternness and serial order from attending to one or another of the multiple serial changes afforded by a pattern as it transforms over time. Often these transformations leave some aspect of the sequence invariant. One important feature of this approach involves its explicit incorporation of both serial change *and* the time period of this change. Rhythm and tempo become integral parts of pattern structure, with the result that the approach offers a different perspective on the problem of structural ambiguity. In particular, this approach relies upon potentially disambiguating effects of timing.

Another significant difference between this approach and that of coding theory involves the relationship between perception and production (or reproduction). In contrast to coding theory, perception and production are not independent in this view. Theoretically, the connecting link between the two involves dynamic attending. Attending over time determines the what and when of perceiving during an encounter with a sequence. To recollect the sequence later, overt recall is guided by an attentional scheme refined during perceiving. That is, while a given sequence does indeed offer different kinds of transformational relations, the perceiver, perhaps guided by tempo and rhythm, focuses upon only one of these and it is this transformational structure which becomes prepotent not only in future serial anticipations (dynamic expectancies), but also in ultimately guiding that person's later reproductions. In short, while coding theory emphasizes the independence of perception and production, the transformational approach assumes a dependence wherein production, in a sense, recapitulates the attentional activities that guided perception.

Transformational theories are relatively new to the serial pattern scene. The one of interest here (Jones, 1976a, 1981; Boltz & Jones, Note 1) merges an emphasis upon contour and rate seen in rate-relational models with more formal accounts of serial rule structure and rhythm developed in coding theory. These ideas appear within a new framework, one that shares some assumptions of the Gibsonian school of perception (Gibson, 1979). Jones (1976a) has proposed that change occurs in time patterns through various serial transformations of the unfolding events and these changes retain, at some level, invariant relationships that define a pattern's form. Thus, in the sequence 123654, the transformation of $1 \rightarrow 2 \rightarrow 3$ by t_1 leaves invariant +1 while that between $6 \rightarrow 5 \rightarrow 4$ leaves invariant −1. These and related transformations may be expressed in terms of mathematical groups. However, in this case transformational groups are used to define psychologically salient symmetries within the pattern itself and not in a memory code in much the way group theory has been used to express invariant shapes of natural patterns (e.g. Hahn & Jones, 1981; Weyl, 1956).

Group properties elegantly simplify theory building here in three ways. First, they provide a basis for explaining different ways in which a pattern can be perceived, in particular *pattern* ambiguity as opposed to *memory code* ambiguity. Second, they provide a basis for formally exploring well-known differences in psychological salience of different serial rules. For example, event repetitions and contour changes differentially affect performance (Boltz & Jones, Note 1; Dowling, 1978; Jones, 1976b; Jones et al., 1978). Correspondingly, in group theory, the identity transforms and rules that change direction (improper rotations) both have special status. Third, the existence of kernel rules that generate the whole group (group generators) provide a basis for speculations about a formal link between perceiving, perceptual learning and production mastery (see Jones, 1981a).

Ultimately this view offers a basis for developing hypothesis about interactive effects of several structural variables. That is, by acknowledging that each transformation is associated with a time period, there exists a potential to formalize interactions of temporal and non-temporal serial relationships. For example, if contour changes in a sequence are especially salient then these points will be more likely to define important accents, and the temporal distribution of transformations that signal contour change (e.g. improper rotations) will identify important accent periods within a pattern's rhythm. Accents are temporal foci of attending. If this is true, one might expect that serial patterns in which contour changes occur regularly in time (that is, isochronous accents) to be easier than those with an irregular accent structure (Jones, 1976a; 1976b; Martin, 1972).

The idea of a contour-by-rhythm interaction was explored by Boltz & Jones (Note 1) using tonal sequences with different contours. Four of these are shown in Figure 2: two possess one contour change and two possess three changes. These represent patterns of a larger set of sequences studied by Boltz & Jones in a reproduction task using musically knowledgeable listeners. All melodies occurred in two different rhythms, one designed to make contour change points temporally predictable, the other designed to space them irregularly in time. A contour-by-rhythm interaction was observed. Patterns with three countour changes were no harder than those with one change only when the rhythm lent temporal regularity to these contour accents. Thus, timed occurrences of salient transformations are important.

Another aspect of this research is also relevant, particularly for the issue of rule recursion. Notice that two of the patterns in Figure 2 are labeled hierarchical (H) and half are linear (L). The H patterns are rule recursive melodies constructed from transformations defined on a familiar musical alphabet (C major scale). The L patterns lack rule recursion but they are carefully matched with H patterns in

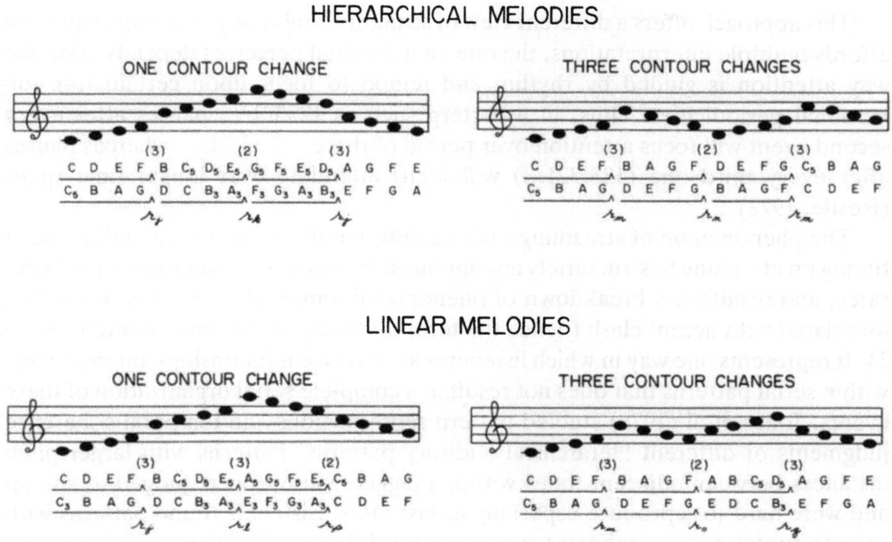

Figure 2. Examples of Hierarchical and Linear melodies with one and three contour changes (Boltz & Jones, note 1). Hierarchical melodies can be recursively rewritten using transformational rules applied to the C major scale while Linear melodies cannot be so rewritten. (Rules, indicated by r_j, operate on immediately preceding phrases to yield a subsequent phrase. The actual rules parallel Restle's Transpose and Mirror Image rules but they form a mathematical group; pitch-skip distances, in scale steps, at transitions are shown above the notes.)

contour and even interval rule properties. The H patterns should be easier to reproduce than L patterns according to coding theory since their codes are the more economical ones. In fact, H patterns were significantly easier than L sequences only when three temporally predictable contour changes were present.

These findings suggest that people respond immediately to dynamic pattern *shape*; the impact of rule recursion appears to depend upon the way pattern shape guides attending. Thus, where contour segments a pattern evenly, rule recursion facilitates performance. Evidently then there are circumstances where rule recursive serial patterns are more memorable than others. Nevertheless, these circumstances qualify current coding theory interpretations. Instead of static codes, it seems that relative timing of certain kinds of transformations that are correlated with rule hierarchies is responsible for much of the lawful performance.

Boltz & Jones suggest that accent isochrony and time ratios involving periods of associated the different (time-nested) serial transformations are important. Patterns are simpler, in that they more efficiently guide attending, if time ratios of these nested transformations are simple integer-ratio subdivisions (or multiples) of salient accent periods (Jones, 1976a). In general, patterns with rhythms that violate such simple constraints are more likely to lead to clashing of accents if they are fast patterns (and streaming) or to, serial fragmentations (and chunking if they are slow patterns).

This approach offers a different view of structural ambiguity. Although a pattern affords multiple interpretations, the one an individual perceives depends upon the way attention is guided by rhythm and tempo to focus upon certain transformational periodicities. Thus, in a pattern such as 123321..., pauses after every second event will focus attention over period of three: 12.33.21..., whereas pauses after every third one (123.321...) will focus attending over longer time spans (Restle, 1972).

The phenomenon of streaming offers a different illustration of the influence of timing on attending to structurally ambiguous sequences. Streaming occurs at faster rates, and results in a breakdown of phenomenal temporal coherence. It is often associated with accent clash (hence the term rhythmic fission, van Noorden, Note 3). It represents one way in which listeners can perceive relationships among events within serial patterns that does not result in a complete serial organization of these events. Jones et al (1978) studied pattern reproductions and temporal coherence judgments of different hierarchical auditory patterns. Patterns with larger pitch distances between adjacent tones within a jagged contour were judged to stream and were hard to reproduce especially at fast rates. Listeners found patterns with smooth contours more coherent and reproduced these more accurately[5]. The point is that all of these sequences, because they are rule recursive (hierarchical), are ambiguous. Yet the *way* they are perceived depends on accent structure (contour), pitch distance and pattern rate. At slow rates people can attend to transformations between adjacent tones to hear trill-like sequences (e.g., $L_1 H_1 L_2 H_2 \ldots$), but at fast rates people are more likely to attend over periods associated with higher-order transformations (i.e., $L_1 L_2 \ldots$ and $H_1H_2 \ldots$). Here streams occurred in fast patterns with large pitch distances, a fact which supported Jones' (1976a) hypothesis about the function of rate-of-change of pitch distance in stream function. This is significant, for it illustrates that even in sequences with rule recursion, streaming occurs as a function of pitch-time tradeoffs. The result is that attending follows the higher order time periods of overlapping streams and serial order information about the hierarchical pattern as a whole is lost.

In summary, under the assumption that people respond to and use dynamic serial transformations that are embedded within time sequences, rhythm and rate become integral parts of the serial organization as it is perceived and reproduced. Supporting research illustrates that rhythm and rate do interact with other aspects of serial structure, often in ways that qualify predictions derived from rate-relational or coding theory models. Testable hypotheses about the role of accent isochrony and the effects of pitch time tradeoffs have been proposed, but specific models that fully encompass the complex interactions of transformational invariants remain to be refined.

Conclusions

Three broad theoretical approaches reflect respectively different approaches to describing the organizational structure of events in time and the way this structure is perceived. Rate-relational theories emphasize the role of certain serial relationships and assume that time places constraints on processing these relationships. They tend to ignore the role of rhythm, however. Coding theories, on the other hand, emphasize the symbolic translation of serial relationships in acquiring (perceiving) and ultimately using codes to reproduce organizational structure. In contrast to rate-relational models, they have failed to consider systematic and selective influences of rate (and/or rhythm) on encoding and decoding processes. Finally, a dynamic transformational approach focuses upon the way in which temporal periodicities associated with various pattern transformations contribute to good or poor rhythms.

References

Allport, D.A. Phenomenal simultaneity and the perceptual moment hypothesis. *British Journal of Psychology*, 1968, *59*, 395-406.

Boltz, M., Marshburn, E., Jones, M.R., & Johnson, W. (in press). Serial pattern structure and temporal order recognition. To appear in *Perception and Psychophysics*.

Bregman, A.S. The formation of auditory streams. In: J. Requin (Ed.), *Attention and Performance VII*. Hillsdale, NJ: Lawrence Erlbaum Associates, 1978, pp. 63-75.

Bregman, A.S., & Campbell, J. Primary auditory stream segregation and perception of order in rapid sequence of tones. *Journal of Experimental Psychology*, 1971, *89*, 244-249.

Butler, D., & Brown, H. Tonal structure versus function: Study of the recognition of harmonic motion. *Music Perception*, 1984, *2*, 6-24.

Collard, R.R.A., & Buffart, H.F.J.M. Minimization of structural information: A set-theoretical approach. *Pattern Recognition*, 1982, *6*, 21-32.

Collard, R., & Povel, D.J. Theory of serial pattern production: Tree traversals. *Psychological Review*, 1982, *89*, 693-707.

Collard, R., Vos, P., & Leeuwenberg, E. What melody tells about metre in music. *Zeitschrift für Psychology*, 1981, *189*, 25-33.

Dannenbring, G.L., & Bregman, A.S. Stream segregation and the illusion of overlap. *Journal of Experimental Psychology: Human Perception and Performance*, 1976, *2*, 544-555.

Deutsch, D. The processing of structured and unstructured tonal sequences. *Perception and Psychophysics*, 1980, *28*, 381-389.

Deutsch, D., & Feroe, J. The internal representation of pitch sequences in tonal music. *Psychological Review*, 1981, *88*, 503-522.

Divenyi, P.L., & Hirsh, I.J. Identification of temporal order in three-tone sequences. *Journal of Acoustical Society of America*, 1974, *56*, 144-151.

Divenyi, P.L., & Hirsh, I.J. The effect of blanking on the identification of temporal order in three-tone sequence. *Perception and Psychophysics*, 1975, *17*, 246-252.

Dowling, D.J. The perception of interleaved melodies. *Cognitive Psychology*, 1973, *5*, 322-337.

Dowling, D.J. Scale and contour: Two components of a theory of memory for melodies. *Psychological Review*, 1978, *85*, 341-354.

Ehrlich, S. Rhythm: Recent French contributions. *Acta Psychologica*, 1960, *17*, 155-176.

Fay, W.H. *Temporal sequence in the perception of speech*. The Hague: Mouton, 1966.

Fraisse, P. *Les structures rhythmiques*. Louvain: University of Louvain, 1956.

Fraisse, P. Rhythm and tempo. In: D. Deutsch (Ed.), *The psychology of music*. New York: Academic Press, 1982, pp. 149-180.

Fraisse, P. Perception and estimation of time. *Annual Review of Psychology*, 1984, *35*, 1-36.

Garner, W.R. *The processing of information and structure*. Potomac, MD: Lawrence Erlbaum Associates, 1974.

Garner, W.R., & Gottwald, R.L. Some perceptual factors in the learning of sequential patterns of binary events. *Journal of Verbal Learning and Verbal Behavior*, 1967, *6*, 582-589.

Garner, W.R., & Gottwald, R.L. The perception and learning of temporal patterns. *The Quarterly Journal of Experimental Psychology*, 1968, *20*, 97-109.

Gibson, J.J. *The ecological approach to visual perception*. Boston: Houghton Mifflin, 1979.

Greeno, J., & Simon, H.A. Processes for sequence production. *Psychological Review*, 1974, *81*, 187-198.

Hahn, J., & Jones, M.R. Invariants in auditory frequency relations. *Scandinavian Journal of Psychology*, 1981, *22*, 129-144.

Handel, S. Perceiving melodic and rhythmic auditory patterns. *Journal of Experimental Psychology*, 1974, *103*, 922-933.

Handel, S., & Oshinsky, J.S. The meter of syncopated auditory polyrhythms. *Perception and Psychophysics*, 1981, *30*, 1-9.

Handel, S., Weaver, M.S., & Lawson, G. Effect of rhythmic grouping on stream segregation. *Journal of Experimental Psychology: Human Perception and Performance*, 1983, *9*, 637-651.

Hirsh, l.J. Auditory perception of temporal order. *Journal of the Acoustical Society of America*, 1959, *31*, 759-767.

Hirsh, I.J. Temporal order and auditory perception. In: H.R. Moskowitz, B. Sharf, & J.D. Stevens (Eds.), *Sensation and measurement*. Dordrecht: Reidel, 1974, pp. 251-258.

Hirsh, I.J., & Fraisse, P. Simultanéité et succession de stimuli, hétérogènes. *l'Année Psychologique*, 1964, *64*, 1-19.

Hirsh, I.J., & Sherrick, C.E., Jr. Perceived order in different sense modalities. *Journal of Experimentual Psychology*, 1961, *62*, 423-432.

Howard, J.H., & Ballas, J.A. Acquisition of acoustic pattern categories by exemplar observation. *Organizational Behavior and Human Performance*, 1982, *30*, 157-173.

Idson, W.L., & Massaro, D.W. Cross-octave masking of single tones and musical sequences: The effects of structure on auditory recognition. *Perception and Psychophysics*, 1976, 19, 155-175.

Johansson, G. Visual motion perception. *Scientific American*, 1975, *232*, 76-89.

Jones, M.R. Cognitive representations of serial patterns. In: B. Kantowitz (Ed.), *Human information processing: Tutorials in performance cognition*. Potomac, MD: Lawrence Erlbaum Associates, 1974, pp. 187-229.

Jones, M.R. Time, our lost dimension: Toward a new theory of perception, attention, and memory. *Psychological Review*, 1976a, *83*, 323-355.

Jones, M.R. Levels of structure in the reconstruction of temporal and spatial serial patterns. *Journal of Experimental Psychology: Human Learning and Memory*, 1976b, *2*, 475-488.

Jones, M.R. Auditory patterns: Studies on the perception of structure. In: E.C. Carterette & M.P. Friedman (Eds.), *Handbook of perception*, Vol. 8. New York: Academic Press, 1978, pp. 255-288.

Jones, M.R. Only time can tell: On the topology of mental space and time. *Critical Inquiry*, 1981a, *7*, 557-576.

Jones, M.R. A tutorial on some issues and method in serial pattern research. *Perception and Psychophysics*, 1981b, 30, 492-508.

Jones, M.R., Kidd, G., & Wetzel, R. Evidence for rhythmic attention. *Journal of Experimental Psychology: Human Perception and Performance*, 1981, *7*, 1057-1073.

Jones, M.R., Maser, D.J., & Kidd, G.R. Rate and structure in memory for auditory patterns. *Memory and Cognition*, 1978, 6, 246-258.

Jones, M.R., & Zamostny, K.P. Memory and rule structure in the prediction of serial patterns. *Journal of Experimental Psychology: Human Learning and Memory*, 1975, 1, 295-306.

Kotovsky, K., & Simon, H.A. Empirical tests of a theory of human acquisition of concepts for sequential patterns. *Cognitive Psychology*, 1973, *4*, 399-424.

Leeuwenberg, E.L.L. Quantitative specification of information in sequential patterns. *Psychological Review*, 1969, *76*, 216-220.

Macar, F. Time psychophysics and related models. In: J.A. Michon & J.L. Jackson (Eds.), *Time, mind, and behavior*. Heidelberg: Springer Verlag, 1985, pp. 112-130.

McNally, K.A., & Handel, S. Effect of element composition on streaming and the ordering or repeating sequences. *Journal of Experimental Psychology: Human Perception and Performance*, 1977, *3*, 451-460.

Martin, J.G. Rhythmic (hierarchical) versus serial structure in speech and other behavior. *Psychological Review*, 1972, *79*, 487-509.

Massaro, D.W. *Experimental psychology and information processing*. Chicago: David MacNally, 1975.

Miller, G.A., & Heise, G.A. The trill threshold. *Journal of the Acoustical Society of America*, 1950, *22*, 637-638.

Oshinsky, J.S., & Handel, S. Syncopated auditory polyrhythms: Discontinuous reversals in meter interpretation. *Journal of the Acoustical Society of America*, 1978, *63*, 936-939.

Povel, D.J. Internal representation of simple temporal patterns. *Journal of Experimental Psychology: Human Perception and Performance*, 1981, *7*, 3-18.

Povel, D.J. A theoretical framework for rhythm perception. *Psychological Research*, 1984, *45*, 315-337.

Povel, D.J. Time, rhythms and tension: In search of the determinants of rhythmicity. In: J.A. Michon & J.L. Jackson (Eds.), *Time, mind, and behavior.* Heidelberg: Springer Verlag, 1985, pp. 215-225.

Povel, D.J., & Collard, R. Structural factors in patterned finger tapping. *Acta Psychologica*, 1982, *52*, 107-123.

Povel, D.J., & Essens, P. Perception of temporal patterns. To appear in *Music Perception*.

Povel, D.J., & Okkerman, H. Accents in equitone sequences. *Perception and Psychophysics*, 1981, *30*, 565-572.

Preusser, D. The effect of structure and rate on the recognition and description of auditory temporal patterns. *Perception and Psychophysics*, 1972, *11*, 233-240.

Restle, F. Theory of serial pattern learning: Structural trees. *Psychological Review*, 1970, *77*, 481-495.

Restle, F. Serial patterns: The role of phrasing. *Journal of Experimental Psychology*, 1972, *92*, 385-390.

Restle, F. Structural ambiguity in serial pattern learning. *Cognitive Psychology*, 1976, *8*, 357-381.

Restle, F., & Brown, E.R. Serial pattern learning. *Journal of Experimental Psychology*, 1970a, *83*, 120-125.

Restle, F., & Brown, E.R. Organization of serial pattern learning. In: G.H. Bower (Ed.), *The psychology of learning and motivation: Advances in research and theory*, Vol. 4. New York: Academic Press, 1970b, pp. 249-331.

Rosenbaum, D.A., Kenny, S.B., & Derr, M.A. Hierarchical control to rapid movement sequences. *Journal of Experimental Psychology: Human Perception and Performance*, 1983, *9*, 86-102.

Royer, F.L., & Garner, W.R. Response uncertainty and perceptual difficulty of auditory temporal patterns. *Perception and Psychophysics*, 1966, *1*, 41-47.

Simon, H.A. Complexity and the representation of patterned sequences of symbols. *Psychological Review*, 1972, *79*, 369-382.

Simon, H.A., & Kotovsky, K. Human acquisition of concepts for sequential patterns. *Psychological Review*, 1963, *70*, 534-546.

Simon, H.A., & Sumner, R.K. Pattern in music. In: B. Kleinmuntz (Ed.), *Formal representation of human judgment*. New York: Wiley, 1968, pp. 219-250.

Stroud, J.M. The fine structure of psychological time. In: H. Quastler (Ed.), *Information theory in psychology.* Glencoe, IL: Free Press, 1955, pp. 174-207.

Vitz, P.S., & Todd, T.C. A coded element model of the perceptual processing of sequential stimuli. *Psychological Review*, 1969, *76*, 433-449.

Warren, R.M. Auditory temporal discrimination by trained listeners. *Cognitive Psychology*, 1974, *6*, 237-256.

Warren, R.M., Obusek, C.J., Farmer, R.M., & Warren, R.P. Auditory sequence: Confusion of patterns other than speech and music. *Science*, 1969, *144*, 586-587.

Weyl, H. Symmetry. In: J.R. Newman (Ed.), *The world of mathematics*, Vol. 1. New York: Simon & Schuster, 1956, pp. 671-724.

Reference Notes

1. Boltz, M., & Jones, M.R. *Does rule recursion make melodies easier to reproduce?* If not, what does? Manuscript under review.
2. Gabrielsson, A. Music psychology - A survey of problems and current research activities. *Basic Musical Functions and Musical Ability.* Publications issued by the Royal Swedish Academy of Music, no. 32, 1981.
3. Noorden, L.P.A.S. van. Rhythmic fission as a function of tone rate. *IPO Annual Progress Report*, 1971, *6*, 9-12.
4. Noorden, L.P.A.S. van. *Temporal coherence in the perception of tone sequences*. Doctoral Dissertation. Eindhoven: Eindhoven University of Technology, 1975.
5. Warren, R.M. Relation of verbal transformations to other perceptual phenomena. In: *Conference Publication 42*. IEE/NPL Conference on Pattern Recognition, Teddington, England: Institution of Electrical Engineers (Suppl. 8 pp), 1968.

Chapter 14. Time, Rhythms and Tension: In Search of the Determinants of Rhythmicity[1]

Dirk-Jan Povel

Introduction

All over the world people enjoy producing and listening to rhythmical patterns. Good rhythms may produce a feeling of tension in the listener or elicit tendencies to move or dance. This psychological attribute of rhythmical patterns will be referred to as ›rhythmicity‹. I will use the adjective ›rhythmical‹ to denote this same meaning; thus pattern X may be said to be more rhythmical than pattern Y. From the viewpoint of the psychologist studying perception it is a most intriguing phenomenon that even simple temporal patterns may evoke strong responses.

Understanding rhythm perception thus, at least, requires understanding why and how rhythms differ in rhythmicity. The aim of this paper is to trace the factors that determine the rhythmicity of a rhythmical pattern. The ultimate goal is to predict rhythmicity responses associated with rhythmical patterns. Recently a theoretical framework for the perception of rhythmical patterns has been proposed (Povel, 1984; Povel & Essens, 1985) which is taken as a starting point for the present research. Basic to this model is the notion that the temporal structure in rhythmical patterns can only be assessed accurately by means of some ›clock‹. By clock I mean a system producing pulses at regular intervals thus allowing the measurement of time. It is assumed that a listener tries to generate an internal clock which is subsequently used to specify the temporal structure in a sequence. Since each sequence can in principle be described in terms of several clocks, differing in time unit and phase, the model predicts the features of the clock that has the best chance to be induced. The tendency to tap a beat at regular intervals along with a rhythmical pattern is seen as a reflection of this internal clock. The model, formalized in terms of a computer program, is described in detail in Povel & Essens (in press) along with a few experiments testing predictions from the model.

Since, to my knowledge, no systematic work in this field has been done, I have adopted an exploratory approach. In a first attempt I developed two hypotheses concerning the cause of rhythmicity in rhythmical patterns which were tested in pilot experiments. The first hypothesis, in part inspired by the ideas of Yeston (1976), states that a pattern will evoke perceptual tension when it simultaneously induces two different clocks. The two clocks will yield two different codings of the pattern thus giving rise to an ambiguous percept which produces the experienced tension. The second hypothesis states that the rhythmicity of a pattern is higher, the

1 The research reported in this chapter was in part supported by a grant from the Netherlands Organization of Pure Research (ZWO).

greater the discrepancy between the accent distribution of the pattern and that of some ideal accent distribution like that proposed by Martin (1972). The results of the experiments (Povel, Note 1) led me to reject both hypotheses.

In this paper I propose a different hypothesis based on a process-oriented approach which seems capable of explaining the majority of the results collected so far. This hypothesis addresses the influence of displaced beats in a temporal pattern.

The Displaced Beat Hypothesis

The hypothesis of interest is based on the assumption that a subject, when listening to a pattern, constantly checks whether the current clock (beat) as inferred from the structure of the previous tones in the pattern still fits. This is the sort of monitoring function that one supposes to be active when following a piece of music in which small temporal changes inevitable occur, or when playing in duet, or even when constructing the specious present (see Michon, 1985, chapter 2 of the present volume).

The following assumptions are made:

1. Rhythmicity arises when the monitoring mechanism, on basis of one or more events in the sequence, registers a deviation from the prior pattern of such a degree that an adjustment of the current clock must be made. This will create uncertainty and ambiguity which is experienced as perceptual tension or rhythmicity.
2. Rhythmicity will be higher when such adjustment must be made more often.
3. Rhythmicity will be higher the stronger the current clock is at the moment that contradictive evidence occurs.

In a few examples I will show how these principles apply to actual temporal patterns. Consider the following pattern A in which tones are indicated with ›I‹, the relative duration of the intervals between tones with the help of dots ›.‹ and the actually induced clock (beats) with ›!‹.

```
1   2   3     4   5     6       1
I . I . I . . I . I . . I . . . !          (A)
!_______!_______!_______!_______!
```

As already stated, the displaced beat hypothesis holds that a subject, when listening to a pattern, constantly checks whether the current clock inferred from previous accented tones still matches. If the next accented tone falls slightly off-beat, either early or late, the subject tends to accept that tone as the actual beat and expects the next accented tone to occur at the regular interval. As said, the fact that the subject must revise his current clock will produce tension. Translated into example B presented:

```
!-------!-------!-------!-------!          (regular beat)
1   2   3     4   5     6       1
I . I . I . . I . I . . I . . . I                           (B)
!_______!_____!_______!                    (displaced beat)
              ^       ^
              !       !
   displaced beat   next beat ?
```

the subject supposes that tone 4 is a displaced beat and consequently assumes that the next accented tone will occur at the indicated location (relative to this displaced beat). Tension will further increase when the next beat does not occur at the expected location. The subject must now again revise his or her expectation and accept the regular clock (indicated above the pattern) as the correct one. Note that tone 5 may be seen as confirming the displacement hypothesis since that tone falls precisely in the middle of the displaced beat interval, as is the case in the first beat interval of the pattern. This view of the process involved in creating rhythmicity nicely explained why subjects report that the tension in the pattern is caused by a lengthening of the interval between tones 5 and 6. According to the displaced beat hypothesis tone 6 is felt to arrive too late since the next beat interval is expected to occur one time-unit earlier.

In the following pattern the hypothetical displacement is disconfirmed earlier by tone 5 which does not fall in the middle of the supposed beat interval but instead in the middle of the regular beat of the pattern (indicated above the pattern). Thus in pattern C counter-evidence with respect to the displaced beat hypothesis occurs earlier:

```
!-------!-------!-------!-------!
1   2   3     4     5   6       1
I . I . I . . I . . I . I . . . !          (C)
!_______!_____!_______!
              ^     ^
              !     !
           beat?   No!
```

I assume that the perceived tension is higher the later the displacement hypothesis is disconfirmed, which leads to the prediction that pattern B is more rhythmical than pattern C. This is one of the predictions that will be tested in the experiment to be reported below.

Another way to test the displaced beat hypothesis is by continuing the pattern in accordance with the hypothesized expectation of the subject and see what happens to the judged rhythmicity of the sequence. According to the displaced beat hypothesis the sequence is expected to continue not as actually occurs in pattern D:

```
I . I . I . . I . I . . I . . . !          (D)
```

but as in pattern E:

```
I . I . I . . I . I . I . . . !
!_______!_____!_______!_______!            (E)
```

(note that the latter sequence is 15 time units long instead of 16)

Pattern E indeed does not seem to be a very rhythmical pattern. This could be taken as supporting the idea that not the irregularity itself is the main cause of the tension but the disconfirmation of the expected continuation of the pattern.

The hypotheses outlined in the preceding paragraphs are tested in an experiment in which predicted rhythmicity is related to rhythmicity judgments about patterns that vary in their degree of disconfirmatory tension.

Experiment

Stimuli

In this experiment 15 rhythms are used as displayed in Table 1. The patterns are formed by systematically varying the configuration of tones in the second and third beat interval. Patterns 1-12 have a length of 16 time units and Patterns 13-15 are 15 time units long.

The time unit is the shortest interval occurring in the patterns, that is, the interval between two dots. The duration of the time unit used in this experiment is 150 ms. The patterns are composed of two types of sounds produced by means of a drum computer (DRUMATIX, Roland, TR-606) controlled by a PDP 11/03 computer. The main pattern is formed with the sound of a *high tom* symbolized in Table 1 with ›I‹. Synchronous with the main pattern an isochronic sequence is generated with the sound of the *closed hihat*, indicated with the symbol ›!‹. This sequence is added to ensure the induction of a beat on the part of the listener (Note that in Patterns 13-15 the *hihat* sequence is not isochronous). The *high tom* sequence is produced at a comfortable loudness level, while the *hihat* sequence is generated at a level just above threshold. In Patterns 1-7 only the length of the intervals in the main Pattern is varied while in Patterns 8-12 also physical accents (by increasing the intensity of the concerning sound) are added. Patterns 13-15 are transformations of Patterns 3-5 in a similar way as Pattern D in the introduction was transformed.

Procedure

Subjects compared pairs of rhythms and indicated which rhythm of a pair they judged as more rhythmical. To this end subjects were provided with a button panel containing two stimulus buttons and two response buttons. Each stimulus button had a response button placed directly below it. If a subject pushed one of the stimulus buttons the corresponding rhythm was presented in a repeated fashion. Presentation could be stopped by pressing the same button. Next the subject could push the other stimulus button to receive the other rhythm. In this way the subject could shift from one rhythm to the other as often as desired. The response was given by pressing the response button corresponding to the rhythm judged as the more rhythmical.

The 16 pairs of rhythms used in the experiment are shown in Table 2. Rather than having all possible pairs compared (120 pairs) I decided to include only those pairs that differ critically according to the hypothesis.

The pairs were presented in a semi-random order such that two consecutive pairs never contained the same rhythm. When 16 comparisons had been made, the experiment was repeated using a different order of presentation. Thus subjects made 34 comparisons in total and judged each pair twice. The experiment (including practical trials) took on the average 20 minutes.

```
I = high tom ;  ! = closed hihat

    1   2   3   4   5       6       1
1.  I . I . I . I . I . . . I . . . !          222244
    !       !       !       !       !

    1   2   3     4 5       6       1
2.  I . I . I . . I I . . . I . . . !          223144
    !       !       !       !       !

    1   2   3     4     5   6       1
3.  I . I . I . . I . . I . I . . . !          223324
    !       !       !       !       !

    1   2   3     4         5       1
4.  I . I . I . . I . . . . I . . . !          22354
    !       !       !       !       !

    1   2   3     4   5     6       1
5.  I . I . I . . I . I . . I . . . !          223234
    !       !       !       !       !

    1   2   3   4     5     6       1
6.  I . I . I . I . . I . . I . . . !          222334
    !       !       !       !       !

    1   2   3   4       5   6       1
7.  I . I . I . I . . . I . I . . . !          222424
    !       !       !       !       !

    1   2   3     4 5   6   7       1
                  <
8.  I . I . I . . I I . I . I . . . !          2231224
    !       !       !       !       !

    1   2   3     4 5       6       1
                  <
9.  I . I . I . . I I . . . I . . . !          223144
    !       !       !       !       !

    1   2   3     4 5 6     7       1
                  <
10. I . I . I . . I I I . . I . . . !          2231134
    !       !       !       !       !

    1   2   3   4   5 6     7       1
11. I . I . I . I . I I . . I . . . !          2222134
    !       !       !       !       !

    1   2   3   4   5 6 7   8       1
                      <
12. I . I . I . I . I I I . I . . . !          22221124
    !       !       !       !       !

    1   2   3     4     5 6       1
13. I . I . I . . I . . I I . . . !            223314
    !       !     !       !       !

    1   2   3     4       5       1
14. I . I . I . . I . . . I . . . !            22344
    !       !     !       !       !

    1   2   3     4   5   6       1
15. I . I . I . . I . I . I . . . !            223224
    !       !     !       !       !
```

Table 1. Rhythms used in the experiment. Of each pattern one complete period is shown which includes the interval between the last tone and the first tone of the next period. In a few patterns accents are added, indicated by < above the tone. The column at right shows the main pattern notated as a numerical sequence of onset-to-onset intervals.

I = high-tom ;   ! = closed hihat

```
Pair        Rhythm   Resp %                         Rhythm   Resp %

01.          (1)      11                             (2)      89

    I . I . I . I . I . . . I . . . !    I . I . I . . I I . . . I . . . !
    !_______!_______!_______!_______!    !_______!_______!_______!_______!

02.          (2)      4                              (3)      96

    I . I . I . . I I . . . I . . . !    I . I . I . . I . . I . I . . . !
    !_______!_______!_______!_______!    !_______!_______!_______!_______!

03.          (2)      27                             (7)      73

    I . I . I . . I I . . . I . . . !    I . I . I . I . . . I . I . . . !
    !_______!_______!_______!_______!    !_______!_______!_______!_______!

04.          (3)      70                             (7)      30

    I . I . I . . I . . I . I . . . !    I . I . I . I . . . I . I . . . !
    !_______!_______!_______!_______!    !_______!_______!_______!_______!

05.          (2)      30                             (11)     70

    I . I . I . . I I . . . I . . . !    I . I . I . I . I I . . I . . . !
    !_______!_______!_______!_______!    !_______!_______!_______!_______!

06.          (3)      62                             (4)      38

    I . I . I . . I . . I . I . . . !    I . I . I . . I . . . . I . . . !
    !_______!_______!_______!_______!    !_______!_______!_______!_______!

07.          (3)      16                             (5)      84

    I . I . I . . I . . I . I . . . !    I . I . I . . I . I . . I . . . !
    !_______!_______!_______!_______!    !_______!_______!_______!_______!

08.          (4)      32                             (5)      68

    I . I . I . . I . . . . I . . . !    I . I . I . . I . I . . I . . . !
    !_______!_______!_______!_______!    !_______!_______!_______!_______!

09.          (8)      55                             (9)      45
                  <                                    <
    I . I . I . . I I . I . I . . . !    I . I . I . . I I . . . I . . . !
    !_______!_______!_______!_______!    !_______!_______!_______!_______!

10.          (8)      18                             (10)     82
                  <                                    <
    I . I . I . . I I . I . I . . . !    I . I . I . . I I I . . I . . . !
    !_______!_______!_______!_______!    !_______!_______!_______!_______!

11.          (9)      16                             (10)     84
                  <                                    <
    I . I . I . . I I . . . I . . . !    I . I . I . . I I I . . I . . . !
    !_______!_______!_______!_______!    !_______!_______!_______!_______!
```

```
12.          (6)      91                            (7)      9
    I . I . I . I . . I . . I . . . |     I . I . I . I . . . I . I . . . |
    |_______|_______|_______|_______|     |_______|_______|_______|_______|

13.          (6)      75                            (11)     25
    I . I . I . I . . I . . I . . . |     I . I . I . I . I I . . I . . . |
    |_______|_______|_______|_______|     |_______|_______|_______|_______|

14.          (3)      62                            (13)     38
    I . I . I . . I . . I . I . . . |     I . I . I . . I . . I I . . . |
    |_______|_______|_______|_______|     |_______|_____|_______|_______|

15.          (4)      73                            (14)     27
    I . I . I . . I . . . . I . . . |     I . I . I . . I . . . I . . . |
    |_______|_______|_______|_______|     |_______|_____|_______|_______|

16.          (5)      84                            (15)     16
    I . I . I . . I . I . . I . . . |     I . I . I . . I . I . I . . . |
    |_______|_______|_______|_______|     |_______|_____|_______|_______|
```

Table 2. **Pairs of rhythms compared with respect to rhythmicity. Percentage of responses that judge one or the other rhythm in a pair more rhythmical are indicated above each rhythm. N = 22. (NB. Rhythm number 12 is not used in the comparisons.)**

Subjects

Only subjects with some musical experience (at least 4 years music lessons) could volunteer for the experiment. Twenty-two subjects, undergraduates in psychology, participated in the experiment.

Results

In Table 2 the results of the paired comparisons are summarized. For each rhythm of a pair the percentage of responses is indicated which judge that rhythm as the more rhythmical of the two. I will now successively discuss the outcome of the comparisons of the 16 pairs of rhythm against the background of the proposed hypothesis.

Pair 1 (1,2). According to the hypothesis rhythm 2 may cause the listener to expect the beat to coincide with tone 4 (compare Pattern A in the introduction). This expectation, however, is immediately shown to be wrong by the occurrence of tone 5 which is subjectively accented. Since a similar expectation does not arise in rhythm 1, rhythm 2 is theoretically more rhythmical than rhythm 1. This is supported by the data.

Pair 2 (2,3). In both these pairs an expectation may arise that the beat coincides with tone 4. This expectation is disproved at a later point in rhythm 3 than in rhythm 2. Therefore rhythm 3 is theoretically more rhythmical, which is confirmed by the data.

Pair 3 (2,7). According to the hypothesis, rhythm 7 will not evoke the expectation of a displaced beat, tone 4 being too early for such an expectation. Moreover, tone 4 fills the second beat interval in the same way as tone 2 fills the first. So I guess that the subject will expect the beat to fall at its proper point in time, that is in the middle between tones 4 and 6. This expectation turns out to be correct when the rest of the pattern is heard. Still there may be some uncertainty in the subject since the beat does not coincide with a tone. In rhythm 2 a displaced beat expectation may arise, but the tension evoked will be small as shown above. For this pair no clear-cut prediction can be made. The data show that rhythm 7 is judged more rhythmical in 73 percent of the cases.

Pair 4 (3,7). Here two rhythms are compared that both were part of earlier pairs. It was argued that rhythm 7 may evoke some uncertainty whereas rhythm 3 will arouse tension because a displaced beat expectation arises. Theoretically it is to be expected that rhythm 3 is the more rhythmical. This is in accordance with the empirical findings: 70 percent of the responses judged rhythm 3 more rhythmical.

Pair 5 (2,11). Here we have two rhythms that differ in an interesting way. Rhythm 2 induces an expectation of a displaced beat which is disconfirmed almost immediately, whereas in rhythm 11 the expectation that the regular beat coincides with tone 5 is immediately disconfirmed by the occurrence of tone 6. Note that tone 6, being the second of a two-tone cluster, is subjectively accented (Povel & Okkerman, 1981). Also in this case no prediction can be made. It turns out that subjects find rhythm 11 more rhythmical in 70 percent of the cases.

Pair 6 (3,4). Both rhythms in this pair evoke an expectation of a displaced beat, caused by tone 4. The only difference between the two rhythms is the duration of the interval before the next tone arrives: in rhythm 3 after three time units, in rhythm 4 after five time units. Tone 5 in rhythm 3 may, in a way, disconfirm the displaced beat hypothesis, albeit that this disconfirmation is not yet decisive. Decisive disconfirmation comes with the occurrence of tone 6. In rhythm 4 nothing happens between the arousal of the displaced beat expectation at tone 4 and its rejection at tone 5. Now one might argue that tone 5 in rhythm 3 adds to the uncertainty of the correct perceptual interpretation thus increasing the experienced rhythmicity. The data support this reasoning only marginally: 62 percent of the judgments indicate rhythm 3 as the more rhythmical.

Pair 7 (3,5). A possible interpretation of rhythm 3 has just been given so I will now look at rhythm 5. Like in rhythm 3 tone 4 will evoke an advanced beat expectation. It may further be conjectured that tone 5 strengthens this expectation because it occurs at the same interval from the assumed beat as tone 2 from the first beat. It may now be supposed that the fact that tone 6 does not occur at the expected location but one time unit later is most surprising because highly unexpected. As shown before in rhythm 3, the occurrence of tone 5 adds uncertainty about the correctness of the displaced beat hypothesis thus weakening it. Hence the main difference between the two rhythms pertains to the *strength* of the expectation of the location of the next beat. The supposition that the tension (rhythmicity) is greater the stronger the displaced beat hypothesis is indeed confirmed by the responses: 84 percent indicate rhythm 5 as the more rhythmical one.

Pair 8 (4,5). Here again two previously judged rhythms are compared. According to the hypothesis rhythm 5 is more rhythmical since the strength of expectation of a displaced beat is stronger for this rhythm than for rhythm 4. Indeed, 68 percent of the responses indicate rhythm 5 as the more rhythmical.

Pair 9 (8,9); *Pair 10* (8,10); *Pair 11* (9,10). These three pairs can be treated together since they are variants of pairs 6, 7 and 8. The difference is that pairs 9, 10 and 11 possess a tone at the location of the second beat while pairs 6, 7 and 8 do not. Furthermore I added physical accents to tone 4 of the rhythms 8, 9 and 10 such that displaced beat expectations may arise analogous to rhythms 3, 4 and 5. Thus we predict similar findings for these two groups of pairs. In looking at the responses we may conclude that the findings are indeed comparable. The preferences are in the same direction, although the distribution of responses differ somewhat. The fact that we do find these similarities may be taken as supporting the hypothesized mechanism.

Pair 12 (6,7). In rhythm 6 the listener will assume the second beat to fall at its regular location but it is disturbing that directly after the beat a (subjectively accented) tone occurs (tone 5) which seems to contradict this assumption. In rhythm 7 tone 5 occurs at a location in line with a similar assumption, namely precisely in the middle of the beat interval. Therefore it is predicted that rhythm 6 will be judged more rhythmical than rhythm 7, which is indeed found: 91 percent of the responses prefer rhythm 6 as the more rhythmical.

Pair 13 (6,11). Here again we have two rhythms previously used. In line with the notions proposed, rhythm 6 should be more rhythmical. Data indeed show that 75 percent of the responses prefer rhythm 6 to rhythm 11 with respect to rhythmicity.

Pair 14 (3,13); *Pair 15* (4,14); *Pair 16* (5,15). These pairs can also be discussed together as they are highly comparable. In fact they are variants of the rhythms 3, 4 and 5 obtained by continuing the patterns in accordance with the assumed displaced beat expectation. This is realized by shortening the penultimate beat interval by one time unit. The resulting rhythms are therefore one time unit shorter. Since in the patterns 13-15 the displaced beat expectation is confirmed I predict that these rhythms will yield less rhythmical tension than the base rhythms they are inferred from. As may be seen in Table 2 this is indeed found: rhythms 13-15 are systematically judged less rhythmical than the rhythms 3-5.

Discussion

I have proposed the hypothesis that the experience of rhythmicity results from the process that updates the internal clock in the light of the incoming stream of events. The results from the experiment reported above support this hypothesis well.

The proposed hypothesis, although rather complicated and in need of further verification, has two attractive features. Firstly, it is completely in line with the generally accepted idea that perceptual tension is related to ambiguity of interpretation. Secondly, it cannot only explain the experienced rhythmicity in pure, that is, abstract or at most mechanically performed rhythms such as I have studied here, but in principle also the rhythmical effects added by a performer. What a performer in producing rhythmical patterns typically does is to play certain notes slightly early or late. These deviations may affect the listeners' clock updating process in the same way as described above for pure rhythms. Thus the custom to play the second beat of the Viennese Waltz (in itself a most uninteresting rhythm) slightly ahead of time may cause the listener to adjust his or her current beat expectation in a similar way as indicated for the first pattern discussed in the introduction. Since the deviations brought about by a performer are small as compared to the displacements in the pure rhythms studied here, the actual perceptual effect will very much depend on the discriminative capacities of the listener.

In future research I will try to relate the presented findings with the concept of syncopation as defined in a recent paper by Longuet-Higgins & Lee (1984). Moreover an attempt will be made to incorporate the notions proposed here in the broader framework of a theory of expectancy (Jones, 1976; see also Jones, 1985, chapter 13 of the present volume).

References

Jones, M.R. Time, our lost dimension: Toward a new theory of perception, attention, and memory. *Psychological Review,* 1976, *83*, 323-355.

Jones, M.R. Structural organization of events in time. In: J.A. Michon & J.L. Jackson (Eds.), *Time, mind, and behavior.* Heidelberg: Springer Verlag, 1985, pp. 192-214.

Longuet-Higgins, H.C., & Lee, C.S. The rhythmic interpretation of monophonic music. *Music Perception*, 1984, *1*, 424-441.

Martin, J.G. Rhythmic (hierarchical) versus serial structure in speech and other behavior. *Psychological Review,* 1972, 79, 487-509.

Michon, J.A. The compleat time experiencer. In: J.A. Michon & J.L. Jackson (Eds.), *Time, mind, and behavior.* Heidelberg: Springer Verlag, 1985, pp. 20-52.

Povel, D.J. A theoretical framework for rhythm perception. *Psychological Research*, 1984, *45*, 315-337.

Povel, D.J., & Okkerman, H. Accents in equitone sequences. *Perception and Psychophysics*, 1981, *30*, 565-572.
Povel, D.J., & Essens, P.J. Perception of temporal patterns. *Music Perception*, in press.
Yeston, M. *The stratification of musical rhythm*. New Haven: Yale University Press, 1976.

Reference Note

1. Povel, D.J. Time, rhythms and tension: In search of the *determinants of rhythmicity.* Internal report 84FU11, University of Nijmegen, 1984.

Chapter 15. Timing in Action

L. Henry Shaffer

Introduction

Skilled action takes definite forms in space and time. If an action is repeated without altering its context then it may be reproduced in precise detail. In particular the ability to reproduce both the temporal pattern and time scale of an action allows us to consider it as a process structuring the time domain. Accordingly we can focus an inquiry on the ways in which time is structured in skilled performances. This does not commit us to suppose that time is independently specified in the control of motor production, and probably it often is not. In any case the performance can serve as a timekeeper for an observer or for the performer.

What are the requirements for a system to act as a timekeeper? It has to be able to produce a series of events, acting as temporal markers, such that the durations between successive events can form a definite pattern on a time scale defined in real time, allowing some degree of random error. By altering one or other of the requirements we can introduce two related concepts. One is a temporal pattern generator, which can produce a definite pattern but need not preserve a time scale; the other is a clock, which can generate only periodic events on a definite time scale. The labeling of concepts is a little arbitrary, because we have been made familiar, through the use of computers, with the idea of a programmable clock that can be programmed to generate temporal patterns. A timekeeper that can produce a variety of temporal patterns must receive an input that programs its parameters and so is a programmable clock. The important consideration then in a flexible system of timekeeping is that it should contain both an information source that can specify temporal patterns and a timekeeper, or clock, that can construct these patterns in real time. A description of timing in motor skill requires us to look at both of these.

Motor Programming

It is easy to show that skilled action is prepared in advance of its execution. If the preparation is in any way inhibited the action becomes slower, less fluent and more prone to error. This was demonstrated for instance in copy typing by restricting the preview of text, so that the typist could read only a few letters ahead of output (Shaffer, 1973). If the action is on a large scale, such as giving a speech or playing a sonata, we have to think of preparation as taking place at different levels of abstraction. The most abstract level is a structural plan, which can be committed to memory before the action or generated at the time the action is improvised. During execution of the plan a continually renewing fragment has to be made sufficiently explicit to control the ongoing motor output. We can call this fragment the motor program. A motor program, then, has to provide an epistemically adequate representation of the current fragment of an intended motor output.

It is supposed that the information represented in a motor program is still quite abstract, in the important sense that it does not specify details of muscle contraction but only the goals of movement, in terms of their space-time coordinates (Bernstein, 1967) or of producing certain sounds, of speech (MacNeilage, 1980) or music. The main reason for supposing this is that skilled action is instantly adaptive to the immediate circumstances in which it occurs, and it is likely that at such short notice the details of movement have to be computed locally by the motor system. On theoretical grounds it is a good strategy in something as complex as a human skill to devolve the processing of information from higher to lower centers of control (Sacerdoti, 1974). Thus we can suppose that there are procedures in the motor system, containing a wealth of background knowledge about movement, which can compute the muscle actions that will achieve goals designated by a motor program.

One of the consequences of supposing a devolution of control is that it allows us to consider timing information being generated at two levels, in the program and in procedures, and being implemented by different forms of timekeeper. This becomes particularly relevant in the context of musical performance (Shaffer, 1982, 1984; Shaffer et al., in press).

Timing Processes

The idea of a timekeeper pacing a performance was developed into a mathematical model by McGill (1962). In this model the timekeeper, or clock, generates pulses at regular intervals, and each pulse triggers an output event which occurs with some random delay. The result is that the durations between successive output events roughly preserve the periodicity of the clock but lose some of its precision, and in the time series there is a negative covariance between successive durations. Wing & Kristofferson (1973) showed that the essential properties of the model are little affected if the clock itself has a random element: the variance-covariance properties of the event durations can be simply stated in terms of the variances of the clock interval and the pulse-to-event delay. In a further generalization, Vorberg & Hambuch (1978) considered pulse intervals forming a recurrent group, generated by a series or hierarchy of clocks, in which the intervals within a group can be the same or different. Such models can thus generate a recurrent rhythm. The expressions for the variance-covariance properties of event duration remain similar to those in the simpler models.

A biological clock can take the form of an oscillator or a counter. An oscillator can produce a pulse at each zero crossing; a counter can produce a pulse and reset to zero each time its count, regular or Poisson, reaches a criterion value (Creelman, 1962; Kristofferson, 1984). It can act as a programmable clock if its oscillatory rate can be altered at a zero crossing, in one case, or if its criterion value can be altered at reset, in the other. A series of clocks in a Vorberg and Hambuch model is equivalent to a single programmable clock given a new criterion value in each cycle (see also Macar, 1985, chapter 7 of the present volume).

A temporal pattern generator lacks the temporal precision of a clock, and we can suppose that it derives its pattern from some form of analogue process. Typically it is assumed to produce a single pattern at a modifiable rate, rather than to be a programmable system. Thus the brain has to store a pattern for each of a set of sequences. This appears to be the assumption made by Kozhevnikov & Christovich (Note 1) for producing words in speech, and by Terzuolo & Viviani (1980) for words in typing. However, these authors have not been clear in distinguishing between the information specifying a pattern and the process that realizes it in real time, or in stating the timekeeping process assumed (see also Jones, 1985, chapter 13 of the present volume).

In the clock models described above it is assumed that a clock pulse triggers a movement with a small random delay. For reasons discussed below it is often unrealistic to suppose that the timed moment in a skilled movement is the moment it begins. It turns out that the mathematics of the models is unaffected if it is assumed instead that a movement is constructed with reference to a clock pulse (Shaffer, 1982).

It should be borne in mind in what follows that the essential function of a timekeeper, or clock, is to give an action a temporal structure or time scale more definite than can be achieved by processes structuring the movements themselves.

Motor Timing

Skilled action can be represented as a sequence of ballistic movements, each taking the limb or articulator to, or towards, a preset goal. This ballistic property is demonstrated in the trajectories of simple pointing or reaching movements, in which velocity increases smoothly to a single peak and then smoothly declines. If such movements are made repeatedly then, in the absence of contrary instruction, movements of the same amplitude have similar velocity curves and tend to have similar peak velocities (Cooke, 1980). An increase in the movement amplitude is often accompanied by an increase in peak velocity. This has the effect of narrowing the bounds of movement duration, but duration is not invariant. In general, movements tend to have a preferred range of speeds outside which they lose their precision. At low speeds they may break up into a succession of smaller movements (Morasso et al., 1983). The constraints on precision are presumably biomechanical, but we shall not inquire into these: what is relevant is that movements can, within limits, have definite time scales, which presumably can be parametrized by the motor procedures that control them.

Another property of skilled movements, apparently, is that they can be located in time relative to different fixed points in their trajectory. This can be illustrated by a few examples. If a person is asked to draw a line, starting in time on the third beat of a metronome, then the movement will be initiated at that moment in time. If asked to tap on the table on the third beat, a hand and finger movement will start earlier and reach completion at the required moment. If the task is to hit an approaching ball with a bat, then a flowing swing will be initiated with the aim of making contact with the ball at the moment of peak velocity, midway through the swing.

In the latter examples the movement has to anticipate a temporal goal. This implies that in situations such as these the skilled performer can construct an internal dynamic model of the world and use this to extrapolate its behavior to a future point in time (Craik, 1943; Poulton, 1957; Bernstein, 1967). The model also has to represent the motor capabilities of the performer if the anticipation is to be effective. The next two examples extend the principle of anticipatory modeling to compound movements.

Timing of Compound Movements

In catching a ball there is a period before contact in which the hand achieves its fine orientation towards the ball and the fingers shape for the grasp, which completes about 10 ms after initial contact. The successful grasp is initiated within about 14 ms of the optimal time (Alderson et al., 1974). The timing of the catch is presumably derived from visual information of the ball's trajectory, which serves to define a spacetime locus for the goal of movement. Note that the solution cannot be defined in either space or time alone.

In the long jump the athlete starts the run-up from a predetermined position and builds up speed in a fixed number of strides. There is an increase in the length of successive strides that is highly stereotyped, with little variation from one occasion to another. Lee et al. (1982) show that in this phase the flight time in each stride is fairly constant, and what appears to be regulated is the vertical component of thrust on the ground. However, a small random component in the stride length will be cumulative over the sequence and this cumulative error must be dissipated in the last few strides, so that in the final stride the foot comes as close as possible to the edge of the board without overlapping it. The good athlete can achieve this final position with an accuracy of a few centimeters, poised for the jump and without sacrificing forward momentum.

Lee et al. argue that in the closing phase of the run-up, optic flow information provides an estimate of time to arrival at the edge of the board and this is used to modify the flight times of the remaining strides, again by regulating the vertical thrust, the effect being to adjust stride length. If their analysis is correct, we see that a parameter derived from the visual input, time to arrival, is translated into a parameter of output, flight time of stride. It seems appropriate that the translation occurs within a single domain, of time, but we should note two things. First, that the independent parameters of input and output are optic flow and muscle force respectively, time being an intermediate parameter. Second, that it is incorrect to suppose, following recent theories of direct perception, that motor output is under direct control of the sensory input. The lengths of the final strides change neither uniformly nor progressively, but show a typical pattern of short-long-short. Perhaps this pattern is adopted to maximize leverage on the final stride, but the point is that these last strides must be controlled by a motor procedure that takes account of a number of factors, only one of which is derived from optic flow information.

In short, we have seen in these sections on motor timing that movements can themselves be parametrized to produce delicate and precise timing in simple and compound actions. In order to achieve this delicacy they often have to rely on internal dynamic models of the world that extrapolate its behavior and the action itself to a future point in time. Thus in the sense of having timing precision, the movements can function as timekeepers.

Rhythm

The concept of rhythm is often taken to entail periodicity, yet this is much too restrictive a criterion to apply to the behaviors that we call rhythmic. Trapeze flying, playing tennis, speaking and playing music are rhythmic activities but they are not periodic or at most only approximately so. Yet the tendency to identify rhythm with periodicity and regular structure is so strong it leads us to inquire whether skilled action is controlled by a central timekeeper, or whether the motor system is itself basically oscillatory in nature. The alternative is that movements are concatenated in time. If each ballistic movement has a temporal reference point then a series of movements can be coordinated in temporal relation using these references. Furthermore a series of movements often performed together may form a single motor trajectory having a definite temporal pattern (Shaffer, 1982).

The ways in which rhythm may enter into a performance will be examined in a variety of skills. It is worth pointing out that at least some of what is characterized as rhythm in a performance is its fluent and unhurried quality, which is a consequence of programming, whether or not there is direct timing control.

Cyclic Action

Bernstein (1967) and others have observed that cyclic repetitive movements, such as walking, running, sewing and filing metal, tend towards periodicity. The evidence relies more on graphical inspection than on analysis of the time series, nevertheless it seems plausible to suppose that there is some form of oscillator controlling the timing of such movements. In a theoretical analysis, Bernstein went on to show that on kinetic grounds this is an efficient way to perform such actions because it optimally conserves energy.

When one turns to other serial skills that do not have such regular movement cycles, the evidence for periodicity is weaker, yet a group based on Haskins Laboratory have argued that all action is basically oscillatory, in which the inherent periodicity may be modulated by contingent factors (see for instance Kugler et al., 1980). They assume that the motor system contains hierarchies, and perhaps heterarchies, of coordinative structures. Each unit structure within a hierarchy functions as an oscillator, with an intrinsic periodicity, controlling some component of the action. These oscillators become coupled within a hierarchy to produce an overall rhythm in the action. It can be seen that extending Bernstein's theory into a universal principle makes it vacuous unless direct evidence of an oscillatory basis can be obtained in non-cyclic action. So far there has been little attempt to test the theory by applying it directly to a variety of skills. Although Fowler (1980) has discussed speech production in terms of the theory, the evidence she gives in support of it is not compelling (Shaffer, 1982; see later in this chapter).

Handwriting and Drawing

Hollerbach (1979) has shown that handwriting can be simulated by modulating the amplitude and the left-to-right displacement of a sinusoidal oscillation, and has successfully implemented this with a robotic arm. Teulings & Maarse (1984) have indeed found a single peak in the power spectrum of the human handwriting signal, reflecting some periodicity of stroke production, but Thomassen & Teulings (1985, chapter 17 of the present volume), in a more detailed analysis show that it is only a tendency conditional upon a number of local factors of production. Recent attempts to provide computational models simulating both the spatial and temporal properties of writing show how such nearperiodicities may arise.

Morasso et al. (1983) describe a theory of trajectory formation in skilled movement, including pointing, drawing and writing, that can account for many of the features of these skills, including their timing. In essence they suppose that the apparently continuous movement in performing such actions is generated by a sequence of discrete ballistic strokes, each having the characteristic property of a bell-shaped velocity profile. The underlying trajectory is thus that of a generalized polygon having straight or curved sides. The final smoothness of curvature arises from a temporal overlap in passing control of production from one segment to the next. The greater the overlap the more smoothing occurs.

The timing properties of drawing or writing thus become a function of the durations of strokes and their sequencing. Morasso et al. assume that there is a preferred speed for a stroke, within a fairly narrow range of speeds, that is reproduced from one occasion to another; below this speed strokes tend to break up into a chain of smaller ones, and above it different strokes may collapse into a single one. The evidence from handwriting samples supports this and shows that the average stroke rate is broadly independent of the overall speed of movement. In addition, the tangential velocity of a stroke tends to vary inversely with the curvature of the segment. Thus the momentary velocity of movement can be under the control of a number of local factors, yet the tendency to keep peak velocity within a fairly narrow bandwidth has the effect of inducing a nearperiodicity in the overall sequence. This research is interesting not only for the light it throws on writing but for its extension to other skills as well.

Typing

In typing, the fingers can move with some degree of independence but are constrained by the keyboard action to produce strokes in sequence. As the degree of skill increases with practice, the timing of the sequence becomes more regular (Genest, 1956) and may achieve a nearperiodicity in highly skilled typists (Shaffer, 1973). The departures from periodicity arise in part from the logistics of moving the hands between rows in typing successive letters with the same hand or of using the same finger. This extra commitment to move the hand or finger during a transition has the effect of delaying the keypress. What this implies, however, is that a hand or finger not involved in a transition can anticipate its next stroke and move towards a canonical position. The implication was tested by presenting a special text that allowed a continuous alternation between hands in typing successive letters. Under these conditions the variance of stroke timing became very small and the overall rate was faster (Shaffer, 1978). Thus when keystrokes can be made from a canonical position their velocities are fairly uniform. There was also a negative covariance of adjacent intervals between strokes, indicating that in a condition of highly regular movement timing the performance may gratuitously assume a clock control.

Rumelhart & Norman (1982) present a model of typing in which the motor commands for successive strokes in a typing sequence are held in readiness in a motor program. Each command in the program has an action potential, but its activation is partially inhibited by each of the commands more immediate in the output sequence. However, the residual action potential for a stroke has the effect of moving the relevant finger and hand towards a canonical position, and at any moment there is a vector of such movement tendencies affecting hand position. The timing of succession is partly arranged by inhibiting any further strokes for a minimum interval during execution of the current one. This model gives a good account of many aspects of timing in fluent typing, including the approximation to periodicity under the condition of continuous hand alternation. One of its limitations is in accounting for some of the timing patterns in a sequence of within-hand transitions. The timing patterns are assumed to arise as a vector resultant of the partial activations of movement towards different keys affecting hand position. This account has the desirable feature of making the timing of a key-press responsive to its sequel but the results, for some typists at least, suggest that the motor system may use a different way of organizing such sequences, grouping them into a compound movement (Shaffer, 1978).

These accounts of writing and typing suggest that their rhythms derive from factors related to the constraints on stroke time and the control of their succession. There is no need to consider clocks or oscillators, except perhaps that clock control may be induced in conditions of highly regular motor output. Nor is any additional mechanism required to account for the reproducibility of timing when a sequence is repeated. This can arise from having stable rules in the control of the timing of movements and their succession and grouping. In particular, there is no need to suppose, as Terzuolo & Viviani (1980) do for typing, that timing profiles for words are stored in the mental lexicon. In any case, what would motivate the storage of such information for a skill like typing? Its rhythms have no expressive function and are not required by the task.

Speech

Speech production combines the kinematic properties of writing and typing. As in writing, an articulator in the vocal system can be involved in a continuous trajectory of movement that approximates a sequence of phonetic goals in a speech fragment. As in typing, the activity of speech is distributed in a number of articulators, the difference being that speech is formed by their parallel and coordinated action. Thus no new principle appears to be involved in the timing of speech, except perhaps in the coordination of parallel movements (Nooteboom, 1985, chapter 16 of the present volume covers some common ground with what follows).

The phonemic structure of a speech act presents a sequence of acoustic goals that have to be mapped into attainable spatial targets. By hypothesis, these are realized under the control of motor procedures. An articulator of the vocal tract can be thought of as executing a sequence of discrete gestures towards its successive targets. If phoneme targets are contiguous then the gestures may overlap in time, producing a ›smoothing‹ of the trajectory. This may result in an incomplete realization of some phonemes, as is observed in fast speech and in the de-stressing of syllables. It may also lead to a loss of vowel quality and, in the limit, to the elision of consonants and whole syllables, so that a sentence like »did you eat yet« may be reduced to »gee chet« (MacNeilage, 1980). The effect of smoothing is not uniform over a speech sequence: there appear to be semantic and syntactic factors that inhibit or limit gestural overlap on certain phonemes or syllables. Rate of speech can be varied not only by controlling overlap but also by controlling the duration of a gesture, and again it is the more sustained sounds, like the vowels, rather than the relatively transient consonants that are mainly affected.

If, on the other hand, an articulator plays no role in the production of one or more phonemes, it may move to anticipate its next target in the sequel. The extent of admissible anticipation is determined by its phonetic consequences (Kent & Minifie, 1977). Thus lip rounding for an open vowel may occur early in a consonantal sequence preceding the vowel, without affecting the consonant sounds, whereas early velar lowering for the production of a nasal sound may occur only if making the earlier sounds nasal is phonologically acceptable. To a first approximation, anticipatory movement can occur as soon as the articulator is free and its movement has no distinctive consequences for intermediate segmental sounds. However one of the most pervasive facts of speech production is that phonetic sounds are modified by their coarticulation.

For many speech sounds there is some degree of tolerance in coordinating the gestures of different articulators. When precise synchrony is required it seems to be arranged by mechanical factors. For example the build-up of air pressure in the oral cavity in the production of stop consonants has the effect of closing the velum and of briefly inhibiting voicing (Ohala, 1983). In general the coordination problem can be solved if the articulators can be given targets in spacetime. Although articulators move with some degree of independence they can receive common timing parameters from the procedures controlling phoneme production.

The remaining timing factor is that of phoneme sequencing. Insofar as it can be determined, phonemes have typical durations that differ from one to another, but these are by no means invariant and appear to be influenced by phonemic context, lexical stress, syntactic position and semantic importance (Huggins, 1978; Nooteboom, 1985, chapter 16 of the present volume). At the phonological level, some of the variables affecting duration are phonotactic, so that for instance vowel duration is shorter when followed by an unvoiced than when followed by a voiced consonant; others are syllabic, so that for instance the durations of all segments get shorter in longer syllables. Timing effects can also occur across syllables, as Sussman & Westbury (1981) have shown: the articulatory compatibility of successive vowels can affect the onset time of the later vowel (the authors describe the result differently!). Thus, it is likely that phoneme sequencing is arranged not on the succession of individual phonemes but according to a hierarchy of structural units, of which the syllable is only one. The syllable seems to be the primary context affecting phoneme duration and overlap, and it seems to be the unit on which stress variables act.

In some oriental languages like Japanese, timing has an active role in making phonemic distinctions, whereas in English this is likely to be a source of confusion only in isolated words or if there is ambiguity about the dialect being used. Timing distinctions occur more at the syllabic level of English, to indicate juncture, so that the proper distinction can be made between, for example, »I scream« and »ice cream«.

It is clear that timing is built into the conventions of the speech code. These timing factors can enter at a number of different levels of the speech structure, and it has to be supposed that a structure of such complexity is represented in the motor program. It does not follow however that a hierarchy of timing factors has to be implemented by a hierarchy of timekeepers, and indeed there need be no timing mechanism other than in the procedures controlling the gestures of the articulators. In particular there is no need for the temporal pattern generators assumed by Kozhevnikov & Christovich (Note 1) to govern the timing of words. The stable timing patterns in speech sequences can arise from phonological rules built into the motor procedures.

This procedural concept of timing in speech productions has much in common with the theory presented by Fowler (1980). Both agree that the timing patterns of speech are achieved by giving rich parametric specification, contained in a motor program, to a motor structure. The theories differ essentially in the nature of this structure and, as a consequence, in what gets timed. A motor procedure is, by assumption, a system that computes a motor output from a set of input specifications. The system assumed by Fowler is a hierarchic coordinative structure operating on an oscillatory principle. She also assumes that the consonantal and vowel sounds of speech are produced by parallel subsystems having some degree of autonomy. Speech timing in this system is arranged by timing a semi-continuous succession of vowel sounds and mapping the timing of consonants onto this. In effect then the oscillatory system for vowels times syllable production.

One of the purposes of assuming parallel subsystems producing vowels and consonants was to explain the fact that in changes of speed or stress, consonant durations in a syllable are less affected than vowel durations. Yet the general theory of coordinative structures assumes that the coupling of oscillators in subsystems leads them to entrain each other, and so it is not clear how it can accommodate this fact. Other weaknesses arise from the timing assumption. If there is a standard duration for each vowel in the sound alphabet, Fowler's theory cannot account for the fact that the duration of a syllable is greater the more consonants occur in the syllable (Huggins, 1978). Nor can it explain how the onsets of consonants following the vowel in the syllable are timed, or why the durations of consonants become compressed when they occur in groups. The computional model has yet to make explicit the phonological rules governing these facts, but the coordinative structure model seems an inappropriate one to account for them at all.

Playing Music

In order to consider timing in musical performance, we first need to describe briefly some aspects of musical structure. With exceptions that can be found in early Gregorian chant, certain folk music and some recent post-serial compositions, the rhythms of music are based upon a meter. A meter takes a recurrent beat, which provides the pulse of the music and organises it into a recurrent group, called the bar or measure, placing major accent on the first beat of a bar. The durations of notes and rests forming a melodic rhythm are obtained by subdividing and combining beat intervals. Though the meter in a piece of music may be regular, the melodic rhythm usually has few regularities or symmetries at the surface level. Thus the processes that generate melodic rhythm are more complex than those described by Jones (1985, chapter 13 of the present volume) or Povel (1985, chapter 14 of the present volume).

Stated in non-musical terms, the score for a piece of music provides a schedule for note production. Its tempo marking specifies a rate for the beat and the notation indicates how notes and rests should be timed in relation to the beat. Although, as will be seen, a player is not required to adhere rigidly to the schedule, the temporal commitment appears to be stronger for this skill than for the others we have discussed, which gives us reason to suppose that the player may make use of an internal timekeeper in controlling the timing of a performance. Given the relationship between melodic rhythm and meter it is also reasonable to suppose that a timekeeper, if used, paces a metrical unit, which is a regular event if not isochronic.

Music contains a number of structures, melodic, harmonic and metrical. Each has its own rhythm and together they define the musical pattern. These structures can also interact to affect musical meaning: harmonically important events usually occur on accented beats and their significance changes if they occur elsewhere;

similarly the interpretation of a melody changes if its groupings begin and end on unaccented beats. It is thus important for the listener to be able to discern the meter in the musical surface, and this is made easier if timing expectations arising from the meter are fulfilled by the performer. This will be a factor affecting the adoption of a timekeeper; others may arise from the technical requirements of performance.

The music written for some instruments, particularly for the keyboard, requires the performer to play several melodies at once. Since the melodies typically have different rhythms this creates special problems of timing. In playing contrapuntal music, such as fugues and canons, which may contain up to four voices or melodic parts (five in the case of organ music), it is usually desirable for the voices to maintain their musical identities. Thus it is important for them to have a rhythmic independence, and this can be better achieved if their rhythms can be constructed relative to a common metrical reference than to each other or, in other words, if separate procedures controlling note timing in each voice can make reference to a common timekeeper. This form of timing control has been demonstrated in performances of a Bach fugue (Shaffer, 1981). The demonstration entails obtaining repeat performances of the piece from the performer and showing, by means of an analysis of variance, that the durations of a particular metrical unit are relatively stable across performances.

Sometimes the music makes use of different so called ›aliquot divisions‹ of the beat in different melodies played simultaneously. For instance one melody may divide the beat into three equal notes and another into four. Though in principle it is possible for a pianist to play the separate rhythms in each hand by counting, in this case, the 1/12 divisions of the beat and placing notes accordingly, this is not feasible in practice at any musical speed of performance. The alternative is to establish the rhythms independently against the beat. Performances obtained of a Chopin study involving the cross-rhythm of three against four showed that the player was not only able to play the separate rhythms convincingly, but could at the same time allow a melody to move away from the beat in a controlled fashion, thus destroying the simple arithmetic relationships that would be obtained by dividing the beat by twelve (Shaffer, 1981, 1984). The rhythmic independence implied by this result supports the idea that the rhythms were produced relative to a superordinate timekeeper.

Sometimes the melodic rhythm is so complex or changes so rapidly that it would be difficult for the player to establish an overall rhythmic coherence other than by reference to a more regular unit, such as a metrical unit. Thus again we should expect the player to adopt timekeeper control to provide this reference, and evidence in support of this has been obtained in multiple playings of a piece by Satie (Shaffer et al., in press).

Finally, if two or more musicians play together they typically play different melodic rhythms. Even if the rhythms were the same it would be difficult to coordinate their timing at the note level by cross reference, given the finite time required to make use of auditory monitoring, and monitoring at this level of detail would tend to inhibit rhythmic independence between players. Thus it is likely that they adopt the more stable reference of a timekeeper pacing the meter and achieve coordination at this level. The quality of coordination in ensembles has been studied by (Rasch, 1979), and an analysis of duet performance supports the idea that the players coordinate their representations of meter (Shaffer, 1984).

In all the analyses of timing in musical performance since Seashore (1938), it has been shown that tempo is not constant but varies continuously, and that the patterns of variation are reproducible across repeated performances. This applies to all the performances described here and it implies both that a timekeeper pacing the meter must be in the nature of a programmable clock, and that the specification of variation is a stable entity across performances. We now have to consider the basis of this stability.

Expression

Speaking and playing music are expressive acts, and the expressive forms they use are not arbitrary but have a definite communicative function. So far timing has been considered in terms of coordinating sequences of graphemes, phonemes and note sounds. It can further be shown that the rhythms of performance can reflect aspects of meaning and structure in a hierarchy of grammatical and musical levels. Hence we are led to suppose that these meanings and structures are represented in the motor programs, and are interpreted by motor procedures.

Speech

English has been described as a stress-timed language and French, in contrast, as a syllable-timed language. Taken as an assertion of isochrony in the production of linguistic units, neither description is correct. What one observes in English speech is that stressed syllables tend to alternate with unstressed syllables, and if there are multiple unstressed syllables between two stressed syllables the durations of all syllables tend to be compressed (Cutler & Isard, 1980). Thus there is some regulation of the stress interval but it does not produce invariance. Also, it may come about not by regulating time, but indirectly through the application of rules governing stress. Liberman & Prince (1977), following Chomsky & Halle (1968), offer a phonological theory of such rules generating stress patterns in words, phrases and clauses. These rules generate binary trees constructed over syntactically marked syllable strings, assigning stress prominence to one of the two branches of each node. The assignments to all nodes dominating a syllable determine its stress level. Whether or not this is the best way to characterise stress structure, the idea that stress structure determines syllable timing seems promising.

Independently of stress timing there is a phenomenon of slowing and pausing at the end of a constituent, the amount varying with the linguistic level of the constituent (Gee & Grosjean, 1983; Huggins, 1978). The gradations of pausing and slowing suggest that a single timing rule may be applied recursively at the different levels of a syntactic hierarchy. Gee and Grosjean argue that the hierarchy is prosodic rather than syntactic, but the main effect of this seems to be to add stress group as a level between word and phrase.

It further appears that the durations of the elements within a constituent – word, phrase or clause – vary inversely with the length of the constituent (Huggins, 1978). However, this phenomenon may be reducible to the ones, already considered, of stress timing and boundary slowing. Certainly, the largest effect of compression occurs when the last element in a constituent is displaced into penultimate or earlier position by an extension of the constituent.

In general it is more constructive to relate prosodic features of speech to strong descriptions of linguistic structure than to pursue each description separately, and as long as it is a two-way exercise it seems the best way to get to understand both the phonetics and phonology of speech. For a further discussion of prosodic timing in speech the reader is referred to Nooteboom (1985, chapter 16 of the present volume).

Music

Having stated that a musical score is a schedule for timing the notes and rests in a performance, this has now to be qualified by adding that both composer and player agree that it should be used flexibly. A notation may only approximately capture the timing intentions of the composer, and these may have to be transmitted orally across generations of music teachers. More important is that a rigid adherence to the score makes the music bland and colorless, whereas carefully shaped departures from notated values not only can improve the melodic shape but can help to convey to a listener the musical structure of the piece.

It is perhaps surprising that the expressive modulations of notated rhythms in performance, which have the appearance of being spontaneous, are highly reproducible in repeated performances, even when the music has been played at sight, that is, without preparation. Modulations in the timing of melodic groups of notes and of the meter, and even departures of synchrony between voice parts, and hence from the beat, are all reproducible (Shaffer, 1981, 1984; Shaffer et al., in press). The precision of reproduction, first demonstrated by Seashore (1938), implies either a remarkable precision of memory for parametric detail or, more plausibly, that the player can use systematic rules to derive these features from an analysis of musical structure.

While on this point, it is worth noting that the expressive forms in playing a piece of music may not be invariant with tempo change. This was observed by Michon (1974) in performances of »Vexations« by Erik Satie, and confirmed by Clarke (1982), who showed that it arises mainly from changes in the lowest level of note grouping in melodies: groups formed at fast tempi tend to split into smaller groups at slow tempi.

As in the case of language, there are many ways of describing musical structure. Of the current methods of description only some are explicit enough to allow us to explore the relations between the structures of a piece of music and the expressive forms in a performance. Perhaps the most explicit is the proposal by Lerdahl & Jackendoff (1983) for a set of generative rules of tonal music. It distinguishes four kinds of hierarchic structure, based on properties of melody, harmony and meter. At each of these hierarchic levels, these properties group the music into thematic, or melodic, units, assign a metrical alternation of strong and weak beats, assign the most salient pitch even within a melodic group, and assign regions of melodic and harmonic tension and relaxation. Unlike linguistic theory there are two kinds of rules, those that generate possible structural descriptions and those that express preferences among these, based on criteria of musical coherence.

Making use of these rules it has been possible to show close relationships between structures of meter and melody and many of the grosser changes of tempo that occur in a performance. Similar to the rhythms of speech, there is a tendency for the performer to slow at the boundary of a melodic group, by an amount that is related to the structural depth of the group and to its coincidence with a metrical boundary (Todd, Note 2). If the music departs from conventional tonality it may lose some of its hierarchic structure, and this in turn may be reflected in a less differentiated use of boundary slowing (Shaffer et al., in press). Many problems of how to characterize expression when the music departs from conventional tonal and rhythmic forms remain to be solved. However, it again seems that the most constructive way to attempt to understand the uses of expression is to relate expressive forms to musical structures.

Summary

The central point in this chapter is a distinction made between the structures of timing in an intended performance, which are represented formally in a plan and more explicitly in a motor program, and the timing and motor processes that interpret these, translating them into the rhythms of performance. These structures can be hierarchic and complex, nevertheless can specify the forms quite precisely; the timing processes are in principle fairly simple and may involve at most two levels of timing, one being a procedural control of the timing of a motor sequence and the other a timekeeper, or programmable clock – used mainly in playing music – to give the rhythms a definite time scale.

There has not been room to discuss relationships between perception and action, but we can end on a single brief comment. Listening to speech or music, one can listen to the surface detail or to the message. In the latter case but not the former, expressive patterns of timing will be heard as regular if they are compatible with one's own conception of expressive form (Huggins, 1978; Cutler & Isard, 1980). The listener and the performer are engaged in communicating meaning (a point stipulated also by Nooteboom, 1985, in chapter 16 of the present volume), and timing plays only an adjunct role in this process. Only when it violates expectations does it intrude on awareness.

References

Alderson, G., Sully, D., & Sully, H. An operational analysis of a one-handed catching task using high speed photography. *Journal of Motor Behavior*, 1974, *6*, 217-226.

Bernstein, N. *The coordination and regulation of movements*. Oxford: Pergamon Press, 1967.

Chomsky, N., & Halle, M. *The sound pattern of English*. New York: Harper & Row, 1968.

Clarke, E. Timing in the performance of Erik Satie's ›Vexations‹. *Acta Psychologica*, 1982, *50*, 1-19.

Cooke, J. The organization of simple skilled movements. In G. Stelmach & J. Requin (Eds.), *Tutorials in motor behavior*. Amsterdam: North Holland Publishing Company, 1980, pp. 199-212.

Craik, K. *The nature of explanation*. Cambridge: Cambridge University Press, 1943.

Creelman, C.D. Human discrimination of auditory duration. *Journal of the Acoustical Society of America*, 1962, *34*, 582-593.

Cutler, A., & Isard, S. The production of prosody. In: B. Butterworth (Ed.), *Language production I*. New York: Academic Press, 1980, pp. 245-269.

Fowler, C. Coarticulation and theories of extrinsic timing. *Journal of Phonetics*, 1980, *8*, 113-133.

Gee, J., & Grosjean, F. Performance structures: A psycholinguistic and linguistic appraisal. *Cognitive Psychology*, 1983, *15*, 411-458.

Genest, M. L'analyse temporelle du travail dactylographique. *Bulletin du Centre d'Etudes et Recherches Psychotechniques*, 1956, *5*, 183-191.

Hollerbach, J. A competence model of handwriting. *Visible Language*, 1979, *13*, 252-264.

Huggins, A. Speech timing and intelligibility. In: J. Requin (Ed.), *Attention and performance VII*. Hillsdale, NJ: Lawrence Erlbaum Associates, 1978, pp. 279-297.

Jones, M.R. Structural organization of events in time. In: J.A. Michon & J.L. Jackson (Eds.), *Time, mind, and behavior*. Heidelberg: Springer Verlag, 1985, pp. 192-214.

Kent, R., & Minifie, F. Coarticulation in recent speech production models. *Journal of Phonetics*, 1977, *5*, 115133.

Kristofferson, A.B. Quantal and deterministic timing in human duration discrimination. In: J. Gibbon & L.G. Allan (Eds.), *Timing and time perception*. Annals of the New York Academy of Sciences, vol. 423, 1984, pp. 3-15.

Kugler, P., Kelso, J., & Turvey, M. On the concept of coordinative structures as dissipative structures. In: G. Stelmach & J. Requin (Eds.), *Tutorials in motor behavior*. Amsterdam: North Holland Publishing Company, 1980, pp. 348.

Lee, D., Lishman, J., & Thomson, J. Regulation of gait in long jumping. *Journal of Experimental Psychology: Human Perception and Performance*, 1982, *8*, 448-459.

Lerdahl, F., & Jackendoff, R. *A generative theory of tonal music*. Cambridge, MA.: MIT Press, 1983.

Liberman, M., & Prince, A. On stress and linguistic rhythm. *Linguistic Inquiry*, 1977, *8*, 249-336.

Macar, F. Time, psychophysics and related models. In: J.A. Michon & J.L. Jackson (Eds.), *Time, mind, and behavior*. Heidelberg: Springer Verlag, 1985, pp. 112-130.

MacNeilage, P. Distinctive properties of speech control. In: G. Stelmach & J. Requin (Eds.), *Tutorials in motor behavior*. Amsterdam: North Holland Publishing Company, 1980, pp. 607-622.

McGill, W.J. Random fluctuations of response rate. *Psychometrica*, 1962, *27*, 3-17.

Michon, J. Programs and ›programs‹ for sequential patterns in motor behaviour. *Brain Research*, 1974, *71*, 413-424.

Morasso, P., Mussa Ivaldi, F., & Ruggiero, C. How a discontinuous mechanism can produce continuous patterns in trajectory formation and handwriting. *Acta Psychologica*, 1983, *54*, 83-98.

Nooteboom, S.G. A functional view of prosodic timing in speech. In: J.A. Michon & J.L. Jackson (Eds.), *Time, mind, and behavior*. Heidelberg: Springer Verlag, 1985, pp. 242–252.

Ohala, J. The origin of sound patterns in vocal tract constraints. In: P. MacNeilage (Ed.), *The production of speech*. New York: Springer Verlag, 1983, pp. 189-216.

Poulton, E.C. On prediction in skilled movements. *Psychological Bulletin*, 1957, *54*, 467-478.

Povel, D.J. Time, rhythms and tension: In search of the determinants of rhythmicity. In: J.A. Michon & J.L. Jackson (Eds.), *Time, mind, and behavior.* Heidelberg: Springer Verlag, 1985, pp. 215-225.

Rasch, R. Synchronization in performed ensemble music. *Acoustica*, 1979, *43*, 121-131.

Rumelhart, D.E., & Norman, D.A. Simulating a skilled typist: A study of skilled cognitive-motor performance. *Cognitive Science*, 1982, *6*, 1-36.

Sacerdoti, E. Planning in a hierarchy of abstraction spaces. *Artificial Intelligence*, 1974, *5*, 115-135.

Seashore, C.E. *Psychology of music*. New York: McGraw-Hill, 1938.

Shaffer, L.H. Latency mechanisms in transcription. In: S. Kornblum (Ed.), *Attention and performance IV.* New York: Academic Press, 1973, pp. 17-41.

Shaffer, L.H. Timing in the motor programming of typing. *Quarterly Journal of Experimental Psychology,* 1978, *30*, 333-345.

Shaffer, L.H. Performances of Chopin, Bach and Bartok: Studies in motor programming. *Cognitive Psychology,* 1981, *13*, 327-376.

Shaffer, L.H. Rhythm and timing in skill. *Psychological Review,* 1982, *89*, 109-122.

Shaffer, L.H. Timing in solo and duet piano performances. *Quarterly Journal of Experimental Psychology,* 1984, *36A*, 577-595.

Shaffer, L.H., Clarke, E., & Todd, N. (in press). Metre and rhythm in piano playing. To appear in *Cognition*.

Sussman, H., & Westbury, J. The effects of antagonistic gestures on temporal and amplitude parameters of anticipatory labial coarticulation. *Journal of Speech and Hearing Research*, 1981, *24*, 16-24.

Terzuolo, C., & Viviani, P. Determinants and characteristics of motor patterns used for typing. *Neuroscience*, 1980, *5*, 1085-1103.

Teulings, H-L., & Maarse, F. Digital recording and processing of handwriting movements. *Human Movements Science*, 1984, *3*, 193-217.

Thomassen, A., & Teulings, H-L. Time, size and shape in handwriting: Exploring spatio-temporal relationships at different levels. In: J.A. Michon & J.L. Jackson (Eds.), *Time, mind, and behavior.* Heidelberg: Springer Verlag, 1985, pp. 253-263.

Vorberg, D., & Hambuch, R. On the temporal control of rhythmic performance. In: J. Requin (Ed.), *Attention and performance VII*. Hillsdale, NJ: Lawrence Erlbaum Associates, 1978, pp. 535-555.

Wing, A., & Kristofferson, A.B. Response delays and the timing of discrete motor responses. *Perception and Psychophysics*, 1973, *14*, 5-12.

Reference Notes

1. Kozhevnikov, V., & Christovich, L. *Speech: Articulation* and *perception* (Publication no. 30543). U.S. Department of Commerce, Washington DC, 1965.
2. Todd, N. A model of expressive timing in tonal music. In preparation.

Chapter 16. A Functional View of Prosodic Timing in Speech

Sieb G. Nooteboom

Introduction

This paper attempts to explain some aspects of prosodic timing in speech. Current explanatory principles such as ›isochrony‹, ›anticipatory shortening‹, and ›time compression of motor commands in short-term storage‹, are blamed for begging the question. An alternative, functional, view is proposed, relating prosodic timing to requirements of efficiency in speech communications. It is suggested that a speaker continuously adapts his articulation to what a listener (supposedly) needs at each point in the utterance. Following this general principle, prosodic timing, including the insertion of well formed speech pauses, is related to the time course of word perception in connected speech. It is attempted to demonstrate that a number of well known regularities in prosodic time follow quite naturally from the functional view taken here, if such factors as the local redundancy of the message and the quality of the communication channel are taken into account.

Some Systematic Aspects of Speech Timing

The timing of speech movements in connected speech shows enormous variability. In speech research we have to come to grips with this variability and find out to what extent variations in speech sound and pause durations are systematic, what their causes are, and how speech timing affects the perception of speech. We may divide the factors that are commonly held responsible for systematic aspects of speech timing in the following three classes:

(1) *›Physiological‹ Factors*. These have to do with the physiology and the mechanics of speech production. For example, more open vowels like /a/ have, under otherwise identical conditions, longer durations than more closed vowels like /i/, simply because it takes more time to open the mouth more than to open it less. There are many interacting physiological factors affecting speech timing. Although quantitative models precisely predicting such systematic variations in segment durations from the physiology and mechanics of speech production are still rather primitive and incomplete, there seems to be nothing mysterious anymore about the basic principles involved.

(2) *Phonemic Factors*. Sound systems of many languages make use of differences in segmental durations for contrasting different word forms. Often the contrastive use

of durational differences can be explained diachronically from a systematized exaggeration of small durational differences that find their origin in the physiology of speech production. A case in point is provided by modern English, where the probably universal and physiologically caused difference in duration between vowels preceding the voiced stops /b/, /d/, /g/ and fricatives /v/, /z/, and vowels preceding the voiceless stops /p/, /t/, /k/ and fricatives /f/, /s/ are generally shorter: the difference in duration between the vowels of BEAD (long) and BEAT (short) can easily be of the order of 80 ms, and this magnitude cannot be explained simply on the basis of the physiology of speech production. In some English dialects all other differences between postvocalic /d/ and /t/ have disappeared completely, and the voiced-voiceless distinction in postvocalic stops is replaced by contrastive use of vowel length. Phonemic control of sound durations is in principle straightforward, although attempts towards quantitative modeling of speech timing show that in actual speech production there are complex and detailed interactions between physiological, phonemic, and other factors which are not easily modeled (See Klatt, 1976; Nooteboom, Note 3).

(3) *Prosodic Factors*. In this chapter all systematic variations in speech timing that are not explained by physiological or phonemic factors, will be termed prosodic. Examples are the shortening of unstressed relative to stressed syllables, the shortening of stressed syllables as a function of the number of unstressed ones following within the same ›prosodic unit‹ (rhythmical foot, word, or short phrase) the lengthening of syllables immediately preceding major syntactic boundaries, and the presence or absence of speech pauses on such boundaries. Current explanations of prosodic timing are less straightforward than those of physiologically or phonemically caused aspects of timing. One explanatory principle, favored by many researchers, is called ›isochrony‹, a supposed tendency to make all units of a certain kind (syllables, rhythmic feet, words, word groups) equally long (See Abercrombie, 1967; Allan, 1975; Lehiste, 1970, 1977; Lindblom et al., Note 2). Because ›isochrony‹ is just a tendency (the feet, etc., are not really made equally long), the validity of the explanatory principle is not easily verified. One recent attempt to find acoustic-phonetic evidence for isochrony in the scansion of metrical poetry failed badly (Loots, 1980). Another difficulty is that it is not at all clear why speakers should strive for isochrony: the explanatory principle itself is begging the question. An alternative explanatory principle, supposedly accounting for the same phenomena, has been proposed by Lindblom et al. (Note 2). They have suggested that there is ›time compression‹ in the short term storage of motor commands in the minds of speakers: consecutive planned motor commands are supposed to come closer together on the mental time schedule for execution, the more motor commands there are yet to be executed in the motor program. The advantage of this explanation is that it not only accounts for the fact that the longer units are generally spoken relatively faster than shorter ones, on the same level of organization, but also for the fact that rate of articulation generally slows down towards the end of units: as pronunciation proceeds, there remain fewer and fewer motor commands in short term storage, and thus there is less compression. Basically, however, we are

faced with the same problem as with the notion of isochrony: where does such time compression come from? Again, the explanatory principle is begging the question.

In this paper I shall leave physiologically and phonemically caused aspects of speech timing aside, and focus on prosodic timing only. In the next section I shall propose an alternative view of prosodic timing, which may be called functional. By trying to relate prosodic timing to requirements of speech communication I hope to avoid the kind of problems posed by the earlier mentioned explanatory principles. The remainder of the paper provides an elaboration of this functional view, relating prosodic timing in speech to the time course of speech perception, in particular to the time course of word perception. I shall attempt to demonstrate that a number of well known regularities in speech timing follow quite naturally from the view taken here.

A Functional View of Prosodic Timing

A plausible question would be: How does speech timing affect speech perception? But we may also turn this question upside down and ask: How does speech perception affect speech timing? The notion of speech perception in this latter question requires some clarification. It must be taken to refer to what a speaker has internalized about perceptual processing by the listener. The question, then, becomes whether and how speech timing is affected by what a speaker explicitly or implicitly knows about what a listener needs for successful perception. I shall answer this question by a simple act of hypothesis in the affirmative, and introduce the following general principle: a speaker continuously adapts his articulation to his estimates of what a listener needs at each point in the utterance.

In order to relate this general principle to actual speech timing we need some additional assumptions:

(a) Precise articulation costs time. Thus, whenever a speaker estimates that the listener needs detailed acoustic information, his speech will slow down.

(b) The need for precise articulation decreases with increasing local redundancy. When segmental information can be confidently predicted from context, the speaker can allow himself to speak sloppily and fast. If not, articulation has to be precise and slow.

(c) Fast articulation, whenever allowed by local redundancy, is an efficient means of minimizing sensory storage problems for the listener. This would suggest that the speaker's behavior is governed by two potentially conflicting tendencies. One is to give all the segmental information needed, and thus articulate carefully and slowly. The other is not to let the listener wait longer than strictly necessary, and thus speak fast. I suggest that actual speech timing strikes a balance between these two tendencies, and is modulated by the continuously changing amount of local redundancy.

(d) Slowing down articulation, accompanied by an appropriate pitch movement, and often also by a speech pause, may be used by the speaker to mark the end of a coherent part of the utterance, which is easy for the listener to recognize

›in one go‹. According to this principle, the stream of speech is segmented into chunks that are suited to the listener's instantaneous processing capacities as estimated by the speaker. The speaker's strategy for inserting well formed prosodic boundaries and speech pauses may thus depend on who the listener is, what the contents of the message are, and how redundant the message is for this particular listener, what the estimated auditory quality of the utterance is as affected by hearing deficits in the listener, unfamiliarity of the listener with the sound structure of the language, environmental noise, a bad telephone line, etc.

If we now seek to explain systematic aspects of actual speech timing from the above principles we need some insight into the systematicity of local redundancy in segmental structure from the standpoint of the listener. This will be discussed below in an attempt to relate speech timing to the time course of word perception.

Relating Prosodic Timing to Word Perception

I shall first concentrate on the production and perception of monomorphematic words spoken in isolation, without contextual cues. Local redundancy is here strongly related to lexical redundancy, that is, to the amount of phonemic information in a word form that is present over and above what is strictly necessary to distinguish this word uniquely from all other words in the lexicon. Lexical redundancy systematically increases with word length expressed in number of phonemes: in general we need all phonemes in their correct order to recognize monosyllables like the word CAN. A polysyllabic word like ELEPHANT, however, is, for instance, uniquely determined by the initial word fragment ELEPH...., the remainder of the word, strictly speaking, being redundant information. This systematic relation between word length and lexical redundancy quite naturally explains, according to the present functional view of prosodic timing, that, other things being equal, speed of articulation increases with increasing word length (Figure 1).

A special note must be made here on the difference between lexically stressed and unstressed syllables. This difference, at least in some languages such as the Germanic ones, has a strong effect on speech timing which does not easily follow from the present line of reasoning. I therefore hypothesize here that in these languages one syllable in polysyllabic words, the stressed one, is singled out to carry the main load of segmental informativeness. The other, unstressed, syllables, for which the unstressed character may be more important than their segmental make-up, are thereby available to be compressed or expanded according to the instantaneous needs of the listener. This explains why unstressed syllables are much more variable in duration than stressed ones.

Although differences in lexical redundancy explain quite nicely why longer words are spoken faster than shorter words, redundancy can not be the only factor involved in the control of speech timing. If we look in more detail at the timing of speech in polysyllabic words, we see that articulation tends to be fastest at the beginning of a word, and then slows down towards the end. At any rate, this is the

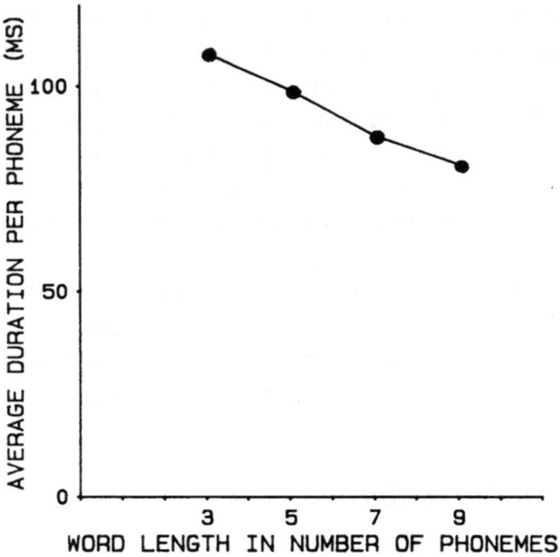

Figure 1. Average duration per phoneme as a function of word length expressed in number of phonemes. Measurements were made in prosodic imitations of real words, consisting of repetitions of the nonsense syllable /mam/. Stress was always on the first syllable. Data from Nooteboom (Note 3).

case when the word is spoken in isolation and thus constitutes a whole utterance by itself. Since we know that polysyllabic words can generally be recognized before the acoustic word ending (see Marslen-Wilson, 1980), we would have predicted the opposite from the distribution of local redundancy: the word beginning is heard before the word can be recognized by the listeners, and thus provides the most important cues for recognition. It is therefore in some sense least redundant. The word ending is often heard, at least in optimal communication conditions, after recognition. It therefore is most redundant and serves at most for confirmation of the selected candidate word. Following this argument it is to be expected that the word beginning is pronounced most carefully, and therefore slowly, and the word ending most sloppily and fast. There are indeed indications that word endings are generally spoken more sloppily and less precisely than are word beginnings. We see for example that disappearance of phonemic contrast is more frequent at word endings than at word beginnings, and, more drastically, that even deletion of whole syllables in the history of a language is more common for word final than for word initial syllables. But we also find that at this microlevel of prosodic timing the distribution of speed of articulation is precisely the opposite of what one would have expected: word beginnings are spoken faster than word endings. It is as if there is another tendency at work, superimposed on the tendency to speak faster when there is more lexical redundancy, namely to provide the listener as quickly as possible with the auditory cues necessary for initial recognition, and to allow him or her to relax after initial recognition has taken place.

Of course, speeding up articulation at the beginning of a polysyllabic word comes into conflict with the need for precise articulation, particularly when this word beginning contains a lexically stressed syllable. Indeed we see that, although the durations of stressed syllables are somewhat affected by their position in the word, getting longer towards the end, this effect is only slight in comparison with the rather drastic changes in the duration of unstressed syllables. These may be very short in initial and medial positions, but about as long as stressed ones in final position (for instance Klatt, 1976; Lindblom, Note 1; Nooteboom, Note 3). These observations can be explained by assuming that speech timing is adjusted to the time course of word recognition, in the sense that the time interval during which the more redundant auditory cues come in and during which the listener can integrate the recognized word in his representation of the ongoing discourse or dialogue, is maximized. The speeding up at the beginning of polysyllabic words is necessarily blocked in segmentally informative stressed syllables, but the slowing down at the end of polysyllabic words can occur freely in both stressed and unstressed syllables.

Summarizing my speculations so far, I suggest that the prosodic timing of speech in the pronunciation of words spoken in isolation and without contextual cues can be explained by:

(a) differences in lexical redundancy, articulation being necessarily slow in nonredundant parts of the message, and either fast or slow in redundant parts;

(b) differences in segmental informativeness of syllables, segmentally informative stressed syllables being spoken more precisely and therefore more slowly than segmentally less informative unstressed syllables;

(c) a tendency to provide the listener as quickly as possible with the cues necessary for identifying the intended word and to provide the listener with ample time for interpreting the recognized word, following initial recognition.

I assume that this line of reasoning can, in principle, be extended to morphologically complex words, word groups, and even whole utterances, where similar regularities can be found in prosodic timing, contextually redundant words being spoken faster than informative words, and articulation slowing down towards the end of units. The one feature in connected speech that we do not encounter in the pronunciation of words in isolation, is the deliberate insertion of well formed speech pauses. This will be discussed in the following section.

Speech Pauses and the Recognition of Words in Connected Speech

Well formed speech pauses are silent intervals, interrupting the flow of connected speech, accompanied by lengthening of the prepausal syllable and often by an appropriate pitch movement, such as a rising pitch in the ›comma intonation‹. When the actual silent interval is omitted but there is prepausal lengthening, the auditory effect is very much the same. It is as if the most important cue for perceiving a speech pause, or prosodic break, is the prepausal lengthening.

It has been observed that speech pauses tend to occur at major syntactic boundaries, and some researchers even suggest that their occurrence is indeed determined by syntax (Cooper & Paccia-Cooper, 1980; Klatt, 1975). This is only partly true, however. Pausing strategies vary enormously from speaker to speaker, and from situation to situation. Syntax is certainly not the only factor determining whether or not a pause will be made. According to the present functional view of prosodic timing a speaker should make a speech pause when, according to his estimates, the listener needs one. When the speech is carefully pronounced and communication conditions are favorable, the listener will hardly need any speech pauses. But this may change drastically when, for whatever reason, the auditory quality of speech is degraded. To elucidate this, let us turn again to the time course of word recognition, this time for words in connected speech. Following Marcus (1984) we may distinguish three phases in word perception: (1) an initial phase during the intake of the onset of the acoustic word form, in which a number of possible lexical candidates are selected (See also Marslen-Wilson, 1980); (2) a recognition phase, in which one candidate is selected as corresponding to the stimulus, and (3) a final phase, used for confirmation (or disconfirmation) of the selected initial candidate, and also for tracking the word ending, and, at the same time, the onset of the next word.

These three phases of word perception are exemplified in Figure 2 for two hypothetical cases in which all phonemes are easily and rapidly identified from the stream of speech. When the to-be-recognized word is a polysyllable, like ELEPHANT, the onset of the next word is available to the listener at the very moment it comes in. However, when the word is a monosyllable like BAT, the listener has often no way of knowing that the word form is complete. The word might have been BATTLE, or BATLY, or BATTERY. It is only after the incoming sequence of speech sounds makes a nonword, like BATIS, that the listener can find out where the one word has ended and the other started. Finding the onset of the next word in such a case, necessarily involves some amount of backtracking in the auditory signal.

The situation worsens for the listener, when the speech is degraded, and the sequence of speech sounds derived from the auditory signal may contain many ambiguities. This is exemplified in Figure 3 for two hypothetical cases. Here, finding the word ending and thus, the onset of the next word, may involve some backtracking, even for lexically redundant words like ELEPHANT. But for monosyllables like BAT, backtracking may become necessary over considerable stretches of speech. As long as the ambiguities are not resolved by further information, the listener may have to keep in mind an ever increasing number of alternative perceptual solutions to the riddle presented by the low quality speech. Of course, when the speech quality is degraded too much, the number of candidate perceptual solutions may become too high and perception may break down altogether.

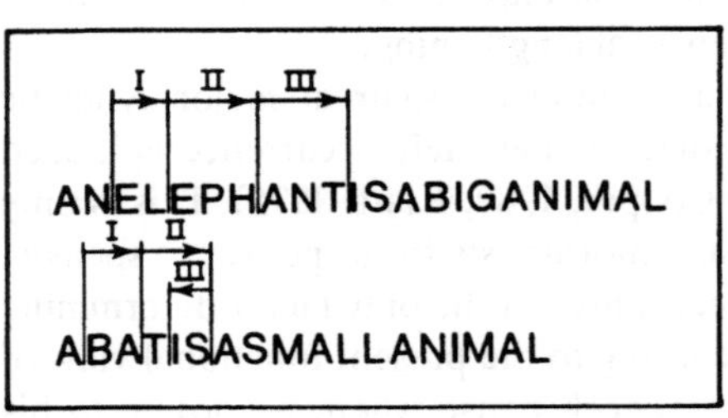

Figure 2. The three phases of word perception in high-quality speech, as suggested by Marcus (1984). In the case of monosyllabic words the recognition phase *(II)* may extend beyond the end of the spoken word, in which case finding the onset of the next word (phase *III*) involves backtracking.

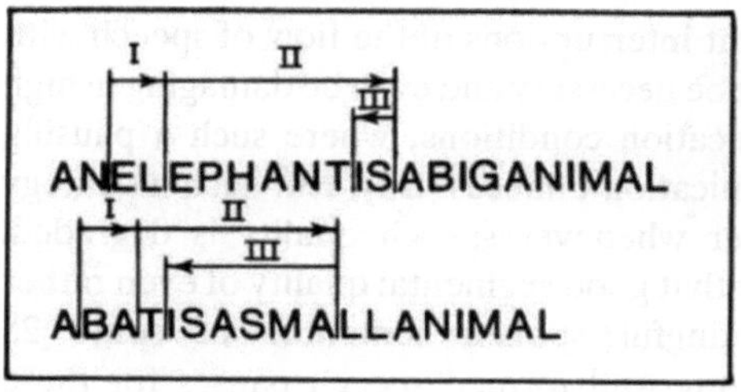

Figure 3. The same as Fig. 1, but for degraded speech quality. Now phase *II* may be stretched considerably beyond the end of the spoken word, thus also increasing the amount of necessary backtracking for finding the onset of the next word.

One may easily imagine several possible ways in which a speaker can help his listener to avoid such troubles when the communication channel is noisy. One way would be to make the message itself more redundant, by using long, polysyllabic words instead of monosyllabic ones, and by using long, syntactically and semantically redundant phrases instead of shorter nonredundant ones. Another way would be to frequently interrupt the flow of speech by well formed pauses at appropriate points in the message. Each speech pause would first of all provide an unmistakable cue to a particular word or word group boundary position, thus reducing the number of alternative perceptual solutions to the puzzle provide by the incoming sound stream. Secondly, it would provide the listener with extra processing time to solve the riddles in the immediately preceding chunk of speech. In extreme cases, the speaker might insert a speech pause after each single word, but this would probably break up the auditory coherence of speech considerably and would also make communication very slow. It is therefore reasonable that, when relatively frequent speech pauses are required, these are made at word group boundaries each time some meaningful and not too long part of the message is completed. It is noteworthy, though, that in actual speech production speech pause positions cannot be easily predicted from syntactic structure alone. Speakers may differ considerably in their pausing strategies, and nonsyntactic factors such as word group length may also be involved (de Rooij, Note 5).

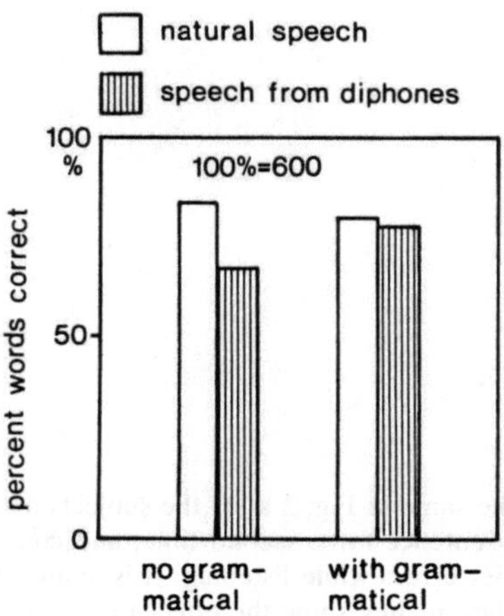

Figure 4. Percentages of words correctly reproduced in an immediate recall task with meaningful sentences consisting of 25 monosyllabic words each, for natural speech and speech synthesized from diphones, both with and without grammatical speech pauses.

According to this viewpoint, frequent interruptions of the flow of speech with well formed speech pauses may even not be necessary and even be damaging in high quality speech and favorable communication conditions, where such a pausing strategy would only slow down communication unnecessarily. But such a strategy would be of great help to the listener whenever speech quality is degraded. Elsewhere I have shown experimentally that good segmental quality of even rather long syntactically correct and meaningful spoken sentences of some 25 monosyllabic words does not require any well formed speech pauses for these sentences to be intelligible. But when these same sentences are presented in degraded speech, in this case by a speech syntheziser of less than optimal quality, intelligibility benefits considerably from the insertion of well formed speech pauses after all syntactically coherent word groups. The results of this experiment are summarized in Figures 4 and 5 (Nooteboom, Note 4).

A Note of Caution

It seems to me that in the traditional views of prosodic timing in speech, notions like rhythm, isochrony, syllable timing versus stress timing, anticipatory shortening, and time compression, are invoked too easily. These notions seem to refer to organizing principles that are supposed to be operative in the act of speaking but, at the same time, seem to be superimposed on speech communication from outside. They are considered in order to explain regularities in speech timing that would otherwise seem inexplicable, but since they are really begging the question, they actually explain very little. In this paper I have deliberately taken the view that prosodic timing in speech is exclusively controlled by requirements of efficiency in speech communication.

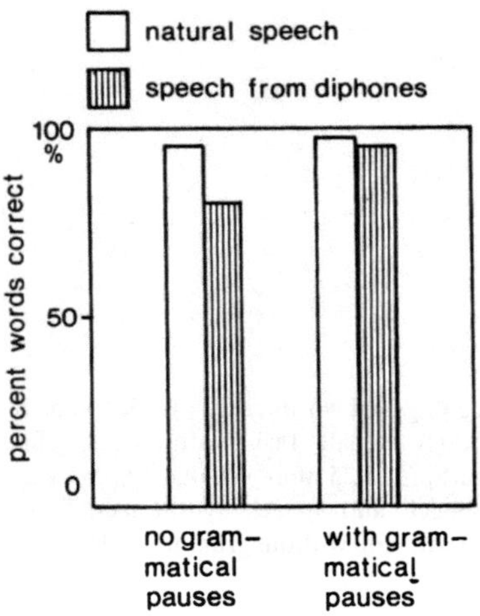

Figure 5. The same as Fig. 3 after the subjects had heard each sentence for a second time, and could correct earlier errors while listening. It is assumed that these data mainly show the effect to intelligibility.

This view has both advantages and disadvantages. One disadvantage is that is obviously too narrow. Speakers are certainly not always optimally efficient in timing their speech movements. Personal clumsiness or stylistic and esthetic factors that escape the notion of efficiency may come into play. It is conceivable that in such esthetic or stylistic variations a desire for rhythmicality, that is external to the lexical, syntactic, and semantic properties of the message, partly controls the timing of speech. After all, the observation that some explanatory principle as such is begging the question does not necessarily imply that the principle is incorrect.

Another disadvantage of the present view is that the explanation of known regularities in speech timing is not always as adequate as one would hope. I am, for example, not entirely satisfied with my attempt to explain the slowing down of speech over the duration of polysyllabic words (or other utterances) from a tendency to minimize the time a listener must wait for the necessary cues for recognition and to maximize the time available for interpretation. The problem is that the facts themselves run counter to my initial assumption that less redundant parts of the stream of speech are spoken carefully and slowly, and more redundant parts are spoken sloppily and fast.

An obvious advantage of the present view is that it accounts easily for the fact that speech timing does not follow mechanistically from the lexical, syntactic and semantic properties of the message as some researchers seem to suggest (e.g. Cooper & Paccia-Cooper, 1980). Actually, speech timing is highly conditioned by the communicative situation. We time our speech differently when we address a large audience in a reverberating hall than when we speak to a single person in a quiet room, and differently again when we speak over a bad telephone line. We have a different timing when we speak to a little child than when we speak to an adult, etc. Such observations support the main idea of this paper, that prosodic timing is to a great extent controlled by adjusting articulation to the instantaneous needs of the listener.

References

Abercrombie, D. *Elements of general phonetics*. Edinburgh: Edinburgh University Press, 1967.

Allan, G.D. Speech rhythm. *Journal of Phonetics*, 1975, *3*, 75-86.

Cooper, W.E., & Paccia-Cooper, J. *Syntax and speech*. Cambridge, MA.: Harvard University Press, 1980.

Klatt, D.H. Vowel length is syntactically determined. *Journal of Phonetics*, 1975, *3*, 129-140.

Klatt, D.H. Linguistic uses of segmental duration in English. *Journal of the Acoustical Society of America*, 1976, *59*, 1208-1221.

Lehiste, I. *Suprasegmentals*, Cambridge, MA.: MIT Press, 1970.

Lehiste, I. Isochrony reconsidered, *Journal of Phonetics*, 1977, *5*, 253-265.

Loots, M. *Metrical myths*. The Hague: Martinus Nijhoff, 1980.

Marcus, S.M. Recognizing speech: On the mapping from sound to word. In: H. Bouma & D.N.G. Bouwhuis (Eds.), *Attention and performance X*. Hillsdale, NJ: Lawrence Erlbaum Associates, 1984, pp. 151-163.

Marslen-Wilson, W.D. Speech understanding as a psychological process. In: J.D. Simon (Ed.), *Spoken language generation and recognition*. Dordrecht: Reidel, 1980, pp. 39-67.

Reference Notes

1. Lindblom, B.E.F. Temporal organization of syllable production, Speech Transmission Laboratory, Royal Institute of Technology, Stockholm. *Quarterly Progress and Status Report*, 1968, *2/3*, 2-5.
2. Lindblom, B.E.F., Lyberg, B., & Holmgren, K. Durational patterns of Swedish phonology: Do they reflect short-term motor memory processes? Unpublished manuscript, Department of Phonetics, Institute of Linguistics, Stockholm University, 1976.
3. Nooteboom, S.G. *The production and perception of vowel duration*. Doctoral dissertation, University of Utrecht, 1972.
4. Nooteboom, S.G. The temporal organization of speech and the process of spoken-word recognition. Eindhoven: Institute for Perception Research, *Annual Progress Report*, 1983, *18*, 32-36.
5. Rooij, J.J. de. *Speech punctuation*. Doctoral dissertation, University of Utrecht, 1979.

Chapter 17. Time, Size and Shape in Handwriting: Exploring Spatio-temporal Relationships at Different Levels

Arnold J. W. M. Thomassen and Hans-Leo Teulings

Introduction

Time, the organizational principle of every motor skill, manifests itself in several ways in the case of handwriting. First, time is systematically involved in the preparation of movements before their onset and during their performance. The study of reaction times and movement times in handwriting (see Thomassen et al., 1984) has revealed temporal features of the processing stages of writing performance; there are relevant analogies with speech and typewriting (see the review by Sternberg et al., Note 1; and Shaffer, 1985, chapter 15 of the present volume).

Furthermore, time is involved in the representation of the various letters, whose correct production depends on the exact phasing of several motoric subsystems (antagonist pairs) (see Hollerbach, 1980). Finally, in the execution of a writing movement, the global and local features of the spatial trajectory, such as size and curvature, determine the absolute amount of time spent on each segment. In correspondence with the other topics discussed in this chapter, only the latter general relationships between the spatial and temporal organization in handwriting are the subject of the present contribution.

Handwriting has a number of typical constancies. There is a striking shape similarity between a word in a person's small handwriting on paper and the same word in his much larger writing on a blackboard (Merton, 1972): across the difference between the muscular and skeletal systems involved there is ›motor equivalence‹ (Bernstein, 1967). Not only are the written shapes similar, but also the relative distribution of time over the segments of a complex writing movement is invariant. Under certain variations in size there is even a tendency for the absolute duration of writing to remain constant (isochrony). Recently, a constancy principle has been described under the term ›isogony‹. The principle implies that angular velocity in handwriting tends to be constant, irrespective of the radius of curvature (Viviani & Terzuolo, 1980, 1982), or equivalently, that instantaneous writing speed (tangential velocity) is proportional to the local radius of curvature. If a letter's size is varied, isogony produces isochrony: the letter will still be produced in the same amount of time. However, it has appeared that overall time constancy is not always accomplished; moreover, there are a number of context effects which locally moderate angular velocity (Viviani & McCollum, 1983; Lacquaniti et al., 1983).

We shall be concerned with the factors determining execution time as a function of size in continuous cursive writing. Since, for the present purpose, we wish to avoid preparation effects (Teulings et al., 1983; Hulstijn & Van Galen, 1983; Van

Galen & Teulings, 1983) and effects due to different letter shapes, we shall focus on the continuous writing of simple repetitive or alternating loop patterns as writing tasks, most often corresponding to the cursive letters *e* and *l* in various sizes. The production of these patterns will be studied at three levels of context, viz., (1) macro-context, that is as a function of the size when specified between ›words‹ (2) meso-context, that is as a function of the size differences between adjacent letters within a ›word‹ and (3) micro-context, that is as a function of the local curvature of strokes within a letter. The main problem that we shall discuss is whether or not the global size of handwriting and the local curvature of cursive loops bear a straightforward and robust relationship to the time required for their production. If such a relationship can be established, this would be indicative of a central control mechanism that is in charge not only of the shape of the writing trace, but also of its performance as a precise function of time. If, on the other hand, the spatio-temporal relationships in handwriting would appear to depend on different situational and performance factors at various levels, the need for postulating such a central timing mechanism would be absent. In order to gain a unifying insight into the size-time relationships, data from the literature and from our own experiments will be expressed by parameters allowing the relationships to be treated uniformly across these levels.

Macro-Context

If the overall size of writing is changed, will there be an effect on movement time, or is such a change (between trials) a condition where isochrony holds? The literature is inconclusive. Some authors observe a tendency towards isochrony under size variation in writing (e.g. Freeman, 1914); others report a lengthening of time intervals over a range of increasing sizes (e.g. Michel, 1971). There are, of course, differences among the conditions in these experiments. Sometimes

Table 1. Exponent values *b* of the relation between global duration *t* and size *s* as a function of macro- and meso-context (that is, $t = ks^b$) as estimated from various publications and those of the relation between local duration (r/v) and local size (r) in micro-context (that is, $r/v = k'r^b$)

Authors	Context		
	Macro	Meso	Micro
Denier van der Gon & Thuring (1965)	0	0.50	—
Wing (1980)	0.90	0.41	—
Greer & Green (1983)	0	0.38	—
Lacquaniti, Terzuolo & Viviani (1983)	—	0.45	0.67
Present authors (simple patterns)	0 to 0.50	—	≈ 0.57
Present authors (complex patterns)	—	0.33	≈ 0.45

Note: b = 0 denotes constant time, or isochrony; b = 0.50 denotes constant force, leading to a time increase; b = 1.0 denotes constant absolute velocity resulting in proportional time with size.)

maximum speed is required, sometimes ›normal‹ speed; also the size range employed may vary. However, at least some macro-context size effects reported recently (see Table 1) were accompanied by isochrony. We assume that the narrow frequency band of the output of the motor system is responsible for this form of constancy. Logically, though, at some point along the size scale, the accelerations required to complete a letter in a given amount of time must reach a ceiling due to the maximum force available; beyond that isochrony must break down. This leads to the hypothesis that writing time will remain constant across size variations in the neighborhood of normal letter heights (approx. 0.5 cm), but will increase at some point where force demands become too high.

To test this hypothesis, twelve students were instructed to write various loop patterns in seven different sizes ranging from 0.25 to 16.5 cm. The instruction was to write at maximum speed. The experiment involved recording the handwriting signal produced with a special pen on a Vector General Data Tablet DTl; its accuracy is better than 0.2 mm. Sampling rates were 200 Hz or 125 Hz while sampling periods were 4 to 6 s, respectively. The signal was low-pass filtered (cosine-shaped transition band from 8 to 24 Hz; see Teulings & Thomassen, 1979 and Teulings & Maarse, 1984).

The results showed that isochrony was indeed accomplished only across the smallest sizes, the stable range running from 0.25 to 1.0 cm. Beyond this range, size increase was accompanied by (moderate) time increase (see Figure 1a). Estimates of the acceleration force levels involved reveal that the force increase which accompanied isochrony across increasing sizes at the lower end of the continuum, leveled off at the high extreme. The force curve reaches the expected plateau just above the tallest size written (16 cm) (see Figure 1b). In order to arrive at these estimates, we assumed that writing size is proportional to the product of acceleration force and time squared. Thus we base our calculations on the relationship $s = fat^2$, where s is size, or stroke length; f is a parameter describing the efficiency of an acceleration pattern (f is virtually constant; see Teulings et al., note 2; under the empirically supported assumption of a sinusoid movement, $f = 0.203$); a is the (absolute) top acceleration in a stroke; t is the execution time of an up or down stroke.

The above findings indicate that across words of increasing size isochrony is maintained as long as the required force is readily available; lack of available force is compensated for by additional time. Thus, writing a pattern with a fixed size appears to involve two mechanisms determining the minimum amount of time per loop, viz., (1) the limited-frequency characteristics of the output of the motor system; and (2) a ceiling of available force. At small sizes (in the order of millimeters) the former mechanism sets a limit to the minimum amount of time; at larger sizes (in the order of centimeters) duration minima are bounded by force. In Table 1 the data are expressed as exponent values in which t is a power b of size s, as follows: $t = ks^b$, where k is a constant. The values of b that we obtained range from zero for the smaller sizes to 0.50 for the largest sizes. If we look at the data in the recent literature, which have also been entered in Table 1, we see that in these other studies the required force was in general not near a ceiling, which is in agreement

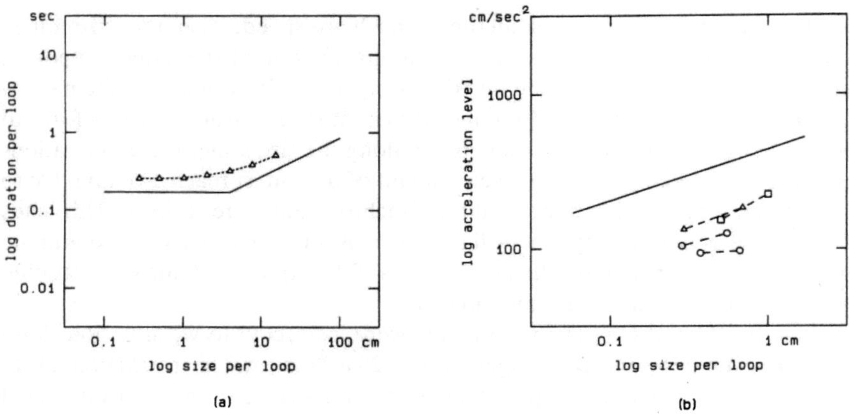

Figure 1. Duration per loop *(a)* and acceleration level *(b)* as a function of loop size if size is varied at the macro-context level. The slopes, i.e., the exponent values of the linear reference lines in panel *a* are 0 und $^1/_2$, and in panel *b* they are 1 and 0; the positioning of these reference lines in the plots is arbitrary.

with the small sizes employed in these studies. We may conclude that the output and force mechanisms just mentioned provide a satisfactory explanation of the macro-context results. An anomaly is found in the macro-context data collected by Wing (1980), where an almost proportional time increase (that is, constant absolute velocity) was found; this finding must be regarded as deviant in writing studies.

Meso-Context

The adjacency of letters of the same or of different size (such as the letters *e* and *l* in a word like pellet) constitutes their meso-context. The specific problem here involves the abrupt size adjustment for each letter, e.g. leading to a size increase for the *l* that was preceded by *e*. This adjustment may be accomplished by an increase of force, of duration, or of both. The data in the literature are again ambiguous. Denier van der Gon & Thuring (1965) report a time increase under equal force for *l* following *e*. In contrast, several authors (Wing, 1980; Greer & Green, 1983) observed a combined increase of both time and force. This was also hypothesized by us on the basis of the observed tendency towards isochrony in normal handwriting and on the evidence suggesting that force increase in the motor system takes more time than is available during the onset of individual strokes in fluent writing (e.g., Denier van der Gon & Thuring, 1965; Stelmach & Teulings, 1983).

In an experiment involving four male righthanded subjects aged 23 to 31, the pseudoword *mehelmen* was written 48 times in different sizes and at different speeds. The apparatus was the same as that in the preceding experiment. The results show that the mean height s and duration t for central *e* and *l* were, respectively, $s_e = 0.293$ cm; $s_l = 0.687$ cm; $t_e = 241$ ms; $t_l = 319$ ms. The power-function describing the size-time relationship of these data is $t = ks^{0.33}$, which supports our hypothesis (see Figure 2a). By assuming that s_l/s_e is proportional to $a\ t\ a_l t_l^2/a_e t_e^2$, we can estimate the force increase accompanying the size increase under the assumption that force is mainly determined by acceleration. We obtain $a = k's^{0.34}$. Thus we observe an exponential increase for both time and force, which confirms our hypothesis. To the extent that these values are similar, they might even imply that there was a balanced trade-off between force and time at the size increase in our experiment (see Figure 2b).

In an analogous fashion we derived the exponents relating time to size for the data reported in the studies by Wing (1980) and Greer & Green (1983). The results are presented in Table 1. There is agreement among these data and ours to the extent that duration increases with an exponent of the order 0.30 to 0.40, which implies a certain amount of force trade-off against time. The same applies to the data reported by Lacquaniti et al. (1983), which stem from a meso-context in which loops of different size were drawn without (rightward) progression; the exponent, converted by us to allow comparison of their time-size relation to ours, was 0.45. Finally, an exponent of exactly 0.50 corresponds to the meso-context observations of Denier van der Gon & Thuring (1965). These figures have also been entered in Table 1. To sum up, we observe that size increase at the meso-context level is always accompanied by some time increase; in most cases, however, there is also some force increase, so that the additional duration of taller letters is less than proportional.

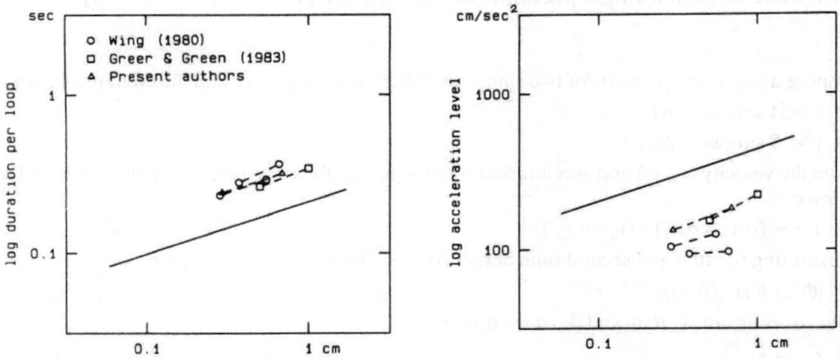

Figure 2. Duration per loop *(a)* and acceleration level *(b)* as a function of loop size if size is varied at the meso-context level (only the range of normal writing size is investigated here). The slope of the reference lines in panel *a* and *b* is $^1/_3$; the positioning of these reference lines in the plots is arbitrary.

Micro-Context

We have discussed the execution times of entire letters or loops as a function of their global size. If we wish to study time and size at the micro-context level, that is at a specific point along the writing trace, we can express local size by r and local duration by r/v, where r is the radius of the circle matching the curve at that point, and where v is the instantaneous absolute, or tangential, velocity. As stated above, the relationship between curvature and movement time has been described as following the isogony principle: curves subtending equal angles are produced in equal amounts of time. Under constant angular velocity, a bigger circle will be produced with a proportionally higher tangential velocity than a smaller circle. For similar patterns, the principle of isogony would thus result in isochrony across sizes. However, we failed to observe this in the meso-context.

Indeed, perfect isogony appears not to be the rule in complex patterns. Velocity increase with radius of curvature is less than proportional and, often unpredictably, constant angular velocity tends to shift to different levels. Lacquaniti et al. (1983) observed that in a large variety of drawing tasks, angular velocity $v/r = (1/k) \cdot (1/r)^{2/3}$ or $r/v = kr^{2/3}$ (the ›two-third power law‹), where k is a constant gain factor which depends on the meso-context; it describes the shifts just mentioned. Insight into the principle underlying the two-third power law is gained by making the assumption that every segment of a writing trace can be described by parts of two orthogonal sinusoids of a single frequency (e.g. Hollerbach, 1980). The presence of such a predominant frequency (5 Hz) in handwriting has been shown by Teulings & Maarse (1984). Thus, over a segment of some duration, components of this frequency (shared by x and y, though with varying phase) will always make the largest contribution to the movement. As Lacquaniti et al. (1983) have indicated, and as the present authors show[1], the applicability of the two-third power law depends on the presence of such a single predominant frequency (see Figures 3 and 4a).

---

1 Suppose a movement consists of two sinusoids of the same frequency and different phases and:

(1) $x = A \sin(\omega t + \alpha)$

(2) $y = B \sin(\omega t + \beta)$

From the velocity (v_x, v_y) and acceleration vectors (a_x, a_y), the curvature 1/r can be estimated as follows:

(3) $1/r = (\alpha_x v_y - \alpha_y v_x) / (v_x^2 + v_y^2)^{3/2}$

Substituting the first and second time derivatives we obtain, after rearranging:

(4) $[\omega^3 A B \sin(\beta - \alpha)]^{-1/3} \cdot r^{2/3}$

Thus for constant A, B, ω and $\beta - \alpha$ we may write

(5) $r/v = kr^{2/3}$

with

(6) $k = [\omega^3 A B \sin(\beta - \alpha)]^{-1/3}$

It may be noted that k is relatively insensitive to variations of $\beta - \alpha$ and of A or B. The values of A, B and ω may vary with either the meso-context or the macro-context. To make quantitative comparisons possible, one may set $1/\omega$ proportional to the duration and A proportional to B and proportional to the size.

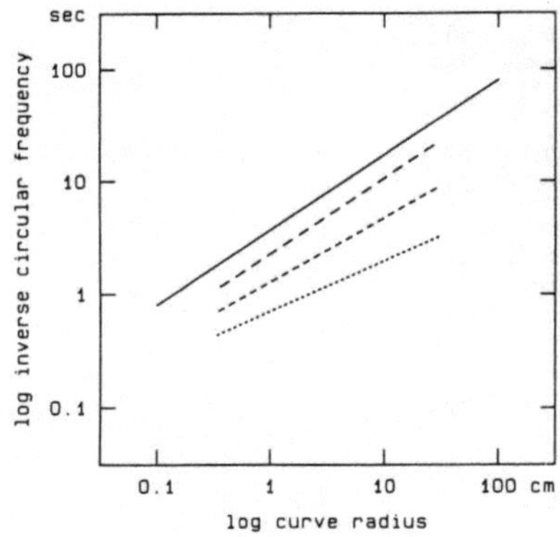

Figure 3. Size-time relationships under varying curvature at micro-context level: local duration (i.e., curve radius divided by tangential velocity) as a function of local size (i.e., curve radius). The slope of the drawn reference line is $^2/_3$, that of the dotted lines represent the values 0.67, 0.57 and 0.45 from Table 1. The positioning of these lines in the plot is arbitrary.

By performing simulations we could study the behavior both of the exponent value b (the power of the function relating r/v to r) and of the correlation measure ϱ (the square root of the fraction of the variance explained by the fitted power relation between r/v and r). Introducing a horizontal translation, transforming ›stationary‹ into ›progressive‹ writing (Thomassen & Teulings, 1983), causes k in the relation $r/v = krf^{2/3}$ to become time dependent (rather than constant) when the translation velocity is of the order of the maximum x-component velocity, that is when ellipses tend to change into cuspids[2] (see Figure 4b, 4e). The latter time dependence is reflected by a departure of the exponent b from 2/3 and by a non-unity correlation ϱ. Other effects result if we introduce a second component with a frequency between zero and that of the main component, their amplitudes being of the same order[3] (see Figures 4c, 4d, 4f). We see that in these instances the relation between r/v and r depends on a number of momentaneous movement parameters, so that it may be concluded that the accuracy of duration predictions on the basis of curvature decreases with increasing complexity.

2 If constant horizontal translation is included we may substitute instead of the x function (1):

(7) $x = A \sin(\omega t + \alpha) + Ct$

Then the expression for k becomes

(8) $k(t) = \left[\omega^3 A B \sin(\beta - \alpha) + \frac{C}{\omega A} \sin(\omega t + \beta)\right]^{-1/3}$

Therefore k becomes time dependent, introducing in general a non-linearity in the relation between log (r/v) and log r with a slope departing from 2/3.

3 If a second frequency component is added, we may substitute instead of the x and y functions (1) and (2):

(9) $x = A_1 \sin(\omega_1 t + \alpha_1) + A_2 \sin(\omega_2 t + \alpha_2)$

(10) $y = B_1 \sin(\omega_1 t + \beta_1) + B_2 \sin(\omega_2 t + \beta_2)$

The expression for k becomes

(11) $k(t) = [\omega_1^3 A_1 B_1 \sin(\beta_1 - \alpha_1) + \omega_2^3 A_2 B_2 \sin(\beta_2 - \alpha_2)$
$- \omega_1^2 \omega_2 A_1 B_2 \sin(\omega_1 t + \alpha_1) \cos(\omega_2 t + \beta_2) - \omega_1 \omega_2^2 A_1 B_2 \cos(\omega_1 t + \alpha_1) \sin(\omega_2 t + \beta_2)$
$- \omega_1 \omega_2^2 A_2 B_1 \cos(\omega_1 t + \beta_1) \sin(\omega_2 t + \alpha_2) - \omega_1^2 \omega_2 A_2 B_1 \sin(\omega_1 t + \beta_1) \cos(\omega_2 t + \alpha_2)]^{-1/3}$

which is difficult to reduce. The number of terms expands further if additionally a constant translation is introduced.

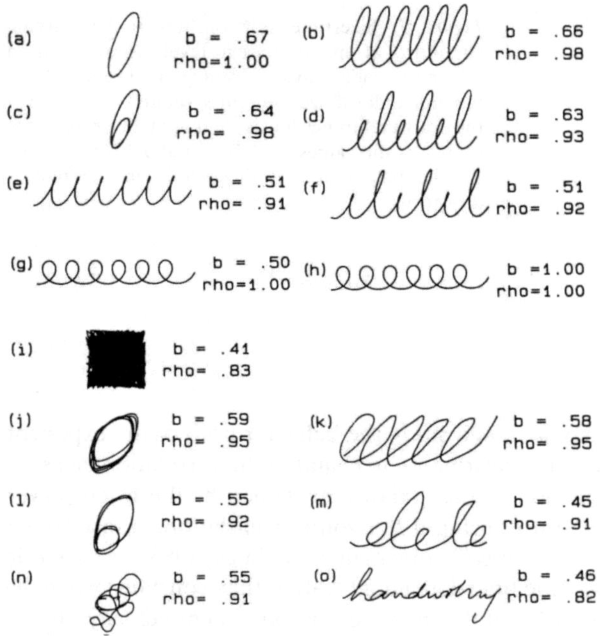

Figure 4. Simulated *(a-i)* and recorded *(j-o)* writing patterns together with the best fitting exponent b satisfying $r/v = kr^b$ and the correlation ϱ between log (r/v) and log r.

When a simulated random pattern was analyzed (Figure 4i), we obtained exponent and ϱ values of the order 0.40 and 0.80 respectively; these values may thus be regarded as a criterion for nonobedience to the two-third power law. The samples of an individual's handwriting that we collected appear to approximate the two-third power law when the patterns are relatively simple (Figures 4j, 4k, 4l, 4n). However, when size differences and translation are combined in a ›word‹, the exponent and ϱ values drop almost to those of the random pattern, which indicates that the two-third power law is no longer valid (Figures 4m, 4o). Different writing strategies, finally, may of course drastically affect the exponent value. It will approach 1/2 under a successful strategy of constant acceleration force (e.g. circular movements with constant progression superimposed; see Figure 4g); and it will approach unity to the extent that absolute velocity approaches constancy (see Figure 4h).

To sum up, at the micro-context level of writing, a lawful two-third power relationship holds between local curvature (size) and execution time in simple drawing movements having a narrow frequency band and little or no translation. Normal handwriting, however, cannot be described accurately by the two-third power law (see also Table 1) and there is no need for a theory of handwriting specifically to account for such a law.

Conclusion

We discussed a number of original experiments and studies from the recent literature on size-time relationships in cursive letter writing. The data of these experiments can, with a small number of assumptions, be recalculated and transformed into power functions which relate the global (duration) and local (velocity) temporal features of letters to their global (size) and local (curvature) spatial features in a uniform manner. This has been done in Table 1 and in Figures 1 to 4. The question which principle is responsible for writing speed must be answered differently at the different levels of writing context. At the macro-context level an increase in ›word‹ size leads to an increase in force but not time, up to a certain size above which force cannot be augmented sufficiently, so that time tends to increase with size up to a power 1/2, a condition where force is constant. At the meso-context level, there is a trade-off between time and force increase in the transitions from small to large letters within a word; time increase with size may vary from a power 1/3 to a power 1/2, that is, from a condition where time and force increase to a similar extent, to a condition where size increase is wholly due to time increase. Finally, at the micro-context level, the relationship between local size (curve radius) and local writing time (the inverse of angular velocity) may also be expressed by a power function; an exponent of 2/3 is found in simple patterns, probably due to the narrow frequency band of these drawing movements. There are serious limitations to the validity of the two-third power law, however, when normal complex words are written.

What do we learn about timing in handwriting from these observations? It appears that such timing is not a matter of time-bound motor programs which are run off irrespective of the required sizes. In contrast, it seems that the programs for letter shapes merely prescribe spatial relationships and general, ordered sequences of movements to achieve the corresponding spatial patterns, or to reach the successive goals within the pattern. To the extent that time constancies are observed, they appear to be largely due to features of effectively and reliably performing output systems, such as fingers, hand and arm. Their physical and physiological characteristics may lead to a ›preferred‹ efficient loop duration at normal writing speed (the 5-Hz peak in the spectrum) and even to time constancy across a certain limited range of sizes if these are fixed within words (small-size macro-context). These effects are probably not due to temporal codes in the stored letter representations or to a central time keeping mechanism, because they give way to adapted timing in an extremely flexible manner. Firstly, writing durations are gradually extended when certain force limits are reached (larger-size macro-context effects). Secondly, writing durations trade off with the sudden force changes required between letters within a word (meso-context effects), even inside the range where between words no such duration effects are observed. Finally, the relationship between local size (curve radius) and local duration (the inverse of angular velocity) can be described less and less in terms of oscillation the more the movement approaches that of normal handwriting (micro-context effects). Indeed, performance durations seem to result from the biomechanical output features of handwriting as much as or more than they do from centrally stored motor programs. As such, force levels and frequency characteristics play an important role. It seems likely, therefore, that the central control mechanism, instead of directly controlling the timing of handwriting, merely takes the temporal consequences of these specific output features into account. Which output system is involved will largely depend on the situational requirements related to factors such as size, friction and writing posture.

References

Bernstein, N. *The coordination and regulation of movements*. New York: Pergamon Press, 1967.

Denier van der Gon, J.J., & Thuring, J.Ph. The guiding of human writing movements. *Kybernetik*, 1965, *2*, 145-148.

Freeman, F.N. Experimental analysis of the writing movement. *Psychological Review Monograph Supplement*, 1914, *17*, 1-46.

Galen, G.P. van, & Teulings, H.-L. The independent monitoring of form and scale factors in handwriting. *Acta Psychologica*, 1983, *54*, 9-22.

Greer, K.L., & Green, D.W. Context and motor control in handwriting. *Acta Psychologica*, 1983, *54*, 205-215.

Hollerbach, J.M. An oscillation theory of handwriting. *Biological Cybernetics*, 1980, *39*, 139-156.

Hulstijn, W., & Galen, G.P. van. Programming in handwriting: Reaction time and movement time as a function of sequence length. *Acta Psychologica*, 1983, *54*, 23-49.

Lacquaniti, F., Terzuolo, C., & Viviani, P. The law relating the kinematic and figural aspects of drawing movements. *Acta Psychologica*, 1983, *54*, 115-130.

Merton, P.A. How we control the contraction of our muscles. *Scientific American*, 1972, May, 30-37.

Michel, F. Etude expérimentale de la vitesse du geste graphique. *Neuropsychologica*, 1971, *9*, 1-13.

Shaffer, L.H. Timing in action. In: J.A. Michon & J.L. Jackson (Eds.), *Time, mind, and behavior.* Heidelberg: Springer Verlag, 1985, pp. 226-241.

Stelmach, G.E., & Teulings, H.-L. Response characteristics of prepared and restructured handwriting. *Acta Psychologica*, 1983, *54*, 51-67.

Teulings, H.-L., & Maarse, F.J. Digital recording and processing of handwriting movements. *Human Movement Science*, 1984, *3*, 193-217.

Teulings, H.-L., & Thomassen, A.J.W.M. Computer-aided analysis of handwriting movements. *Visible Language*, 1979, *13*, 219-231.

Teulings, H.-L., Thomassen, A.J.W.M., & Galen, G.P. van. Preparation of partly precued handwriting movements: The size of movement units in handwriting. *Acta Psychologica*, 1983, *54*, 165-177.

Thomassen, A.J.W.M., Keuss, P.J.G., & Galen, G.P. van (Eds.), *Motor aspects of handwriting: Approaches to movement in graphic behavior.* Amsterdam: North Holland Publishing Company, 1984.

Thomassen, A.J.W.M., & Teulings, H.-L. Constancy in stationary and progressive handwriting. *Acta Psychologica*, 1983, *54*, 179-196.

Viviani, P., & McCollum, G. The relation between linear extent and velocity in drawing movements. *Neuroscience*, 1983, *10*, 211-218.

Viviani, P., & Terzuolo, C. Space-time invariance in learned motor skills. In: G.E. Stelmach & J. Requin (Eds.), *Tutorials in motor behavior.* Amsterdam: North Holland Publishing Company, 1980, pp. 525-533.

Viviani, P., & Terzuolo, C. Trajectory determines movement dynamics. *Neuroscience*, 1982, *7*, 431-437.

Wing, A.M. The height of handwriting. *Acta Psychologica*, 1980, *46*, 141-151.

Reference Notes

1. Sternberg, S., Knoll, R.L., Monsell, S., & Wright, C.E. Control of rapid action sequences in speech and typing. *Invited Address. Division of Experimental Psychology, Annual Meeting, American Psychological Association*, 1983.
2. Teulings, H.-L., Thomassen, A.J.W.M., & Galen, G.P. van. *Invariants in handwriting: The information contained in a motor program*. In preparation.

Part IV. Notions: The Concept and Meaning of Time

Chapter 18. Semantics of Time

Johan van Benthem

The Philosophical Setting

Philosophical speculation about the nature of time has had a long tradition, with contributions by many major philosophers, from Saint Augustine to Kant (e.g.. Smart, 1979). In this century, a more systematic philosophy of time has developed along two lines, one within the philosophy of science (e.g.. Sklar, 1977), the other within logic (e.g.. Prior, 1967). Many specific issues have fueled research in these areas, of which the following four set the scene for our topic.

(1) What is the status of time in the cluster: time, space, spacetime, causality, modality? The prevalent approach has been to treat time as analytically separable from this context; but there are interesting, more ›holistic‹ relativistic or modal variants, in which time only emerges at a secondary stage (e.g. Winnie, 1977; Burgess, 1979), as a construction out of spacetime locations or ›world states‹. And of course, whether separable or not, our views of time and space seem structurally similar – witness the ubiquity of spatial metaphors in describing time, and temporal indicators when describing space (in terms of imagined movement).

(2) Is there a unique structure of time, or rather a bundle of temporal ontologies? Here, the term ›ontology‹ is used in the philosophical sense of ›possible structure of what exists‹. A more pluralistic perspective has become predominant, for various reasons. For instance, Russell (1926) distinguishes between ›common sense‹ conceptions of time, based upon extended intervals or events, and ›scientific‹ ones, in terms of durationless mathematical instants. And even among the latter, there is a variety of possible formal ordering patterns. For instance, temporal order might be discrete or dense, bounded or unbounded, etcetera.

(3) What are the basic patterns structuring time, and how do they arise? There is a distinguished line of philosophical and psychological accounts of the basic temporal structure, often intertwined. In particular, attempts have been made to explain the emergence of such basic ordering notions as ›precedence‹ or ›betweenness‹ as *invariants* under certain movements through time, or equivalently, under changes in temporal perspective. (see Beth (1935) or Weyl (1963) on the Helmholtz-Klein tradition; but also, e.g., Bohm (1965) on the work of J.J. Gibson.) Another relevant example is Michon's (1985) analysis of Fraser's five levels of temporal structure in chapter 20 of the present volume.

(4) What are the temporal aspects of other ontological categories? For instance, there are more or less radical views of *individuals* as physical manifestations through time; and also the different temporal structures of various types of *event* (instantaneous, durative or repetitive) have been a matter of study (see Vendler

1967, Mourelatos 1978). Mainstream philosophy has been biased towards ›eternal‹ ontological pictures, not changing over time, but more dynamic alternatives are now gradually coming to the fore.

These four broad questions play a role whenever choices are to be made concerning formal modelings of time, in particular when the aim is to give an account of temporal discourse and temporal reasoning as expressed in natural language.

Mathematical Ontology

Even without any specific language in mind, one may at least investigate plausible candidates for possible temporal structures with their properties. This enterprise is called ›mathematical ontology‹ in recent literature (see van Benthem, 1982). A choice of such a structure involves selection of a ›structure type‹ consisting of temporal *entities*, ordered by temporal *relations*, satisfying certain *conditions*. For instance, one can think of numerical points, with binary precedence satisfying the axioms of linear order, or open intervals with both precedence and inclusion, again subject to suitable postulates.

Recurring specific examples are the common mathematical structures of integers, rationals or reals, in their standard order. But there are good reasons for preferring a more sensitive conceptual analysis that does not presuppose the validity of the particular choices embodied in such standard scientific models of time. For instance, as for the temporal entities themselves, the priority of durationless over extended items should not be taken for granted. Furthermore, given one type of temporal entity, various ordering patterns may be reasonable. For instance, it has been suggested that binary precedence codifies a notion of irreversible temporal order (›time's arrow‹), whereas a truly basic notion – say ›betweenness‹ ought to be unbiased on this score (Needham, Note 3). Likewise, when extended temporal items are considered fundamental, a whole range of possible ordering relations has been imposed in addition to (total) precedence. One notable example is the pair ›co-begin‹, ›co-end‹ (Thomason, Note 4), which, interestingly, also plays a prominent role in Montangero (1985, Chapter 19 of the present volume). Perhaps, psychological considerations can cut through the thicket of mathematical options.

Finally, even given a selection of entities and relations, various structural conditions may still be imposed. Notably, point patterns can still be either discrete or dense, interval pictures either atomic or infinitely ›descending‹ (with intervals arranged like nested Russian dolls). This plurality of admissible temporal structures may actually be just what is needed in sorting out our rich - and not necessarily consistent - intuitions concerning time. For instance, one way of visualizing time is as a *family* of ever finer-grained interval structures, perhaps even starting at the top with one single undivided chunk of ›Time‹. (Notice that, in this perspective, each separate level might have smallest atoms, whereas the union of all levels may have become descending.) Another relevant example comes from the philosophy of science: in particular, the foundations of Measurement Theory. There exists an interesting account of the genesis of global scientific time as the minimal mathematical receptacle needed for embedding all possible ›local times‹ arising from experimental reports (see Manders, 1982).

Mathematical articulation of various philosophical intuitions about the structure of time still continues - from rather straightforward ordering properties to more esoteric ones, such as ›homogeneity‹ or ›isotropy‹ of time. In this way, various temporal ontologies arise, some of them tolerant of curious temporal patterns, others virtually restricted to the mathematical standard examples of integer, rational or real time. The virtue of our precise mathematical setting is that philosophical speculation can often be turned into concrete results. One case in point is the theorem in van Benthem (1982) stating that, on countable linear orders, homogeneity has the effect of ruling out all but the structures of integer time, rational time and one other intriguing possibility: dense from a global perspective, but discrete at a local level. (The latter picture has actually been suggested for our physical universe.) Other questions in this area have led to results about the existence of *limits* of sequences of temporal structures, classification of *invariants* under certain operations on temporal structures, or structural *connections* between various temporal ontologies (notably, between those based on intervals and those based on points).

Results like those referred to usually employ the abstract theory of order relations, with a sprinkling of logical ›model theory‹. Another mathematical connection which should be mentioned here is Measurement Theory (Suppes & Zinnes 1963). Considerations of measurement can be brought to bear upon the earlier issues, but some caution is needed. For instance, much attention is paid in measurement theory to the standard structure of the real numbers, with its obvious mathematical hierarchy of $\{R\}, \{R, <\}, \{R, <, +\}, \{R, <, +, \cdot\}$, passing from mere topological to metric structure. (Concomitantly, one may consider corresponding groups of automorphisms: bijections, monotonic maps, linear maps, narrowing down to mere identity; see Michon, 1985, chapter 20 of the present volume.) But, this obvious mathematical structure is not necessarily a temporal one: R in its role of a model for temporal order must be distinguished from R as a measurement scale for temporal duration. For instance, in the former role, the next reasonable layer of structure would seem to be the introduction of an ›equidistance‹ predicate, on top of the topological order. Even so, there remain several relations between the two viewpoints, to which we cannot do justice here.

We now turn to another angle upon time, viz. its reflection in natural language. Afterwards, the interplay of these two perspectives will form our central topic, the semantics of time.

Temporal Expressions in Natural Language

A pervasive phenomenon in natural language is the temporal flavor of many expressions occurring in them. The world as experienced by us is in flux, and language has to cope with this, one might say. Various broad linguistic mechanisms enable us to describe events as taking place in time: tenses (with temporal auxiliaries), aspects, temporal adverbs and connectives. For instance, tenses relate events to our present moment in various ways; witness »Mary sings« (present), »Mary sang« (simple past), »Mary will sing« (simple future), »Mary is singing« (present progressive), »Mary has sung« (present perfect), »Mary had sung« (past perfect) – and higher combinations occur too (»Mary will be singing«, »Mary would have sung«).

In addition, aspects give information about temporal features of events, such as durative/repetitive, perfective/imperfective. Under this heading also, one usually finds the Vendler's (1967) classification of verb phrases into four temporally distinct kinds: states (»Mary loves John«), activities (»Mary sings«), accomplishments (»Mary wrote a letter«) and achievements (»Mary died«). For instance, activities are »descending«: their temporal subparts are also instances of the same activity whereas, e.g., achievements describe almost atomic periods.

Then there is an overwhelming variety of temporal adverbial expressions, such as »always«, »often«, »seldom«, »never«, »now«, »yesterday«, »at six o'clock«, »for an hour«, »in an hour«, etc. And finally, temporal connectives are coordinating phrases such as »while«, »during«, »after«, »before«, »since«, »until«.

In addition, various other linguistic constructions have temporal overtones, witness the case of conditional statements. (»If« often means »whenever«.) Notice that we are concerned here with the ›low key‹ ways in which temporal reference enters natural language, not with the special vocabulary used in discussing time as such. Still, it may be useful to note in passing, that explicit reference to ›time‹ or ›times‹ can be very tricky. For instance, in »Peter betrayed his Lord three times« reference is made not so much to temporal items as to *occasions*. Notice also the duality between ›discrete‹ and ›continuous‹ uses, as in »Mary will come sometime« versus »Mary took some time getting dressed«. In the latter case time is a kind of stuff which can be consumed poured, and wasted. The duality between these two viewpoints, and possible shifts from one to the other, is a very general phenomenon in natural language.

Of course, the above examples are all taken from English. As usual in semantic research, one has to keep an eye on general cross-linguistic features eventually. Moreover, the linguistic behavior of any one of the above constructions may have its erratic aspects, irrelevant to our concern, which is to investigate the temporal picture (or pictures) presupposed in natural language use. As always, the art of science is to give individual facts a fair hearing – not to let them obscure general insight.

One immediate linguistic concern is the formal description of the above mechanisms, preferably through some kind of illuminating scheme of classification. For tenses one wellknown scheme was proposed in Reichenbach (1947); for aspects – that is, indications of duration, termination or frequency of events – one may consult Verkuyl (1972) or Dowty (1979). A more dynamical theme in these studies is how we may shift our aspectual presentation by various means, such as ›opening up‹ an event description by forming a progressive. In line with current methodological wisdom, such phenomena are often investigated by considering a whole fragment of natural language including some temporal constructions, and then trying to provide an explicit description of the meaning of every sentence in it. Two more theoretical issues are generated by these descriptive concerns. One would like to understand why there is such a small, and relatively stable repertoire of linguistic temporal constructions. But also, one would like to have a clearer view of the roles of eventual temporal ontology versus preliminary temporal representation in the above uses of language. That both are needed, and may be combined in one elegant theory is the contention of Kamp (1979), to which we return below.

For a more psychologically oriented approach to this whole area, the reader is referred to Miller & Johnson-Laird (1976).

Linguistic Semantics

In principle one could try to describe linguistic temporal constructions purely syntactically. But throughout the history of the subject a *semantic* approach has been prevalent, taking into account both linguistic form and temporal reality described, as well as the nature of the interpretation linking one to the other. Thus, a combination arises of linguistic and mathematical concerns which might be called ›logic‹. Our semantic scheme looks as in Figure 1.

In linguistic semantics one tries to be explicit about all three aspects, giving a grammar, a mathematical ontology, and a systematic recipe for relating expressions in the former to statements in the latter. As a miniature example we shall consider a regimented logical, rather than a natural, language, viz. Prior's ›tense logic‹. Its changing fortunes will also illustrate how, in the course of research, all three components of the above scheme are continually modified.

Prior's language is a propositional logic of statements, to which operators have been added for past tense (*PAST*) and future tense (*FUT*). Temporal reality is modeled by simple point orders (T, $<$), with, at each moment, a ›snapshot‹ of the then true atomic statements. The basic interpretative idea is that sentences are evaluated at points in time lying within a larger temporal structure to which the tenses give access thus:

A is true in (T, $<$) at point t.

For instance, past tense functions as follows:

PAST B is true in (T, $<$) at point *t iff*

B is true in (T, $<$) at some point *t' t*.

And a symmetric clause serves for *FUT B* (with t' after, rather than before t). These simple stipulations turn out to be very useful in explaining typical temporal inferences and non-inferences. Indeed, a whole research program has emanated from them (Prior, 1967).

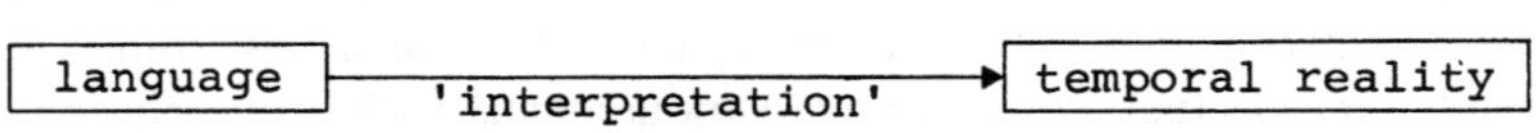

Figure 1.

Nevertheless, all three elements in the above picture can be criticized as faithful renderings of natural language. For instance, morphologically there are just two *tenses* (present, past) while ›future tense‹ is really a combination of present tense with an *auxiliary* - and this point holds generally, with auxiliaries »have«, »will«, »be-ing«. A more sensitive grammar should take this into account.

Next, the above interpretative scheme fails to account for the deictic character of many uses of tenses. To take a standard example, »Lucas did not tease Wieke« seems to mean neither PAST *not* (Lucas tease Wieke), which is too indefinite, nor *not* PAST (Lucas tease Wieke),which is too sweeping. It rather refers to some specific past point of time which the speaker has in mind. A more recent scheme, then, is to let present and past tense pick out contextually supplied present or past points, while the quantificational explanation (»at some«) is reserved for temporal *auxiliaries* (»have«, »will«, »be-ing«). This strategy is also in closer accord with natural language forms (Barwise & Perry, 1983).

Finally, the third component of the picture has been challenged too. The above scheme says very little about the actual temporal structure presupposed in language use. And indeed semantic cues from natural language are often indefinite, not enforcing one unique mathematical option. Nevertheless, there has been a shift towards moving from linguistic observations to formal modelings. Consideration of tenses such as the progressive, but also of the aspects and indeed of verb types themselves, has led to a competing interval or »period« picture of time as coming primarily in extended chunks. This makes for a scheme whose basic notion is »*B* is true in *I* at interval *i*«, where *I* is some suitable temporal interval structure, endowed not only with the already available relation of temporal precedence but also with, among others, inclusion, meet, overlap.

Thus, the delicate issue is evaded of what it means to say that the domestic activities expressed in natural language ›happen‹ at durationless instants *t*. Moreover, using various types of interval (open or closed), various verb types can now be distinguished (say, »dancing« versus »finishing the letter«). Further suitable explanations of adverbial constructions (»at three o'clock«, »for an hour«, »usually«) then become possible, explaining how the temporal type may change by such tags.

Even so, this new semantic picture may still lack discriminatory power. For instance, there are tenses such as French *passé simple* and *imparfait* (or even English simple past and past progressive, to some extent) which describe exactly the same structural situation, but in a different ›perspective‹. For example, »Louis vint« is an ›instantaneous‹ description of Louis' past coming, while »Louis venait« presents it as an open interval, allowing for further specification of what went on inside.

To account for such differences one needs an *enrichment* of the above semantic scheme – adding a component representing a perspective with respect to the real-world events it describes, with certain temporal relations specified. Truth of the discourse will then consist in such a picture having a suitable relation to the actual world: in simple cases in its being ›embeddable‹ therein as a kind of small scene. Thus, one now obtains a triangular semantic format, as in Figure 2.

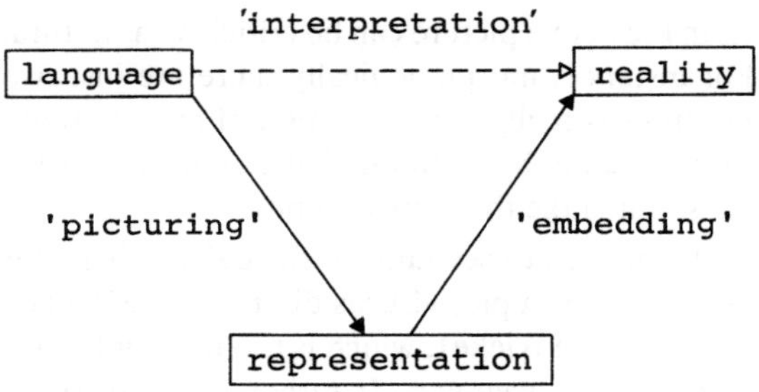

Figure 2

There is additional descriptive power in this scheme since different instructions for representation of discourse may exist on top of the same set of potential truth conditions. Thus, the *passé simple* tells us to picture a certain event as point-like (or ›atomic‹) *in the representation* (unlike the *imparfait*, which carries no such suggestion) – even though it may acquire duration in the richer structure of total temporal reality in which it is eventually embedded. A detailed account of this and other examples may be found in Kamp (1979), one of the seminal sources for the present perspective. In a sense the above scheme reflects a very old idea, dating back to De Saussure and beyond. But, Kamp has provided a detailed technical implementation for a complete fragment of natural language, which has sparked off much subsequent work.

Other refinements in the semantic picture are needed when the interaction is studied of tense with other grammatical constructions. For instance, the combination of tense with quantification (»every princess will have been married«) leads to the familiar picture of worlds of individuals on a temporal string. Thus, the earlier philosophical question of temporal aspects of other categories of object is reflected in language. Likewise, the analogies and interactions between temporal and spatial subsystems of natural language revive the issue of the ›separability‹ of time in our semantic modeling.

Research in this area is still only beginning, and large areas remain uncharted. For instance, and perhaps surprisingly, not much attention has been paid in the literature to the obvious temporal quantifiers themselves: »always«, »mostly«, »often«, »seldom«, »sometimes«, »never«. One reason is that these adverbs have very recalcitrant meanings. For instance, what is expressed by »Wanda always teased her brother«?

One preliminary conclusion may be drawn, however, for our larger theme. The interpretative needs of natural language *constrain*, but do not fully *determine* our structural temporal picture. In particular, they still seem to tolerate various ordering patterns (discrete or dense), and even to require the presence of both point-like and extended paradigms. Of course, these are the effects of surface text analysis. Deeper psycholinguistic studies may add additional constraints, however, such as the earlier-mentioned ›higher‹ intuition of homogeneity: the formal structure of time is the same from each temporal vantage point.

Logical Queries

Starting from the above semantic schemes, various types of theoretical question arise in temporal logic, of which we present only a sample. (see van Benthem, Note 2, for a broader survey.)

Why is there a relatively stable repertoire of temporal constructions? The earlier-mentioned question of the small number of actual tenses is studied in van Benthem (Note 1), where possible tenses are viewed as operations on sets of times (the ›substrata‹ or lifetimes of statements in the model). Actual tenses are then analyzed as such operations which are invariant for certain transformations of the temporal order.

Another, more central example concerns the earlier-mentioned temporal content of types of statement. Is there a systematic basis for aspectual classification in terms of ›inheritance behavior‹ on intervals? Various types of connection are possible between truth of a statement at an interval and truth at (some, all of) its subintervals. Thus, the Vendler classification into states, activities, accomplishments and achievements can be given a more formal semantic backing (see ter Meulen, 1984).

Now we turn to the central concern in logic, viz. the systematic study of reasoning, temporal reasoning in this case. We have various intuitions about validity and non-validity of temporal inferences, and these should be systematized, as well as reflected in the semantics. One immediate topic, then, is the correspondence between certain desired validities and particular features of our temporal structures guaranteeing this validity. For instance, in Prior's scheme, validity of the inference from *PAST PAST B* (»Lucas had lamented«) to *PAST B* (»Lucas lamented«) turns out to amount to *transitivity* of the temporal order, while its converse would enforce *density* (between every two points there is a third). Thus, given a partial picture of temporal entities and their relations inspired by general linguistic considerations, specific inferences impose additional conditions – providing further steps towards a more determinate mathematical ontology. But also conversely, logicians have studied if (and if so, how) certain structural properties manifest themselves in natural inference. The latter direction is noticeable in recent studies of interval semantics (see Humberstone, 1979).

But, complete theories of temporal inference are also of interest. For instance, given some temporal structure (say, the real or the integer line), one may axiomatize the set of all valid inferences in some formal language interpreted on it. Such completeness theorems constitute the core of the traditional subject of tense logic. Actually, most major structures have been investigated in this way, including relativistic spacetime.

Again, there is a converse direction of research too. Given an already existing theory of inference, one may search for a suitable matching class of temporal structures whose logic it is. This direction is found in philosophical studies of temporal reasoning in the tradition (cf. the analysis of Diodorus' so called Master Argument in White, 1984), but also in such practical areas as the semantics of programming languages, when describing temporal behavior of program execution

(Pnueli, 1977). One striking feature of many studies in the latter vein is the interplay between considerations of temporality and modality , especially when the future is concerned (already linguistically, the temporal status of the auxiliary »will« has always been in debate). Accordingly, one important recent trend in tense logic is the development of ›branching structures‹ for time, modeling a certain modal uncertainty (Burgess, 1980, Thomason, 1984). Poetically, this whole development was foreshadowed in Jorge Luis Borges' story ›The Garden of the Forking Paths‹ (Borges, 1981).

Points, Periods and Events

To conclude this survey of mathematical, linguistic and logical concerns, here is a topic in which they are all intertwined.

Whereas the picture of time as a set of discrete durationless points has been predominant in science, there has always been a more continuous picture too, of time as coming in extended items (underlying real events, which have spatio-temporal extent). There is a philosophical motivation for recovering such a ›lost paradigm‹ but also a linguistic one, as we have seen. Various authors have therefore reconstructed an exact mathematical ontology of intervals, ordered by various relations and subject to various constraints. Actually, there is quite some diversity here: authors have disagreed on the proper patterns (precedence, inclusion of overlap, various forms of ›co-begin‹ or ›co-end‹). Outside logic, where redundancy in a set of basic predicates may be a virtue rather than a vice (being a sign of a rich definitional structure), such conflicts tend to evaporate. Likewise, conflicting stands have been taken on the admissibility of ›branching‹ intervals. Both options are being developed.

Once such decisions have been made, however, it becomes possible to compare competing temporal paradigms. In the event, temporal interval and point pictures turn out to be closely related. In a precise mathematical sense, point structures induce interval structures (using ›convex segments‹) while, conversely, the former can be recovered as limits of descending sequences of the latter. Moreover, this connection also creates correspondences between further features of the two ontologies. For instance, it may be studied precisely how limited point structures become enriched when additional periods enter a period structure (for instance, because knowledge is obtained about new events). Thus, rivalry turns into peaceful co-existence.

Such formal studies not only have a dual motivation, but also dual applications. For instance, Thomason (Note 4) gives an interesting reconstruction and extension of Russell's philosophical views concerning the interplay of ›private‹ and ›public‹ time. On Russell's account, our private experiences form small stocks of events (or purely temporally, intervals), which are then collected into one large public experience. Private experiences induce private times, by the above construction; while public time arises in the same way out of public experience. Thomason then investigates the exact connection between private times and public time (see also van Benthem, 1984). Interestingly, the common sense world has extended temporal items, without coming into conflict with the eventual point-like scientific standard picture (which is that of public time).

Thus, both temporal paradigms are needed for an adequate reconstruction of certain philosophical views of time. A similar observation holds for the semantics of natural language; Kamp (1979) employs interval structures at this representational level, but point structures for the eventual semantic modeling of the real world. For instance, an ›instantaneously‹ presented event (such as *passé simple* »Louis vint« or an achievement »Louis died«) comes with a representation instruction that the corresponding interval not be subdivided by further intervals introduced for other events in the discourse. Thus, its interval is ›atomic‹ in the representation, and hence it would be point-like in the corresponding ›private time‹ – even though, in the eventual embedding, this interval will have a certain real extent.

The analogy with the earlier example runs even deeper. Kamp has a dynamic view of developing discourse, creating ever richer event representations with concomitant enrichments of the relevant embeddings. Nevertheless, the information in a text may still remain *partial*, in the sense that not all precedence, inclusion or other relevant relations between events reported need be decided. Thus the event models involved are a kind of partial trees, with every new bit of text ›orienting‹ further events with respect to certain already existing markers. It may be of some interest to display a few constraints which do govern these pictures. The following table has some basic conditions on *precedence* ($<$) and *overlap* (o) by themselves, as well as two conditions where they interact:

$x<y<z$	$\rightarrow$	$x<z$	(transitivity)
never:	$x<x$		(irreflexivity)
$x \,o\, x$			
$x \,o\, y$	$\rightarrow$	$y \,o\, x$	
$x \,o\, y$	$\rightarrow$	$not\,(x<y)$	
$x<y \,o\, z<u$	$\rightarrow$	$x<u.$	

As may be imagined, this topic generates a wealth in theoretical logical questions.

This format of description has found many applications beyond the original tense examples. For instance, the earlier-mentioned distinction between activities (›descending‹) and achievements (›atomic‹) has created a problem. For, upon closer inspection achievements may well describe physical processes which take time. What happens is rather that they are being *presented* as point-like, in a small linguistic context. As before, this phenomenon can be accommodated smoothly within Kamp's account. Presumably, these perspectival mechanisms of natural language play a certain useful role in terms of efficiency of information transfer. At present, formal tools are still lacking to make such conjectures amenable to logical analysis.

Cognitive Science to the Rescue?

Perhaps additional light will be shed upon these issues from a new angle in the field of artificial intelligence, where concerns of feasability, complexity and efficiency of representation and reasoning are more prominent than in theoretical logic. At least, this seems to be the promise of several publications in the area, of which a few deserve mention here because of their continuity with our previous concerns.

What kind of temporal picture must be implimented in a planning or a problem solving system to enable it to represent its information advantageously? And what axioms or rules of temporal reasoning should go with it? The answers are not necessarily found in the earlier ›tense logics‹: parsimony of primitive notions may create unwarranted inferential exertions and complete coverage of validities may be more than is needed for a powerful reasoner. These points may be found in McDermott (1982), who constructs a branching modal-temporal picture on the basis of some plausible intuitive axioms. It may be of interest to see which topics arise out of these practical concerns: causality, continuous change and persistence of facts. The latter topic leads to the issue of so-called ›non-monotonic reasoning‹ in temporal contexts, employing stability presuppositions (allowing one to assume that facts, once introduced, remain in force through time until explicitly canceled) not deductively warranted by one's premises. As yet, there is no definite logical theory in this area.

At least in its basic outline, McDermott's picture is still the point-based one of standard science. An interesting revival of Russellian concerns is found, however, in Hayes (1979), pleading for a reconstruction of a ›naive physics‹, that is, a system of notions and axioms directly reflecting our ordinary reasoning about space, time, motion, etc. Such theories might conceivably model our actual problem solving reasoning more closely and efficiently than those making the detour via scientific representations. The one case study in this area which has been worked out in some detail is the interval ontology of time, quite parallel to the logical research already reported here. A common hurdle, both here and in the earlier philosophical setting, appears to be the extension of such relatively simple static cases to more dynamic ones, involving frequency and motion. A good qualitative common sense theory of motion is not yet available – and it remains to be seen if one can be found in the whirlpool of our kinematical intuitions.

While the various temporal paradigms have arrived at some kind of peaceful co-existence in the logical pantheon, computational concerns might still motivate choices that philosophers cannot make. This thesis is defended by Allen (1983), who gives an interval representation of temporal knowledge which is claimed to be more efficient than any point-based competitor. There are interesting analogies with Kamp's partial event structures here, including the use of orienting markers in a tree-like structure. (see also Estes' (1985) plea for ›semantic networks‹ of this kind in Chapter 10 of the present volume). Owing to the elementary nature of the example (some simple temporal relations between events are being recorded and manipulated), Allen's thesis remains inconclusive, though intriguing. Whether in this manner or otherwise, the area of temporal logic seems ripe for a solid dose of empirical concerns.

Epilogue

It has not been the purpose of this survey to integrate the above semantic research program with current research in the psychology of time. But, a few obvious connections may be marked. At a fundamental level, speculation about the genesis of the temporal structure has often been a joint affair of philosophers and psychologists. Moreover, there may be an interesting link measurement theory. (For instance, some of our final semantic schemes invite comparison with the standard measurement approach.) And, of course, there is the obvious borderland of psycholinguistics.

Perhaps psychologists can profit from the long tradition of extensive and exact models of temporal description outlined in this chapter. Even if they were to reject specific proposals made, the spirit of the enterprise might be worth absorbing. On the other hand, most formal semanticists (like their linguistic colleagues) cherish the belief that eventually some psychological grounding for their constructs will emerge – a very quiet love affair thus far, not often resulting in actual encounters. Perhaps the present volume will help to change that situation.

References

Allen, J. Maintaining knowledge about temporal intervals. *Communications of the Association for Computing Machinery*, 1983, *26*, 832-843.

Barwise, J., & Perry, J. *Situations and attitudes*. Boston: MIT Press, 1983.

Benthem, J. van. *The logic of time*. Dordrecht: Reidel, 1982.

Benthem, J. van. Tense logic and time. *Notre Dame Journal of Formal Logic*, 1984, *25*, 1-16.

Beth, E. *Rede en aanschouwing in de wiskunde*. Groningen: Wolters, 1935.

Bohm, A. *The special theory of relativity*. Reading, MA: Benjamin/Cummins, 1965.

Borges, J. *Labyrinths*. Harmondsworth: Penguin Books, 1981.

Burgess, J. Logic and time. *Journal of Symbolic Logic*, 1979, *44*, 566-582.

Burgess, J. Decidability for branching times. *Studia Logica*, 1980, *39*, 203-218.

Dowty, D. *Word meaning and montague grammar.* Dordrecht: Reidel, 1979.

Estes, W.K. Memory for temporal information. In: J.A. Michon & J.L. Jackson (Eds.), *Time, mind, and behavior.* Heidelberg: Springer Verlag, 1985, pp. 151-168.

Hayes, P. The naive physics manifesto. In: D. Michie (Ed), *Expert systems*. Edinburgh: Edinburgh University Press, 1979.

Humberstone, I. Interval semantics for tense logics. *Journal of Philosophical Logic*, 1979, *8*, 171-196.

Kamp, H. Instants, events and temporal discourse. In: R. Bäuerle, E. Ugli, & A. von Stechow (Eds.), *Semantics from different points of view.* Berlin: Springer Verlag, 1979, pp. 376-417.

Manders, K. On the space-time ontology of physical theories. *Philosophy of Science*, 1982, *49*, 575-590.

McDermott, D. A temporal logic for reasoning about processes and plans. *Cognitive Science* 1982, *6*, 101-155.

Meulen, A. ter. Events, quantities and individuals. In: F. Landman & F. Veltman (Eds.), *Varieties of formal semantics*, Vol.3. Dordrecht: Foris, 1984, pp. 259-279.

Michon, J.A. Temporality and metaphor. In: J.A. Michon & J.L. Jackson (Eds.), *Time, mind, and behavior.* Heidelberg: Springer Verlag, 1985, pp. 288-296.

Miller, G.A., & Johnson-Laird P. *Language* and *perception*. Cambridge, MA.: Harvard University Press, 1976.

Montangero, J. The development of temporal inferences and meanings in 5- to 8-year old children. In: J.A. Michon & J.L. Jackson (Eds.), *Time, mind, and behavior.* Heidelberg: Springer Verlag, 1985, pp. 279-287.

Mourelatos, A. Events, processes and states. *Linguistics and Philosophy,* 1978, *2*, 415-435.

Pnueli, A. The temporal logic of programs. *Proceedings of the 18th IEEE Annual Symposium on Foundations of Computer Science*, Providence, NJ: 1977, pp. 46-57.

Prior, A. *Past, present and future.* Oxford: Clarendon Press, 1967.

Reichenbach, H., *Elements of symbolic logic.* Berkeley: University of California Press, 1947.

Russell, B. *Our knowledge of the external world.* London: Allen & Unwin, 1926.

Sklar, L. *Space, time and space-time.* Berkeley: University of California Press, 1977.

Smart, J.C.C. (Ed.), *Problems of space and time.* New York and London: McMillan, 1979.

Suppes, P., & Zinnes, J. Basic measurement theory. In: *Handbook of mathematical psychology,* Vol. 1. New York: Wiley, 1963, pp. 1-76.

Thomason, R. Combinations of tense and modality. In: D. Gabbay & F. Guenthner (Eds.), *Handbook of philosophical logic,* Vol. II. Dordrecht: Reidel, 1984, pp. 135-165.

Vendler, Z. *Linguistics in philosophy.* Ithaca NY: Cornell University Press, 1967.

Verkuyl, H. *On the compositional nature of the aspects.* Dordrecht: Reidel, 1972.

Weyl, H. *Philosophy of mathematics and natural science.* New York: Atheneum Press, 1963.

White, M. The necessity of the past and modal-tense logic incompleteness. *Notre Dame Journal of Formal Logic*, 1984, *25*, 59-71.

Winnie, J. The causal theory of space-time. In: J.S. Earman, C.N. Glymour, & J.J. Stachel (Eds.), *Foundations of space-time theories.* Minneapolis: University of Minnesota Press, 1977, pp. 134-205.

Reference Notes

1. Benthem, J. van. *Tenses in real time*, 1983, research report 83-28. Vancouver: Department of Mathematics, Simon Fraser University. To appear in *Zeitschrift für Mathematische Logik und Grundlagen der Mathematik*.
2. Benthem, J. van. *A manual of intensional logic.* Stanford: Centre for the Study of Language and Information, Lecture Notes, 1984a.
3. Needham, P. *Temporal intervals and temporal order.* Manuscript, Department of Philosophy, University of Uppsala, 1979.
4. Thomason, S. *Possible worlds, times and tenure.* Vancouver: Department of Mathematics, Simon Fraser University, 1979.

Chapter 19. The Development of Temporal Inferences and Meanings in 5- to 8-Year Old Children

Jacques Montangero

Developmental psychology can at least partially answer one of the fundamental questions asked by van Benthem (1985) in chapter 18 of this volume: What are the basic patterns structuring time as it is conceptualized and how do they arise? Our own studies of the development of time judgments and related parameters have, more specifically, been aimed at answering two major questions. The first question asks what the possibilities and limitations of reasoning about time in children are and how they develop. This immediately raises several issues which I shall mention briefly.

(a) Do conceptual representations of time develop with age? Some authors tend to think that development takes place only in aspects extraneous to time itself, whereas I believe, in accordance with Piaget's findings, that both the quality of reasoning about time and its meaning develop in children.

(b) If time concepts develop, one should endeavor to define the prelogic initially applied to time by the child and the meaning of duration at different levels of development in more precise terms than Piaget's ›time intuition‹.

(c) When compared with adults' answers, children's judgments of duration are sometimes erroneous. One aim of psychological research is to explain these errors and any unexpected responses. Piaget (1969) referred to the absence of an operational system and to the incapacity of the child to take the relation between velocity and time into consideration. Other explanations rely on the idea that children are more likely to use misleading cues (Levin, 1977) or rules that are applied to irrelevant cues (Siegler & Richards, 1979). However, none of these explanations can account for all observed error categories or for the fact that correct answers co-exist with wrong ones in the same individuals.

(d) One specific aspect of children's time judgments is their great variability, between children if we study them in the same situation or in the individual child if we test it in different situations (see Levin, 1977; Montangero, 1977). In contrast with Siegler & Richard's model which postulates one rule only at each level of development, a satisfactory model should include several potentially available rules in order to explain the variability of judgments.

The second major question addressed by our research is: What are the significant components of the concept of duration? It can be subdivided into three further questions.

(a) Can duration be defined exhaustively on the basis of the three parameters time, speed and distance only, as in the model of time in classical physics and in that of Piaget? From the results of a number of psychological experiments, we know that other parameters must be involved. Fraisse (1967) for example stipulates the importance of frequency and number in duration judgments. Some of my early experiments showed that apart from having to relate velocities and distances, children also had to connect the relative orders of succession of events to be able to grasp the concept of duration.

(b) Has duration one meaning or several ones, as is suggested in some other chapters of the present volume? It seems evident that time in general, but also time in a more specific sense and in particular time interval or duration, can convey several meanings which should be incorporated in a model of time conceptualization.

(c) If the study of reasoning about time, the inferential aspect, is to take into consideration the meanings on which the inferences bear, one should endeavor to understand the relations between meanings and inferences. Three solutions seem possible: meanings are constituted by inferences and depend on them, or inferences are determined by the meanings to which they relate, or there is a more or less reciprocal interaction between the two.

A Model of the Development of the Concept of Duration

For several years I have conducted research in order to shed more light on the problems outlined above. My experimental results helped me to elaborate a model (in the broad sense of the term) of the developmental levels of knowledge related to duration, in children aged 5 to 9 years. In the framework of genetic epistemology, such a developmental model may serve as an aid to understand a concept also at the level of adult thinking.

Before giving examples which summarize the main findings of the research program and illustrate the features of the model, I shall define six basic features of the model.

(1) The understanding of duration is explained in terms of the relations and meanings involved in three subsystems (see Figure 1). Each of these subsystems is composed of three subjective variables or ›meanings‹, corresponding to the representations of three physical parameters. One of these, time interval or duration, is common to all three subsystems. The first of these subsystems relates three temporal variables, viz. time interval ($\triangle t$), relative starting order (t_1: started before, after or at the same time) and ending order (t_2: ended before, after or at the same time). This subsystem works for judgments such as: »lamp *A* was lit up before lamp *B* and switched off together with lamp *B*; therefore lamp *A* was on for a longer time.« The second subsystem relates duration ($\triangle t$) with velocity (v) and distance (d). For instance; »runner *A* ran twice as fast and covered a distance twice as long as runner *B*; therefore *A* and *B* ran for the same length of time.« The third subsystem concerns discontinuous activities. It relates duration ($\triangle t$) with frequency (f) and a discontinuous quantity which may possibly be expressed in terms of number (n). Example: »*A* jumped the same number of jumps as *B*, but he jumped faster; therefore *A* jumped for less time than *B*«.

Figure 1 represents the three subsystems of relations between ›meanings‹, or represented cues, at an integral, complete developed stage when the within system relations are triadic. This is the case in adults and, at least for certain types of problems, in children of 8 years and older. As is clear from the examples above, the relative duration of two events can be judged by considering the two other ›meanings‹ of a subsystem. Judgments may be given starting from any cue represented in the subsystems, provided the value of the second cue of the appropriate subsystem is also taken into consideration.

In order to refer to the inferences of the subjects, I use a notation which consists of letters, signs (+, −, =) and arrows. The letters refer to the representation of the different cues. The signs + (more), − (less) and = (equal) indicate the relative value of the cues. For instance, v^+ means faster. In the first subsystem the relations before, after and together (at the same time) are indicated by the abbreviations *bef*, *aft* and *tog*. The arrow (→) indicates a functional dependency between meanings. For example, $f^+ t^-$ means that higher frequency (v→) entails less time.

In this notation the three examples given above to illustrate each subsystem can be symbolized as follows:

Subsystem 1: t_1^{bef} and $t_2^{tog} \rightarrow t^+$
Subsystem 2: v^+ and $d^- \rightarrow t^=$
Subsystem 3: f^+ and $n^= \rightarrow t^-$

(2) Depending on the situation presented or on the level of development of the subject, one subsystem may dominate the other two. This means that duration judgments will be derived from the predominant subsystem and that judgments of cues represented in the other subsystems are erroneously derived from the predominant cues.

(3) In young children, the basic form of the relation between ›meanings‹ is dyadic, rather than triadic as it is in older children and adults. The ›meanings‹ of the subsystems are related pairwise (see Figure 2a) as, for instance, in Crépault's (1978) model of time reasoning in adolescents.

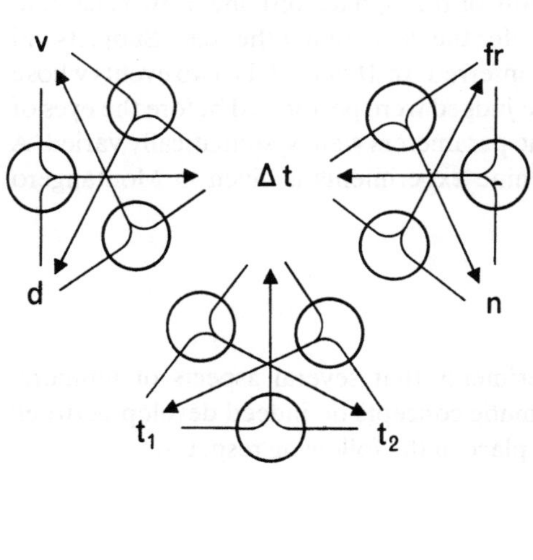

Figure 1. Three subsystems of 'meanings' or cues related to duration.

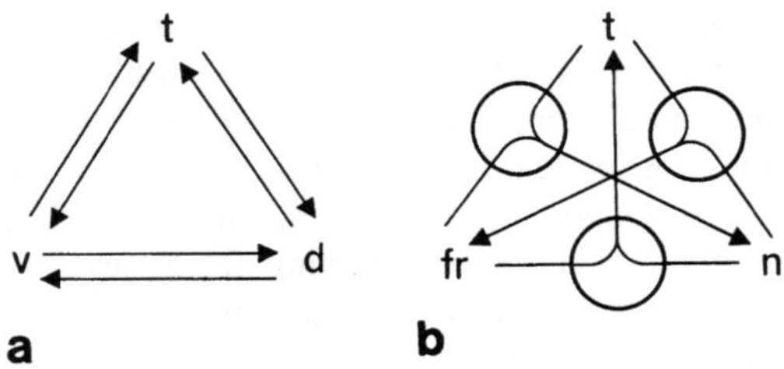

Figure 2. Forms of relations within subsystems: *(a)* dyadic relations; *(b)* relations of three elements at a time.

(4) The different possible pairs of cues have an unequal degree of importance: some pairs are more prominent than others. This means that subjects have a strong tendency to relate the two ›meanings‹ involved in such a pair and that dissociating the two ›meanings‹ is difficult. As a result a cue is sometimes related with an inappropriate cue with which it forms a prominent pair, rather than with another more appropriate cue which is part of the less prominent pair.

(5) The ›meanings‹ or cues represented in the subsystems are part of the meaning of time. Each dyad or triad of represented cues constitutes a partial meaning of the concept of duration.

(6) The relations established between cues or ›meanings‹ (more specifically the relative ease to establish them) are determined by the meanings related. The inferential aspect of time representation is thus determined by the intensional aspect. However, the development of new inferential capacities enriches in turn the meanings.

A Review of Some Experimental Results

In the series of experiments to which I shall now turn children (aged 5 to 9 years) were required to give relative judgments (in terms of ›more‹, ›less‹, ›before‹, etc.) about the duration, speed, order of succession, etc., of two events. Each subject was presented with several situations and intrasubject, intersubject (group of age) and intergroup analyses of the answers were performed. Two main methods were used. *Method 1* consisted of presenting (verbally or through action) one or two cues and asking the subjects whether they could infer the value of another cue. Subjects did not actually see the cues which had to be inferred. In *Method 2* the two events whose duration and other parameters had to be judged were performed before the eyes of the subject. In both methods some of the parameters were systematically varied. A summary of the results of a series of nine experiments is given in Montangero (1984).

Development

It is clearly borne out by these experiments that several aspects of temporal reasoning and of the meanings of kinematic concepts do indeed develop between the ages of 5 and 8. Development takes place in the following respects.

The Number of Cues Related. Experiments with Method 1, designed to study within-subsystem relations, showed that at 5, 6 and 7 years of age most answers imply covariation of two cues. For example, in an experiment studying the first (relative order) subsystem (Montangero, 1981) 80 percent of the answers related to two cues only (e.g., $\triangle t^+ \rightarrow t_2^{aft}$). Among 7-year old children, 42 percent of the subjects (against 5 percent and 18 percent in 5- and 6-year old children respectively) felt on occasion that one should know the value of two parameters in order to infer the third one. Moreover, the majority of 7-year olds did, in fact, explicitly relate three cues at a time for some situations. At the age of 8, subjects refused to make an inference when the value of one parameter only was given.

An experiment with Method 2 (Montangero & Gurtner, 1983) aimed at studying the interference between subsystems 2 (speed-distance) and 3 (frequency-number). The first situation consisted of continuous movements, viz. displacements of two toy bicycle riders. The second situation consisted of ›pure‹ frequency without displacement: jumping on the same spot. The third situation combined displacement and frequency (see Figure 3): two dolls with magnets jumped forward and picked up a coin at each jump. In the latter situation there was a conflict between the cues related to the second subsystem (doll *A* moved faster and further and consequently stopped nearer the goal than doll *B*) and the cues of the third subsystem (doll *A* jumped less quickly and consequently made fewer jumps and picked up fewer coins than the other doll). In this situation 3-, 5- and 6-year old children's judgments centered on the cue represented in one subsystem (e.g. v and d), whereas 8-year olds related the cues of different subsystems (e.g., »He went further (subsystem 2) but he picked up fewer coins (subsystem 3) and both stopped together (subsystem 1)«.

On the whole inferences on duration are first limited to dyadic relations, though sometimes one of the terms of a dyad is a global undifferentiated concept like ›slow and long‹. Toward the age of 8 temporal inferences can deal with three cues at a time; moreover, correct correspondences between elements of different subsystems can be established.

The Hierarchy of Pairs. If we consider the frequency of relations established between two cues or meanings in the answers obtained, we see that this frequency varies according to the meanings related and the sense of the relation (for example, $\triangle t^+ \rightarrow n^+$, viz. a longer duration implies a larger number, is more frequent than $n^+ \rightarrow \triangle t^+$). The proportion of relations established and qualitative signs of the facility to establish them permit us to define a hierarchy of covariations of two meanings within each subsystem from the ages 5 to 7.

Figure 3. Situation 3 from Montangero & Gurtner (1983). Two dolls *A* and *B* start and stop simultaneously. *A* moves forward faster, *B* jumps with a higher frequency and picks up more coins.

The Proportion of Correct Duration Judgments. In experiments using Method 2 the proportion of correct duration judgments for displacements usually increased with age and was always higher, in terms of number of answers and number of subjects, in 8- to 9-year old children. We sometimes observed, however, a regression in the proportion of correct answers. For example, in situations 1 and 3 of the second experiment described above, viz. the ›pure‹ displacement and the ›mixed‹ displacement-frequency situations, correct answers were more frequent in 5-year old children, reaching 31 percent for situation 1 and 50 percent for situation 3, than in 6-year old children, for whom the corresponding percentages were 6 and 27, respectively. This apparent regression is explained by the fact that 5-year old children did not differentiate relative stopping orders (t_2^{tog}) and duration ($\Delta t^{=}$). Therefore, the subjects who could perceive the simultaneity of stops answered that durations were the same, which is the correct answer. At the age of 6, duration and stopping orders were differentiated. Subjects who judged (correctly) that both bicycles or dolls stopped together could still give a different answer for duration. Consequently, the decrease of correct judgments does not necessarily reflect a regression of temporal reasoning. This example clearly shows the necessity of distinguishing answers from underlying processes.

The Meanings of Duration and Speed. In younger subjects there was a maximum of indifferentiation of the meaning of duration with that of distance covered, order, etc. Duration was not considered as an interval with two boundaries but as an extended chunk of time which precedes or follows an event, which implies that duration and succession are not (yet) dissociated. As far as speed was concerned, when both velocity and frequency were present in a situation (as in Montangero & Gurtner, 1983), 5-year old children considered only velocity as speed, and those 6 years of age tended to center on frequency and older subjects considered both velocity and frequency as speeds.

Errors or Unexpected Answers

Many errors can be explained in terms of relative strength of some cues and relations which entail a detour effect. For example, when the covariation $v^+ \rightarrow d^+$ is strong, subjects asked to relate v and t produce an indirect relation: $v^+ \rightarrow (d^+) \rightarrow t^+$. A similar process does occur within other subsystems (see Figure 4).

Figure 4. Judgments explained by a 'detour' process: (a) »The faster one took more time«. $V^+ \rightarrow (d^+) \rightarrow t^+$; (b) »The one who stopped after the other started first«. $t_2^{aft} \rightarrow (t^+) \rightarrow t_1^{bef}$.

Some errors are due to the predominance of a one particular subsystem. For example, in the situation represented in Figure 3, young subjects often derived their judgment of number of coins picked up (situation 3) from the distance traveled ($d^+ \rightarrow n^+$). Subsystem 2 dominates subsystem 3 in this case. The experiments revealed a dominance of subsystem 1 over subsystems 2 and 3 when, for the same experimental situation 3 a number of subjects centered on simultaneity and left the difference of velocity out of consideration.

Variability of Answers

Cases of transfer of one type of duration judgment to different situations (subsystems) occurred only in a minority of subjects. Judgments varied according to the character of the situation presented and the type of question. Some variations seemed to stem from endogenous factors. In an experiment using Method 1 (Montangero, 1979), for example, several 6-year old children alternated the related cues systematically in their successive answers, regardless of the question asked. On the whole, however, there does not seem to be one rule or strategy for judging duration that is characteristic for a particular subject or level of development. Instead several options appear to be available, depending on the subsystem or subsystems, to which the subject actually refer. It is possible, however, to determine what are the most probable tendencies at each development level.

Possibilities and Limitations of Younger Children and Subsequent Development

In 5- and 6-year old children, understanding of duration can be characterized in the following way. Inferences are limited to the relations between pairs of ›meanings‹ (Figure 2a). Sometimes a pair of ›meanings‹ may acquire a global meaning. Thus, for instance, ›quickly done‹ refers both to velocity and time. Most covariations of two elements correspond to the adult model of time when the third element is supposed to be held constant (and equal). For example, ›started before‹ entails ›longer duration‹, as would be the case when stops are simultaneous.

At this age inferences about situations have some features of actions: they tend to be step by step relations and often have a purposeful character. Thus an activity which started after another is supposed also to stop after the other activity »in order to do the same amount of things«.

Also, in young children the predominance of some pairs within a subsystem or of the elements of one subsystem over those of another subsystem entails at least some correct covariations, but also causes indifferentiation, that is the difficulty to dissociate the elements of a couple, and exclusions, that is the difficulty to relate one of the two meanings to a third one. For example in an experiment using Method 1 (Montangero, 1981) 6-year old children had great difficulty answering the question concerning the relations between t_2 and t_1, since t_2 was strongly related to Δt.

In summary, the development toward a correct understanding of duration is found to be crucially dependent on three fundamental factors, to wit, the ability to successively consider all dyads within the three subsystems; the capacity to relate three cues or ›meanings‹ at a time, which implies the possibility to dissociate any couple (see Figure 2b); and finally the establishment of correspondences between elements of different subsystems and of correspondences between each subsystem and a class of situations.

Questions About the Components of the Concept of Duration

By now it will be evident that duration is certainly not definable with the three parameters time, velocity and distance only. There are various representations of duration and patterns of reasoning about time for events that have nothing to do with speed or distance, as was shown by the results of several experiments. Duration can be said to have several meanings, each corresponding to a dyad or a triad of elements from the three subsystems. Duration has therefore any and all of the following denotations. It is an interval delimited by initial and final boundaries (instants). As such it is related to the concept of succession of instants ($\triangle\, t - t_1, t_2$). This corresponds to the interrelation of point structures and interval structures mentioned by van Benthem (1985, chapter 18 of the present volume). Being a succession of instants, duration is also an interval fragmented by a periodicity or rhythm ($\triangle\, t - f$). Duration may be treated as flow measurable by a constant velocity ($\triangle\, t - v$), as a quantity of discrete unities that can be counted ($\triangle\, t - n$) or, finally, as a quantity which can be evaluated by spatial dimensions ($\triangle\, t - d$).

As we saw earlier, these meanings of duration develop and so do temporal inferences and judgments. It seems that any progress in the quality of these inferences (e.g. the possibility to dissociate a dyad, to relate three parameters) entails an advance in the notion of duration, that is, in the potential to consider more aspects of duration at a time, to conceive of it as an interval, or to dissociate it from other parameters. And in turn, it is likely that any progress in the meaning of time entails new developments of temporal inferences.

Conclusion

The issues raised in the introduction of this chapter found answers in the above description and interpretation of some results from an extensive research program. This research also sheds some light on the question of the status of time compared with that of space (a further discussion of which can be found in van Benthem (1985, chapter 18 of the present volume). My model of duration is certainly not isomorphic with a model of space for it specifically implies the possibility to dissociate distance and time. From a psychological point of view, time is analytically separable from space and from other (psychophysical) dimensions, as soon as its relations with these dimensions are correctly established.

Note finally that our research deals with the concept of time interval or duration, which is only one aspect of the concept of time. In common sense representation, a clear conceptualization of time interval seems to occur later in development and to be less basic than the concept of succession, which indeed constitutes a form of ›prototemporality‹ to use one of J.T. Fraser's terms as they are discussed by Michon (1985) in chapter 20 of this volume.

References

Benthem, J. van. Semantics of time. In: J.A. Michon & J.L. Jackson (Eds.), *Time, mind, and behavior.* Heidelberg: Springer Verlag, 1985, p. 266-278.

Crépault, J. Le raisonnement cinématique chez le préadolescent et l'adolescent, I. Esquisse d'un modèle théorique: Concepts de base. *Archives de Psychologie*, 1978, *178*, 133-183.

Fraisse, P. *Psychologie du temps*. Paris: Presses Universitaires de France, 1967 (2nd edition).

Levin, I. The development of time concepts in young children: Reasoning about duration. *Child Development*, 1977, *48*, 435-444.

Michon, J.A. Temporality and metaphor. In: J.A. Michon & J.L. Jackson (Eds.), *Time, mind, and behavior.* Heidelberg: Springer Verlag, 1985, p. 288-296.

Montangero, J. *La notion de durée chez l'enfant de 5 à 9 ans*. Paris: Presses Universitaires de France, 1977.

Montangero, J. Les relations du temps, de la vitesse et de l'espace parcouru chez le jeune enfant. *l'Année Psychologique*, 1979, *79*, 23-42.

Montangero, J. Les relations entre durée et succession: Etude d'une prélogique enfantine appliquée au temps. *l'Année Psychologique*, 1981, *81*, 287-308.

Montangero, J. Perspectives actuelles sur la psychogenese du temps. *l'Année Psychologique*, 1984, *84*, 433-460.

Montangero, J., & Gurtner, J.-L. Vitesse-fréquence, vitessedéplacement et jugements de durée chez l'enfant. *Archives de Psychologie*, 1983, *51*, 368-384.

Piaget, J. *The child's conception of time*. London: Routledge & Kegan Paul, 1969.

Siegler, R.S., & Richards, D.D. Development of time, speed and distance concepts. *Developmental Psychology*, 1979, *15*, 288-296.

Chapter 20. Temporality and Metaphor

John A. Michon

Introduction

Time has been discussed mostly in spatial terms. The use of expressions like »we left this period *behind* us« or »there is a difficult time *ahead* of us« may be no mere incident. It could well indicate the existence of »a thoroughly spatial metaphor, a complex cognitive system that space and time expressions have in common« as was suggested by H.H. Clark (1973; p. 62). Physics has also ›spatialized‹ time, as is pointed out in Park (1985, chapter 3 of the present volume). Conceptually time tends to be represented in common speech as a straight line, passing through the body of the speaker (the present) from behind (past) towards the front (future). It will be clear, however, that this conceptualization does in no way pay sufficient tribute to the richness of temporal expressions that most languages appear to have. More is needed. Psycholinguistically a considerable step forward was made by Miller & Johnson-Laird (1976) in their account of procedural semantics. They too favor a quasi-spatial linear representation of time, but in addition they attribute the complexity of temporal expressions, of ›time talk‹, to the conceptual difficulties people encounter when they try to project their thoughts and experiences on that linear representation of time.

Van Benthem (1985), in his contribution to the present volume (chapter 18) also suggests that, indeed, part of the difficulties may derive from the restrictions imposed by the spatial metaphor in its simpler form. As he points out, however, there are many other possible representations. The formal properties of these alternatives are a matter of mathematics, but their applicability is constrained by semantic limitations. And he proceeds to outline several alternative ›pictures‹ of time. Montangero (1985, chapter 19 of the present volume) also argues for a richer set of temporal forms than the simple spatial metaphor can accommodate. In the present paper I shall discuss several descriptive frames for temporal relations as they have been proposed by Fraser (e.g. 1978, 1982).

Fraser's Levels of Temporality

J.T. Fraser, founder of the International Society for the Study ofTime and author of several ›chronosophical‹ studies, entertains an evolutionary view of time. »Time«, he says, »had its genesis in the early universe, has been evolving, and remains developmentally open-ended (...) The notion of time as having a natural history is difficult to assimilate with received teachings or even to express in non-contradictory statement; (...) Yet, a detailed inquiry reveals that the evolutionary character of time is already implicit in the ways time enters physical science in particular and natural science in general.« (Fraser, 1982, p. 1).

This position implies no less than a straightforward status for time as a *real*, evolving attribute of the material universe. But it is not my intention to discuss this aspect of his unconventional conception of temporality. Instead I wish to suggest a relation between the various forms of time distinguished by Fraser and some basic metaphors that seem to play an important role in the cognitive organization of the world around us. It could be that Fraser's system of ›temporalities‹, irrespective of its significance as a physical theory, *at least* specifies - or instantiates - a rather general complex of cognitive structures. These structures, or basic metaphors will be understood in the present context as generative structures, that is, as sets of symbolic rules which can produce dynamic internal representations or ›images‹ of the world. Metaphors in this sense are not just analogies. They are like scientific theories and models, the principal difference being that they do not *literally* refer to the object and relations they represent (e.g. Black, 1962; Boyd, 1979).

Fraser (1978, 1982) holds that there are five stable, hierarchically related ›natural‹ levels of temporality, each of which has emerged at a particular stage of development of the universe. In this context hierarchical means that properties which emerged at a lower level will remain ›available‹ at the next higher level.[1]

(1) The first stage represents the domain of elementary particles with zero rest mass, viz. photons. All photons move at the (constant) speed of light and, for that reason, any ›temporal‹ relation other than *co-presence* is meaningless. The characteristic form of time is the chaos of *atemporality.*

(2) The second stage is represented by the elementary particles with non-zero rest mass, such as protons, electrons, etc. This is the realm of quantum mechanics and, according to Fraser, the appropriate time form is *prototemporality* which allows determination of *relative order* of events, if perhaps only locally or statistically.

(3) The domain of aggregated matter, from molecule to milky way, determines the third stage. It is described by mechanics, classical as well as relativistic. Its proper time form is called *eotemporality* (after *Eos*, the Greek goddess of dawn), which is the conventional time of clocks: continuous, regular and reversible. The latter concept, reversibility, is of crucial significance in discussions about the true nature of time (Davies, 1981; Prigogine, 1980; see also Michon, 1985, chapter 2 in the present volume and Park, 1985, chapter 3 in the present volume): reversibility implies that the choice of what is to be considered as past or future for both tails of an event sequence is an arbitrary one. Eotemporality does, however, allow temporal distance or *duration* to be specified. This description does indeed hold for a purely mechanistic, clockwork universe.

(4) At the fourth stage, especially but not exclusively (see Prigogine (1980) for exceptions) represented by living organisms, the ›natural‹ time form is *biotemporality.* In biotemporality a ›physiological present‹ is defined which serves

1 Occasionally Fraser refers to a sixth level, sociotemporality, which lies on top of the hierarchy.

to establish the interfacing, or tuning, between processes inside the organism (or system) and events in the environment. Biotemporality moreover, is *irreversible*; »time's arrow« has a meaning here because evolution and development cannot be undone. By this token biotemporality permits the absolute order between events to be specified.

(5) The domain of (human) knowledge, or noosphere, finally determines *nootemporality* in which events have a definitive position on an individual time scale, a personal history which defines a conscious present, a privileged *now*, but also »beginnings and endings«.

Basic Metaphors

Leaving the question unanswered whether Fraser's conceptualization of time forms contains the germs of an empirically testable physical theory of the genesis and evolution of time, I shall now try to establish the plausibility of considering the levels of temporality as instantiations of so called ›basic metaphors‹ or ›world views‹. As I already suggested, basic metaphors are like naive system theories, that is, they are dynamic conceptual structures or schemata that are »general enough to be of use in representing information from distinctly different events that are similar to the extent that they share elements of the same structure« (Schank, 1982, p. 222). Such metaphors permit us to encode and comprehend objects and events in terms of a coherent and reasonably consistent world view. In this sense, as was pointed out in a recent discussion of the role of the metaphor in scientific theory, »the use of metaphor is *one of many devices available* to the scientific community to accomplish the task of accommodation of language to the causal structure of the world« (Boyd, 1979, p. 358; my emphasis).

I wish to extend this point of view by proposing that such metaphors are not arbitrary but rather derive from structural properties of the human mind. This actually amounts to the position that such metaphors cannot take just any conceivable form, but, instead, display considerable morphological stability. This position is, in one form or another, prominent in many recent theories of cognition (e.g. Chomsky, 1980; Fodor, 1981, 1983; Gardner, 1983; Haroutunian, 1983, Goodman, 1984). It is also emerging from several of the chapters in the present volume, in particular Jones (1985, chapter 13) and van Benthem (1985, chapter 18).

Which are these basic metaphors, and how can we know if they constitute a fundamental (exhaustive, stable) set? Several authors have, for different purposes, stressed a group of four basic or root metaphors (e.g. Pepper, 1942; De Mey, 1982; see also Tyler, 1981).

The first view, called formistic or monadic, provides an interpretation of the world that is strictly fact oriented. It seeks to order and classify objects and events, according to the presence or absence of certain attributes, and in terms of sub- en superordination. Mediaeval science was dominated by this view of the world, culminating perhaps in the elaborate ›memory theaters‹ described by Yates (1966; see also Hajdu, 1936), where one would sit and contemplate the known facts about the world, symbolized on walls and ceiling, in order to achieve illumination and mystic unification with God and the Universe.

The second view recognized as a stable basic metaphor centers around mechanical interaction. It specifically recognizes causal relations between objects and the predictability of events. It was, of course, the dominant world view of Europe in the 18th century and is most concisely exemplified by the mechanical clock.

A third view is generally known as contextualism. It differs from the mechanistic view by accounting for the dependence of a system's functioning upon its control of the environment. Contextualism is exemplified by the thermostat, and its science in cybernetics.

Fourthly and finally, what is called organicism (Pepper, 1942) or cognitivism (De Mey, 1982) acknowledges both the structural and functional interdependence of a system and its environment. It accounts for development and learning, recognizes the uniqueness of personal history and centers around the concepts of self-organization and development.

Although this is neither necessary, nor even the intention of the authors who initially outlined them, I prefer to consider these basic metaphors as hierarchically related (and perhaps parts of an exhaustive set as well). The higher levels demonstrably retain aspects of the lower ones. Thus, for instance, cybernetics implies the principles of mechanistic world view. It should also be clear that the choice of basic metaphor will occasionally be prescribed by the nature of the facts that are to be dealt with in a particular representation: *natura dictare delectat*. However, more often than not the choice of basic metaphor will be determined by the needs of the observer rather than by the facts under concern.

A close analysis of Fraser's views about the various states of nature and their intrinsic temporalities reveals that these states may in fact be considered as instantiations or exemplifications of the basic metaphors (for a more detailed discussion see Michon, 1985). By implication therefore the various forms of time – *atemporality, prototemporality, eotemporality, biotemporality* and *nootemporality* – must have their origin in these basic metaphors.

On this account these five levels of temporality should, in summary, be considered as a particular – namely time oriented – set of cognitive representations derived from, or generated by, a well-established set of generative basic metaphors. Each level of temporality does represent the proper (temporal) ordering principle that derives naturally from each of these basic metaphors.

Measurement Structures and Temporalities

Among the various possible formalizations of time – cf. van Benthem's (1985, chapter 18 of the present volume) ›mathematical ontologies‹ – the one that is consistent with the properties of Fraser's five levels of temporality would seem to be the structure of linear measurement theory. If this is so each basic metaphor must imply a natural or best fitting measurement structure, that is, a necessary or at least a preferred way of quantifying relations between events within the context of that metaphor. *Measurement scales* are the actual representations of these quantitative relations, and consequently a scale type should be identifiable with each level of temporality as distinguished by Fraser (1982).

Interestingly enough the formal ordering operations and properties which are naturally allowed in each of the basic metaphors coincide with the measurement operations and scales derived from the abstract theory of measurement. A useful set of scales has been around for quite some time (Campbell, 1928; Stevens, 1951). It includes the familiar nominal, ordinal, interval, ratio and absolute scales (Figure 1). As Narens (1981) pointed out, however, this classification is based on ad hoc considerations and it has perhaps survived mostly because it has proved nearly impossible to extend the set in a meaningful way. Narens showed in his paper that the same scales can be derived from a mathematical argument and indeed constitute a canonical set. The following is the briefest summary of Narens' arguments.

Automorphisms are mathematical entities (mappings) that can be used to specify what transformations a measurement scale may undergo before it will loose its type identity and no longer preserve the established relations between numbers (or ›objects‹ or ›events‹). Thus the nominal scale may be subjected to any transformation that will leave the individual identity of the scaled objects or events unchanged. Order scales may undergo any monotonic transformation, including sign reversal. Interval scales permit transformations that preserve distance relations while ratio scales permit only such transformation as leaves distances and ratios unchanged. The absolute scale finally must remain identical and therefore accepts only the identity transformation.

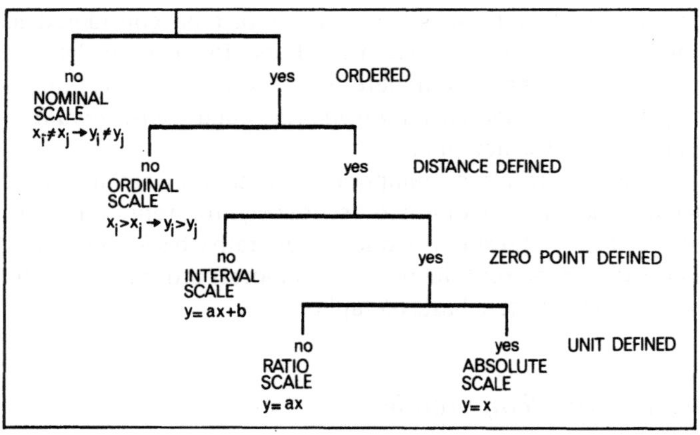

Figure 1. The conventional hierarchy of ordered linear scale types. The nominal scale is not an ordered scale in the proper sense and preserves only the individual identity of its elements. An ordinal scale preserves the rank order of the elements, an interval scale preserves distances, a ratio scale preserves ratios of distances and the absolute scale, permitting only an identity transformation, preserves the absolute identity of all elements on the scale and all distance relations between them.

Narens was able to prove that the automorphisms characterizing various scale types can be exhaustively classified by means of two properties, *homogeneity* and *uniqueness*.

Homogeneity may be understood as an indication of the number of points on a measurement scale that may be varied independently without destroying the properties of that scale. This number is, for instance, larger for an ordinal scale than for a ratio scale. For the absolute scale homogeneity must be equal to zero since this scale permits no variation of any point at all. Uniqueness specifies the number of points on each of two scales that must ›fit‹ in order to determine whether these two scales are automorphically identical. This number is, for example, equal to 2 for an interval scale. Narens demonstrated that the (canonical) measurement scales which have equal degrees of homogeneity and uniqueness are by far the most important ones and coincide with the conventional set of scales of Figure 1. Although it is possible to have scales with degree of homogeneity lower than degree of uniqueness, such scales are rare (relativistic summation of velocities perhaps being the most interesting exception). Figure 2 illustrates the relation between the various scale types.

If we now consider Fraser's levels of temporality again, they appear to represent the temporal ›interpretation‹ of this well-established set of canonical scale types, to wit, the nominal scale (which actually does not count as a proper, ordered, scale type and as such is not a part of Narens' considerations) which corresponds with atemporality (which actually does not count as a proper temporality!) and only allows bijective mappings; the ordinal scale which corresponds with prototemporality and accommodates relative order (A before B); the interval scale corresponding with eotemporality and allowing metric, but reversible relations of distance and duration; the ratio scale which shows the hallmarks of biotemporality, in particular irreversibility; and lastly the absolute scale corresponding with nootemporality, each event having its own unique place defined on the time scale of personal history.

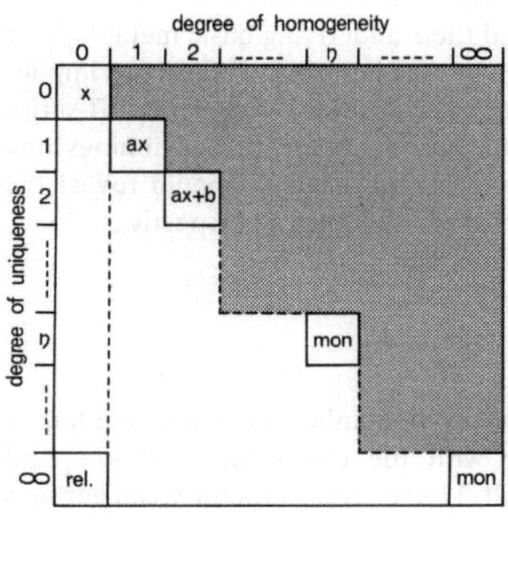

Figure 2. The classification of ordered linear scale types, according to Narens (1981). The 'canonical' or 'natural' scale types appear on the main diagonal, their degrees of homogeneity and iniqueness being equal. Increasing degrees of homogeneity and uniqueness determine the absolute scale $(y = x)$, the ratio scale $(y = ax)$ and the interval scale $(y = ax + b)$, plus any number of ordinal scales *(mon)*, the position of the latter on the diagonal depends on the number of elements on the scale. Since the degree of homogeneity is at most equal to the degree of uniqueness the upper right half of the matrix is void. In the lower left half below the main diagonal one will find spurious, but sometimes important, scale types. An example is the addition theorem of relativity theory $x \circ y = (x + y)/(1 + xy/c^2)$, which is characterized by an invariant constant c (the velocity of light).

Discussion

Each of the five scale types determines one particular way of assigning numbers to empirical variables. The formal properties of these scale types have been given one particular interpretation in Fraser's system of temporalities. As such they do indeed represent, to repeat Clark (1973) once more, »a thoroughly systematic spatial metaphor, suggesting a complete cognitive system.« However, in my view this metaphor is spatial in an abstract (measurement theoretical) rather than a concrete sense. Van Benthem (1985), in his contribution to the present volume (chapter 18), states some caveats about measurement theory as a vehicle for formalizing time. For one thing, he points out that there is a difference between the use of the number system as a model for temporal (order) relations and its use as a measurement scale for duration. It will be clear that in the present discussion the scale types are being used in the former sense, that is as models of time. Van Benthem (personal communication) has also pointed out that from a logical perspective other hierarchical or nested ›scales‹ may be derived that are similar but not quite identical to the measurement structures discussed in the preceding pages. As an example he proposes a ›hierarchy‹ of the following sort:

(a) $[T, =]$
(b) $[T, =<]$
(c) $[T, =, <, +]$
(d_1) $[T, =, <, +,$ ›now‹$]$
(d_2) $[T, =, <, +,$ ›history‹$]$
(d_3) $[T, =, <, +,$ ›event pattern‹$]$

In these formulas T represents the domain of temporal discourse (events, instants, intervals), the symbols $=$, $<$, $+$ are given their temporal meaning of simultaneity, precedence and concatenation (addition). It should be evident that (a), (b) and (c) do indeed correspond to the ›lower‹ three levels of the scale hierarchy presented in the present chapter, but that (d_1), (d_2) and (d_3) represent structures that add qualitative, or rather, cognitive differences which might be found to fit Fraser's levels of temporality and their underlying basic metaphors or stable states of nature in a different, and perhaps a better manner. Thus (d_1) implies the well known distinction between A-series (past, present, future) and B-series (earlier than, later than), whereas (d_2) can accommodate the beginnings and endings characteristic for ›personal history‹. Further analysis should reveal the relative merits of the measurement perspective and the logical perspectives.

Conclusion and Summary

Temporal relations can be conceptualized in a number of ways. Finding a satisfactory conceptual structure to cope with the complexities of temporal experience has turned out to be quite difficult. The required structures can perhaps be expressed most adequately in terms of either temporal logic or measurement theory, as long as the functional aspects of temporal processing are not lost from sight. Both temporal logic and measurement theory suggest that rather more conceptualizations of time are feasible than the simple spatial metaphor that has played such a dominant role in the past. J.T. Fraser's idea of time as a hierarchical set of levels of temporality may well serve as a guide for the direction in which to proceed with at least one fruitful way of analyzing the conceptualization of time. However, as the various chapters in the present volume have shown, this approach is by no means exhausting the ways of representing time that are open to the human mind. The upshot of the present chapter is that the canonical time scales as described here derive indeed – in Van Benthem's words – from a straightforward »mathematical ontology«. While other curious temporal patterns (such as time in relativity theory) may arise, these do not have, in my view, the mundane but stable properties that cognitive representations of time should have, if they are to perform a useful function in everyday life.

References

Benthem, J. van. Semantics of time. In: J.A. Michon & J.L. Jackson (Eds.), *Time, mind, and behavior.* Heidelberg: SpringerVerlag, 1985, pp. 266-278.

Black, M. Models and metaphors. Ithaca, NY: Cornell University Press, 1962.

Boyd, R. Metaphor and theory change: What is »metaphor« a metaphor for? In: A. Ortony (Ed.), *Metaphor and thought.* Cambridge: Cambridge University Press, 1979, pp. 356-408.

Campbell, N.R. An account of the principles of measurement and calculation. London: Longmans Green, 1928.

Chomsky, N. *Rules and representations.* Oxford: Blackwell, 1980.

Clark, H.H. Space, time, semantics, and the child. In: T.E. Moore (Ed.), *Cognitive development and the acquisition of* language. New York: Academic Press, 1973, pp. 27-63.

Davies, P.C.W. Time and reality. In: R. Healy (Ed.), *Reduction, time and reality: Studies in the philosophy of the natural sciences.* Cambridge: Cambridge University Press, 1981, pp. 63-78.

Fodor, J.A. *Representations: Philosophical essays on the foundations of cognitive science.* Brighton, Sussex: Harvester Press, 1981.

Fodor, J.A. *The modularity of the mind.* Cambridge, MA: Bradford-MIT Press, 1983.

Fraser, J.T. *Time as conflict: A scientific and humanistic* study. Basel: Birkhaeuser, 1978.

Fraser, J.T. *The genesis and evolution of time: A critique of interpretation in physics.* Amherst, MA: University of Massachusetts Press, 1982.

Gardner, H. *Frames of mind: The theory of multiple intelligences.* London: Heinemann, 1983.

Goodman, N. *Of mind and other matters.* Cambridge, MA: Harvard University Press, 1984.

Hajdu, H. *Das mnemotechnische Schrifttum des Mittelalters.* Leipzig: 1936.

Haroutunian, S. *Equilibrium in the balance: A study of psychological explanation.* New York: Springer, 1983.

Jones, M.R. Structural organization of events in time. In: J.A. Michon & J.L. Jackson (Eds.), *Time, mind, and behavior.* Heidelberg: SpringerVerlag, 1985, pp. 192-214.

Mey, M. de. *The cognitive paradigm.* Dordrecht: Reidel, 1982.

Michon, J.A. The compleat time experiencer. In: J.A. Michon & J.L. Jackson (Eds.), *Time, mind, and behavior.* Heidelberg: SpringerVerlag, 1985, pp. 20-52.

Michon, J.A. *J.T. Fraser's ›levels of temporality‹ as cognitive representations.* (To be published in J.T. Fraser et al. (Eds.), The Study ofTime V. Amherst, MA: University of Massachusetts Press, 1985.

Miller, G.A., & Johnson-Laird, P.N. *Language and Perception*. Cambridge: Cambridge University Press, 1976.
Montangero, J. The development of temporal inferences and meanings in 5- to 8-year old children. In: J.A. Michon & J.L. Jackson (Eds.), *Time, mind, and behavior.* Heidelberg: Springer Verlag, 1985, pp. 279-287.
Narens, L. On the scales of measurement. *Journal of Mathematical Psychology*, 1981, *24*, 249-275.
Park, D. Brain time and mind time. In: J.A. Michon & J.L. Jackson (Eds.), *Time, mind, and behavior.* Heidelberg: Springer Verlag, 1985, pp. 53-64.
Pepper, S.C. *World hypotheses: A study in evidence*. Berkeley: University of California Press, 1942.
Prigogine, I. *From being to becoming: Time and complexity in the physical sciences*. San Francisco: Freeman, 1980.
Schank, R. *Dynamic Memory.* Cambridge: Cambridge University Press, 1982.
Stevens, S.S. Mathematics, measurement, and psychophysics. In: S.S. Stevens (Ed.), *Handbook of experimental psychology.* New York: Wiley, 1951, pp. 1-49.
Tyler, L.E. More stately mansions: Psychology extends its boundaries. *Annual Review of Psychology*, 1981, *32*, 1-20.
Yates, F., *The art of memory.* London: Routledge & Kegan Paul, 1966.

Author Index

Subject Index

MIX
Papier aus verantwortungsvollen Quellen
Paper from responsible sources
FSC® C105338

If you have any concerns about our products,
you can contact us on
ProductSafety@springernature.com

In case Publisher is established outside the EU,
the EU authorized representative is:
Springer Nature Customer Service Center GmbH
Europaplatz 3, 69115 Heidelberg, Germany

Printed by Libri Plureos GmbH
in Hamburg, Germany